D0312651

WISER IN BATTLE

WISER IN BATTLE

★ ★ ★

A Soldier's Story

LT. GEN. RICARDO S. SANCHEZ

with Donald T. Phillips

HARPER

An Imprint of HarperCollins*Publishers*

www.harpercollins.com

HarperCollins books may be purchased for educational, business, or sales promotional use. For information, please write: Special Markets Department, HarperCollins Publishers, 10 East 53rd Street, New York, NY 10022.

FIRST EDITION

Designed by Renato Stanisic

Library of Congress Cataloging-in-Publication Data is available upon request.

ISBN: 978-0-06-156242-6

08 09 10 11 12 ID/RRD 10 9 8 7 6 5 4 3 2 1

To my family, for all of their love and
support during our journey,
To my fellow soldiers, for all of their sacrifices,
To my Lord, God, for all things.

In every circle, at every table, you will find someone to lead the Army in Macedonia, who knows where the camp should be made, what port held by the territory is best entered, where magazines should be established, how provisions moved; by land or sea, where the enemy should be engaged, and when to hold back. And these people not only tell us how the campaign should be conducted, but what is wrong with the actual campaign, accusing the Consul as though he were on trial . . . If therefore, anyone thinks himself qualified to give me advice, let him come with me to Macedonia.

LUCIUS AEMILIUS PAULUS IN THE ROMAN SENATE, 169 BC

Contents

Preface

I grew up in one of the poorest counties in the United States of America. At the age of fifteen, when the Army Junior Reserve Officer Training Corps came to my high school, I immediately signed up. I loved the military from the very beginning. And by the time I was a junior in high school, I knew it would be my vehicle out of poverty. Today, after thirty-three years in the U.S. Army, I can look back with pride on a period of active duty that spanned from the Vietnam era to ground combat in Desert Storm, Kosovo, and Iraq. But my military career ended at a place called Abu Ghraib. In 2006, I was forced to retire by civilian leaders in the executive branch of the U.S. government. I was not ready to leave the soldiers I loved. The Army was my life. Service to my nation was my calling.

In the aftermath of the terrorist attacks of September 11, 2001, I watched helplessly as the Bush administration led America into a strategic blunder of historic proportions. It became painfully obvious that the executive branch of our government did not trust its military. It relied instead on a neoconservative ideology developed by men and women with little, if any, military experience. Some senior military leaders did not challenge civilian decision makers at the appropriate times, and the courageous few who did take a stand were subsequently forced out of the service.

From June 14, 2003, to July 1, 2004, the period immediately following major combat during Operation Iraqi Freedom, I was the commander of coalition forces, responsible for all military activity in the Iraq theater of war. I was there when Saddam Hussein was captured. I was there when the prisoner abuse scandal at Abu Ghraib occurred. And I was there when low-level enemy resistance expanded into a massive insurgency that eventually led to full-scale civil war.

During that first year of our nation's occupation of Iraq, I observed intrusive civilian *command* of the military, rather than the civilian *control* embodied in the Constitution. I saw the cynical use of war for political gains by elected officials and acquiescent military leaders. I learned how the pressure of a round-the-clock news cycle could drive crucial decisions. I witnessed those resulting political decisions override military requirements and judgments and, in turn, create conditions that caused unnecessary harm to our soldiers on the ground.

After our carefully planned and successfully executed invasion of Iraq, I arrived in the country and was stunned to find that there had been a complete lack of Phase IV post-invasion planning by the administration and the military. Not only was there no strategic vision of what to do next—there was a shocking lack of resources and proper training for our troops. To make matters worse, the combatant commander quickly ordered a massive withdrawal of American forces and redeployed the crucial high-level command centers. Instead of embracing a joint interagency approach, our government and military refused to abandon an outdated Cold War mentality. I find it ironic that I was later criticized for being the youngest and least experienced three-star general in the Army when I was actually one of the most experienced general officers in combined joint interagency operations at all levels of war.

Having fought in Desert Storm and Kosovo, I was well aware of the fundamental responsibilities of a commander in a warfighting environment. Among the most sacred: to take care of subordinates and never send them into harm's way untrained. However, because of our rush to war and the need to mobilize rapidly, some units were deployed without proper training. This fact manifested itself across the board—among active-duty forces, the Reserves, and the National Guard. Some general officers chose to cut corners and certify units as "combat ready" when, in fact, they were not. Throughout my tenure in Iraq, and right up to the day I took off my uniform to retire, I stood up on multiple occasions and adamantly refused to deploy soldiers who were not ready to fight.

When circumstances on the ground required the need for establishment of standards and guidance for soldiers in a variety of areas (such as detention and interrogation procedures, among many others), we immediately acted. I made repeated requests to higher headquarters for help. But when the Pentagon refused to help us with interrogation procedures, I issued the guidelines myself. We knew it would become contentious to do so, but as a leader, I knew that without standards, an army loses its discipline and chaos inevitably ensues. I had to take action for the good of our soldiers, our Army, and the mission.

The sad saga of Abu Ghraib encapsulates the essence of America's failure in Iraq. Abu Ghraib also represents America's initial abandonment of its commitment to human rights and the Geneva Conventions—and an eventual return to reason. In 2004, I refused to sweep Abu Ghraib under the carpet. I still refuse to do so. It remains the personal undertow of my story.

WHEN I BECAME A soldier, I was a nonpartisan, nonpolitical individual who believed in the constraints of civilian control of the military. I also understood that, while on active duty, the Uniform Code of Military Justice precluded me from speaking out against my superiors while in uniform. If I valued my oath—and I did—I had to comply. Since leaving the service, however, I have been encouraged by both civilians and retired four-star military officers to write about my life, my career, and what really happened on the ground in Iraq. I believe now is the right time.

In this book, I will relate key life events and circumstances that have made me the person I am today. I will discuss key events of my military career with special emphasis on experiences and lessons learned that prepared me for my future leadership role as coalition commander in Iraq. The basis for that part of the story, including re-created conversations, is drawn from personal notes, diaries, official reports, an extensive chronological record of the activities, discussions, decisions, and issues I encountered during that time period.

Over the fourteen months of my command in Iraq, I witnessed a blatant disregard for the lives of our young soldiers in uniform. It is an issue that constantly eats away at me. During that time, 813 American soldiers lost their lives, and more than 7,000 were wounded. I cannot do, say, or write anything that would dishonor them. But to *not* set the record straight would, I believe, dishonor the legacy of their service.

There is a camp of commanders who feel that retired generals should not stand up and voice their views on any policy, much less against a policy gone awry. I am now making camp with those who believe our voices must be heard in order to help America prepare for the future battles it must win—so that democracy itself survives.

IN PLUTARCH'S *Moralia*, DAREIOS, the father of Xerxes, said that during battle, he became more levelheaded. I know the feeling. Whether on the field of battle where the wolf rises in the heart against a determined combatant, or in the hallways and back rooms of political safe houses where reason and truth are sometimes nowhere to be found, I strived to remain calm, keep a clear head, and make proper decisions. I had to. The lives of countless soldiers under my command depended on it.

With each battle or crisis, I resolved to gain more wisdom in order to do better the next time out. My ability to do so was inspired, in part, by Psalm 144, which hung on the wall of every office I occupied since I was a major. I also carried it with me into the field wherever I went. I carried it in my heart.

Praise be to the Lord, my rock,
who trains my hands for war,
my fingers for battle.
He is my Loving God and my Fortress.

WISER
IN BATTLE

Prelude

Thursday, May 20, 2004
The White House, Washington, D.C.
Midmorning

I'm Lieutenant General Ricardo S. Sanchez."

"Yes, General Sanchez, the President is expecting you and General Abizaid," replied the guard at the security gate. "Please proceed."

In the van with me was General (GEN) John Abizaid, my friend and superior officer. We had just completed a closed-door Iraq operations update briefing to the Senate Armed Services Committee, which had been a breeze compared to the lengthy grilling we'd gone through during the previous day's public hearing.

Once inside the White House, we were met in the waiting room off the Oval Office by Secretary of Defense Donald Rumsfeld, who had asked us to proceed directly from Capitol Hill for this meeting. A minute or two later, National Security Advisor Condoleezza Rice opened the door and invited us in. President Bush, who was already standing, stepped forward and shook my hand. "Hi, Ric," he said. I barely noticed as a photographer snapped our picture.

There were several other people in the room, including a few presidential advisors. Everybody exchanged greetings and sat down. Abizaid and I sat on the couch to the left of Bush. Rumsfeld was across from us, and Rice was seated to our left directly across from the President.

GEN Abizaid began the conversation. "Mr. President, Ric's convoy was hit by an IED about ten days ago," he said. "When I called him up to ask how he was, his immediate response was, 'Hey, sir, no big deal. A couple of our vehicles were disabled, but none of our soldiers were wounded. Everybody is okay.' That's who this guy is, Mr. President. He doesn't think about himself. He thinks about his soldiers."

President Bush smiled and nodded. "That's good, that's good," he said.

Secretary Rumsfeld spoke next. "Mr. President, I just received a close-hold memorandum from Ambassador Bremer requesting that two additional divisions be deployed to Iraq."

Then turning toward Abizaid and me, he asked, "Have you guys seen this?"

"No, sir."

"Never heard of it."

Bush then addressed Condoleezza Rice, to whom Bremer reported. "Did you know about this?" he asked.

"No, sir," she responded. "I'm not sure why Jerry's doing this."

"Well, why didn't he go through the military chain of command?" asked Bush. "What are we going to do about it?"

"Mr. President, you ought to be glad he didn't send it to you, because now you don't have to respond," said Rice. "Bremer is ready to leave. He'll be writing his book. He needs to go."

"Well, this is amazing," said Rumsfeld, shaking his head negatively. "Mr. President, you don't have to do anything. He addressed it to me. I'll take care of responding to him."

Over the next hour or so, a variety of subjects were discussed, including how our testimony went on the Hill and possible political implications. GEN Abizaid gave a broad review of his theater of operations, including Afghanistan. I spoke about the current ground situation in Iraq with emphasis on the previous months fighting in Fallujah and Najaf. President Bush, as usual, was up to date, quite supportive of our efforts, and very attentive to what I was saying. At one point, he brought up the ongoing turmoil surrounding the Fallujah Brigade. "Are we really making any progress with this?" he asked.

"Well, it doesn't look good right now, sir," I replied. "But it's the most viable alternative we have at the moment."

"Well, be sure to keep us informed."

"Yes, sir. We will."

When the meeting finally broke up, Bush stood up, grasped my hand warmly and said, "Good job, Ric. Thanks for everything you're doing."

"You're welcome, Mr. President," I replied.

As we walked out of the Oval Office, Secretary Rumsfeld asked us to wait for him in the Situation Room. "I'm going back in to see the President for just a second, and then I need to talk to you," he said.

"Any idea what this is about, sir?" I asked Abizaid when we got downstairs.

"No. Not a clue."

It couldn't have been more than a minute or two before Rumsfeld came into the Situation Room and closed the door behind him. He was all business. "The President has approved the following personnel moves," he said. "He can't send General Craddock to Iraq, because it would formalize the shadow chain of command links that the Democrats are trying to prove. Therefore, the choices are Abizaid, Casey, and McKiernan. McKiernan would be a good choice if it were a warfighting assignment. Abizaid has to stay focused on the CENTCOM theater as a whole. Therefore, he's sending General Casey to Iraq."

Then the Secretary of Defense looked directly at me. "Ric, he's afraid to send your nomination [for a fourth star] forward at this time, because it's likely to get mired in the ongoing political debate. He's decided to keep you in V Corps and send General Craddock to Southern Command [SOUTHCOM]. We'll keep you on hold, let this thing die down, and renominate you later. So hang in there."

John Abizaid immediately spoke up. "Mr. Secretary, I don't understand why we're doing this," he said. "Ric has been told he was going to command SOUTHCOM."

"Well, the political conditions are not right," replied Rumsfeld. "We've got to let this blow over."

"But, sir, this is wrong!" protested Abizaid.

"Timing isn't right. We just can't go forward."

I was stunned. After a cordial meeting with the President of the United States, I was being replaced in Iraq and sent back to Germany, and my nomination for a fourth star had been rescinded. I decided, however, to display no emotion. "I understand, Mr. Secretary," I said. After Rumsfeld ended the meeting by saying, "That's all I have for you," I departed. But GEN Abizaid remained behind for a few minutes to speak privately with the Secretary.

While exiting the White House, I ran into Deputy Secretary of Defense Paul Wolfowitz, who was just arriving. Under the awning, he came up to me, looked me in the eye, and shook my hand. "Ric, you're a great American hero," he said in a sincere, almost regretful tone. "It's been nice knowing you."

When John Abizaid joined me in our van a few minutes later, I mentioned Wolfowitz's remark. "My instinct tells me that a much broader decision has been made," I said. "Something is just not right."

"Ric, you're reading too much into it," he replied with a nervous laugh.

John was trying to make me feel better, but I knew something was up. Just how bad it was would not become completely apparent for another several months. At that moment, however, I was deeply disappointed.

Within four days, a well-coordinated press release was launched from the Pentagon in which several national newspapers featured articles setting the stage for my departure from Iraq, and leaving the impression that I was being relieved. A few days before, I had arrived in Washington believing I enjoyed the support of both the administration and my military chain of command. But as I now headed to the airport, it was obvious that I was leaving without any clear knowledge of where I stood or who really supported me. In boarding the airplane for my return flight to the theater of operations, I turned to my aide. "Am I glad to be leaving Washington," I said. "At least in Iraq I know who my enemies are and what to do about them."

Waiting on the runway for our turn to take off, I jotted down a few thoughts in my notebook:

> Putting two and two together, this decision was probably made before meeting with the president . . . What a setback! There must be some reason for it. I will have to place my trust in the Lord. But it is awfully difficult to accept after being told that the nomination was at the level of the president for approval. People will speculate and talk. Few, if any, will know. Many will demand perseverance. The Hispanic kids of America and other Latin American countries deserve my continued struggle. Can't let a political issue defeat me. I have overcome too many obstacles to get to this point.

As the plane took off, I looked out the window and saw the Potomac River flowing below. I was struck by the fact that the Capitol and the White House were on one side, the Pentagon on the other—symbolic, perhaps, of how the civilian government and the military should interact: close civilian control, but clearly separated with regard to command of warfighting forces.

The flowing waters of the river itself also reminded me of my roots in the Rio Grande Valley of South Texas. And my thoughts drifted back to a different world and a simpler time.

PART I

THE SHAPING OF A SOLDIER

★ ★ ★

CHAPTER 1

The Rio Grande Valley

My soul is anchored in a poverty-stricken town on the desolate banks of the Rio Grande River—an international boundary that separates a superpower from a country still struggling to make its way out of the Third World. Less than a hundred miles down the road are the Texas cities of McAllen, Harlingen, and Brownsville. But just on the other side of the river, some 1,200 meters to the south, is Mexico. The flowing water, itself, provides an oasis of life in the dusty, desert landscape—nourishing the plants, animals, and people gathered along its meandering path.

Rio Grande City, where I was born in 1951, is one of the oldest settlements in South Texas. It flourished around Fort Ringgold, a military outpost established in 1848 in the wake of the United States–Mexico War. Occupied by Confederate forces during the Civil War, and the federal cavalry afterward, Fort Ringgold was eventually closed, but was reactivated for brief periods of service during World War I and World War II. Despite the continuing presence of the U.S. military, Rio Grande City had a checkered history marked by ethnic hatred and racial intolerance.

I grew up in a Hispanic community among people who possessed little of material value. My own family was among the worst off in the neighborhood. But our poverty was balanced by a tight-knit network of extended relatives steeped in faith, tradition, and the strong values of honesty, integrity, and respect. The adults in our family were *Rectos*—Spanish for those whose very frames stand erect with honor and pride.

My father Domingo Sanchez was the son of a baker who had emigrated to Rio Grande City from Camargo, Mexico (directly across the river), at the turn of the century. His first marriage produced two sons, Ramon and Domingo Jr. (Mingo). During World War II, Dad was exempt from military service because of his critical skill job as a welder who built airplanes at Laredo Air Force Base. After the war ended, he returned to Rio Grande City because, as he said, "Laredo was too far away from home." It was there where in 1948 he met and married my mother, Maria Elena Sauceda, who was seventeen years his junior.

Mom also had deep Mexican roots. Her family emigrated to Rio Grande City around the turn of the century. Her grandfather was a Yacqui Indian, native to northern Mexico, who wore all whites, a sash, and sandals, and carried a machete with him wherever he went. He and his wife went off to fight in the Mexican Revolution and never returned. It's believed they were killed in battle. Their young son, Carlos Sauceda, was raised in Rio Grande City by his maternal grandparents. Eventually, he married Elena Morales, who gave birth to my mother in 1927.

Soon after they married, my parents' family began to grow. Roberto was born in 1949, then me in 1951, then Leonel three years later. After that, Magdelena de los Angeles, David Jesus, and Diana Margot came along spaced evenly about eighteen months to two years apart. We lived among the dingy, dusty houses that decorated the bedraggled Roosevelt Street. Right across the street from us was the Benito Gonzalez family. They had more than a dozen kids and were migrant workers. Because their family needed money, the Gonzalez kids left school around the age of twelve or thirteen and went to work full-time.

The first house we lived in was an old military barracks that my dad bought dirt-cheap from one of the former World War II camps. I remember them hauling it onto our property and setting it up on concrete blocks. It was only one room, fifteen feet wide by twenty-five feet long, without doors, windows, a bathroom, plumbing, or electricity. We never owned a television. For heat in the winter, my parents would gather mesquite branches from the woods and build a fire. They would place the burning embers into an old aluminum tub and bring it into the house for us to gather around. There was an outhouse on the back part of the lot and, in the far corner, we had a little wooden shack that we used for bathing. The water line on the property stretched from the front to the backyard. It was nothing more than a pipe sticking out of the ground with a faucet on it. So we filled a pail there, hauled it to the shack for bathing or into the house for cooking.

We lived in that one-room house for four or five years, until my father saved up enough money to build a small brick house on the same lot. It had only a tiny living room (about eight by ten feet) and two small bedrooms, but the brick exterior made a big difference in holding the warmth in on cold winter nights. As our family grew, Roberto and I slept in one bedroom in our own bunk beds—and the living room became a third bedroom. On a bunk bed there, my younger brothers Leo and David slept together in the top bunk, and my sisters Maggie and Diana slept in the bottom one. When that house was built, all the necessary water pipes were run in, but my father couldn't afford to buy sinks, bathtubs, or toilets, so we continued to use the outhouse and the shack in the backyard.

Because my dad earned very little as a welder, we were on welfare during most of my youth. I remember standing with my mother in the welfare lines every Thursday waiting to receive allocations of pork, beef pieces, applesauce, cheese, flour, and rice, the bulk of which was gone in a day or two. Then we were back to our normal staples of beans and rice.

While beef was a rarity and a treat, the biggest treat of all was goat, or *cabrito*. Every once in a while, my father would buy one and I'd help him slaughter it in the backyard. We'd prepare it in a variety of ways, such as barbecued or stewed. Sometimes we'd wrap the goat in a sack and place it on top of mesquite embers in a hole in the ground. Eight or ten hours later, we'd have the best food in the world. At Christmas time, there were usually no presents under our tree. Actually, most of the time, we didn't even have a tree. Dad welded some stars out of metal and we placed them on the windows or over the carport. And that's how we usually prepared for the holidays.

School itself was very important to the poor people of the Rio Grande Valley—especially to my mother who, from my earliest memory, placed tremendous emphasis on getting a good education. In grade school, 99 percent of the kids spoke Spanish as a first language. Back then, there was no requirement for us to speak English in the classroom, so everybody fell back on their natural inclinations, including the teachers. They'd try to teach us English, but just as soon as we walked out of class, we reverted back to our native language.

As I think back on my early days in the Rio Grande Valley of South Texas, I am struck by the influence of the military way of life that constantly surrounded me. Our first house came off an old military post. I went to school at Fort Ringgold, which had been purchased from the government by the school district and turned into a learning campus. My classrooms were in Army buildings. And the fort's history was ever present—the Mexican War, the U.S. cavalry, the Civil War, and two world wars. And then there was the constant influence of my older half brothers, Mingo and Ramon, both of whom served in uniform.

By the time I was five years old, Mingo had left the house and enlisted in the Army. After his initial tour, he joined the U.S. Air Force. Whenever he had a scheduled leave, he sent word that he would arrive on a certain day and the first thing he was going to do was inspect us. So my older brother, Robert, and I would scramble to get ready. We'd get haircuts, bathe, polish our shoes, and make sure our clothes were clean. And when Mingo walked in the house, he'd say, "Okay, line up, men." Robert and I then got in formation and stood at attention while Mingo walked up and down looking carefully at our hair, our hands, our clothes, and especially our shoes. "Good. Good. You look good, men," he might say. Or, "Private Sanchez, there's a smudge on your left shoe. Have that taken care of the next time I see you." "Yes, sir. I will, sir."

My other big brother, Ramon, had served a short stint on active duty in the Army and then came home to Rio Grande City and was a member of the Reserves for twenty-five years. One weekend each month, I would see him in his uniform when he reported to the Reserve Center near Fort Ringgold for his regular meetings. And once a year, his unit would deploy to Fort Hood, north of Austin, for mandatory training. Even though they were going only four hours away, all the families would gather for the departure ceremony. The reservists, dressed in uniform and carrying duffel bags, would line up in formation. A few appropriate remarks would be made by the commanding officer and then, as the buses were boarded and pulled away, people cheered, held up small flags, and waved goodbye. And when they came home two weeks later, we would greet them all with the same emotion and fanfare.

I was imbued with this very real love of country early on. It wasn't just Ramon, Mingo, and their commitment to military service. My uncles, Leonel and Carlos Sauceda, had served during World War II, and I would often play with some of the war trophies that my grandmother kept for them. My father was also very proud that he had been designated an essential civilian working for the Air Force during World War II. My dad was a loving, giving father, not prone to straying from his roots. But he also had a problem with alcohol. Whenever Dad had extra money, he'd buy a bottle of mezcal, drink it down, and throw the bottles out in front of the house. As far back as I can remember, he carried a flask in his back pocket everywhere he went. Every day, Robert and I would go downtown with him to visit a couple of bootleggers. Sometimes he'd stay in the car and send us down an alley and into the backyard of a house where we'd tap on the window and give the bootlegger a quarter, who'd then fill up Dad's flask. He also liked to gamble. On weekends, we'd go with him and *Tio* [Uncle] Raul to *la jugada*, a local gambling spot on the northwestern outskirts of town, close to the cemetery. It was just a little corral with a tarp cover where the men would shoot craps and drink. Robert and I usually stood to the side and watched, hoping they'd win, because that meant we'd get a few pennies from the winnings.

My father's drinking and gambling problems eventually took a toll on our family. My parents grew angry with each other and trouble in the house created a stressful environment that lasted for years. My way of coping was to leave early in the morning and not come back until late at night. I just had to take care of myself and keep busy. My father finally moved out of the house when I was about ten. Even though my parents' divorce was a very traumatic event in my life, I still got to see Dad every day, because he went to live with his older brother, Raul, in their childhood home.

Tio Raul ran a little laundry and tailor shop in the middle of town. The building itself wasn't much more than a wooden shack that looked like it was about to fall apart. He recruited me to work for him before I entered the first grade. Customers dropped off clothing in the mornings and Tio Raul would clean, press, and do his tailoring during the day. I brought him his breakfast early in the mornings, and then as soon as school let out, I'd go right back and get to work. I'd empty out the washer, get rid of the water, and sweep out the front of the shop. And then I would be his delivery boy—walking all around town to take the cleaned and tailored clothes back to his customers.

My Tio Raul was a very kind and gentle man. He never married, and spent much of his spare time drinking and gambling with my father. But he almost never missed a day of work. And all through my childhood, he encouraged me to work hard, stay in school, and follow my dreams. He was very much a second father to me.

Over the years, I picked up a couple of other jobs, but they were always secondary, because Tio Raul never let me slack off from my duties for him. Among my other jobs: stocking shelves and sweeping out the local drugstore for fifty cents a week, and selling newspapers for a nickel a piece and keeping one penny for myself. When I wasn't working, I roamed around town with my buddies, David Saenz, and Jose de Jesus (Chuy) Trevino. Chuy's dad was a mechanic and David's dad owned an electronics store down the street from my uncle's tailor shop. I remember very distinctly walking by that shop, looking in the window, and seeing a television set for the first time. Chuy and I would often hang out at David's house, because that's where we got to watch TV. The three of us were almost inseparable. To scrounge up some money, we wandered through town collecting soda and long-neck beer bottles and then sold them back to the bars for a few pennies. Our playground was the plaza, the alleyways and town streets, the railroad, the woods, and the arroyos.

As I grew older, I became attuned to the culture and values of my Hispanic community. Our deep faith in God pervaded everything we did, whether it was in church, in school, or in plain everyday conversation. For instance, the appropriate response to "*Hasta mañana* [Until tomorrow]," was always, "*Si Dios quiere* [If God wills]." We owed everything to God and we would be okay if He wished us to be. *Si Dios quiere.*

The people of the Rio Grande Valley also believed in curses that were tied to the Indian beliefs of northern Mexico. The belief was that children had to have the curse removed by chasing away its evil spirits. I very clearly remember my mother taking us to faith healers, or *curanderas*, for the *curarte de susto* (curing of your fears) ceremony. I was led into a room with lighted candles all around and pictures of saints adorning the walls, and then lay down on a table. The *curanderas* would do some chants, lay hands on me, and pass a broom over me several times to sweep away the evil spirits. And when I came out of that little room, I felt great. I had been healed. And my mother was greatly relieved.

I can remember going through these ceremonies up to the age of twelve or thirteen. Mom put all of us through one every six months or so, whether we had been cursed or not.

Among the values imbued in the Hispanic community and my family, in particular, three stand out. The first was hard work. Not only was I never allowed to slack off, it was drummed into me that if I worked hard, I could achieve anything I wanted in life. Of course, from grade school on, I was self-sustaining to a large degree. I continued to work for Tio Raul until he died of cancer when I was a senior in high school.

The second value involved the handshake. My father and uncle imbued this in me at a very young age. "If you say you're going to do something, don't back away from it," they told me. "You better make sure you know what you're committing to, because once you shake hands, by God, that's it. Your word is your bond."

The third value was simply to tell the truth no matter what. I was taught to live by this principle. Lying never, ever went unpunished in our house. "You better tell the truth," my mom would say, "because if you don't, and I find out later, you'll get an extra whipping." I remember one time when my parents found out that I'd been in a rock fight. When I got home, they confronted me right away. "Okay, did you do it?" they asked. "We'll let you off lighter if you tell the truth." I knew I was going to get my butt whipped. I just had to figure out how to get the least number of whacks. So I told the truth. Mom and Dad never spared the whip.

Interestingly enough, the same woman who took us to the *curanderas* to sweep away evil spirits was also very practical about school work. (And she led by example, because after the divorce, she made a determined effort to obtain her high school equivalency degree.) When I was thirteen, my brother and I came home one day during the summer and announced that we weren't going to school when it started up again in the fall. When she asked us why, we pointed out that the Gonzalez kids were migrant workers and they were not only gone in the springtime, but they dropped out of school as they got older. "Why can they get out of school and not us?" we asked.

My mom paused for a moment and then said, "Okay, you can go pick cotton." Then she went right out and arranged for a truck to come by the next morning at five thirty and take us out about thirty miles to La Gloria, a local ranch, to pick cotton with about thirty other workers. So there I was out in this field with a huge sack on my back getting paid fifty cents for every hundred pounds I picked. Well, I couldn't pick a hundred pounds of cotton in a week. So not only could I not make any money, I was breaking my back and working harder than I ever had in my life. To top it all off, when I got home in the evenings, I still had to go to my uncle's tailor shop, clean up, and make all his deliveries around town.

After about two or three weeks working in the fields, Robert and I told Mom we wanted to go back to school. "Okay," she said, "you boys can go back to school when it starts up again. But you're going to have to keep picking cotton until the summer ends." My mother was an incredibly smart woman.

In the end, all six of her children went to college and earned at least a bachelor's degree. Robert became an RN and director of the operating room technician program at Texas State Technical College; Leo grew up to be a teacher and a coach; Maggie, a high school principal; David, a power systems technician for a power and light company; Diana, a pharmacist. And after college, I started my career in the U.S. Army.

ON MARCH 5, 1966, Private First Class (PFC) Joel Rodriguez, age twenty-two, was killed in the jungles of Vietnam. He had been a popular student and football player at Rio Grande City High School and, after graduation, had joined the Marine Corps. It seemed as though the entire town went into a state of suspended animation to honor and bury its fallen Marine. People moved slowly through the tiny Rodriguez home day and night—passing the flag-draped coffin, pausing to pay respects to his parents and discussing his sacrifice and how he had died.

On the day of the funeral, as the casket was being brought out the front door, I was standing among the townspeople. We all walked behind the hearse, six blocks from his home to the Catholic church, and then another mile to the cemetery. I will never forget the splendid-looking Marines in full dress uniform, the twenty-one-gun salute, the playing of taps, and the folding of the flag. While the sadness and grief was heart-wrenching, the display of honor and affection paid to PFC Rodriguez affected me deeply. I was only fifteen years old, but I knew that *this* was right, that *this* was a noble cause, and that the military was a worthy profession. At that moment, I became convinced that service to my country was a calling I had to answer. I was going to be a soldier.

Less than a year later, the JROTC (Junior Reserve Officer Training Corps) came to Rio Grande City, and David Saenz and I immediately signed up. We were committed and proud to wear our uniforms in high school. During those formative years, the military became an integral source of hope for me. I joined the color guard, raised the flag each morning at school and football games, and marched, trained, and attended classes in teamwork, citizenship, and leadership. Special emphasis was placed on learning about Hispanic heritage in the military, particularly in 1968 when Congress dedicated a special week for the celebration of Hispanic contributions to the United States. I was also proud to learn that, in our nation's history, there were a total of thirty-eight Hispanic Medal of Honor recipients.

During these years, I began to observe and think deeply about things I had never noticed before. Racial prejudice, for one, was something that I never really worried about. After all, I was pretty isolated in a 99 percent Hispanic community. But when we went to the bigger cities of McAllen, Harlingen, and Brownsville, I started to wonder why my family was not allowed in some restaurants, theaters, and other public places. I speculated that this might be part of the reason that my father never strayed very far from Rio Grande City. The issue also came to a head in the late 1960s when the United Farm Workers showed up in our town and started organizing. Their demonstrations resulted in some severe and very disturbing confrontations with law enforcement authorities, which I'll never forget.

I also began to think about the intricate relationship and linkage my family had with Mexico. Most people in the United States viewed the Rio Grande as an international boundary. But to us, it was just a river. The concepts of separated and sovereign nations never entered our minds as we crossed back and forth on the river's primitive pontoon ferry. We regularly visited the small towns of Camargo and San Pedro (now Miguel Aleman) to shop and to receive medical and dental care, all of which were much less expensive in Mexico than in the United States. More important, though, we had relatives on the other side of the river. My sister-in-law, Ramon's wife, Tina, for instance, had grown up in Camargo, and when her father died, the entire family crossed over to attend the funeral. I remember very distinctly going to his wake in the little shack where she had grown up. In a plain wooden casket, the body rested on blocks of ice that were melting and draining into a big bucket below. It was the only way they could preserve the body for two days so they'd be able to have the wake and say the rosary.

As I got older, I realized that my relatives in Mexico were dirt-poor peasants living a Third World–type of existence. They were even poorer than we were, and they had little if any hope for a better life, because the Mexican government didn't care about them. But on our side of that river, the United States offered promise. The government provided food, some medical care, and vehicles to give us a hand up the ladder. Most important, we had hope for a better future.

I gradually realized that I did not want the life offered by Rio Grande City. The depth of my family's poverty weighed heavily on me. The cardboard I had to place in my shoes was embarrassing. I didn't have nice clothes to wear to social engagements. I couldn't participate in extracurricular activities, because I had to work. And I became dejected when I was selected to attend the Presidential Classroom for Young Americans in Washington, D.C., because I knew I would not be able to afford the clothing to wear at the event. My Tio Raul, however, came to the rescue. He bought me a new pair of shoes and fashioned an outfit for me (slacks, a jacket, some shirts, and a tie) from clothes that had been abandoned in his shop. In the end, I went to Washington, D.C.—and I will always remember and love my Tio Raul for his kindness.

During my sophomore year, a senior in our school was awarded a four-year ROTC scholarship—and I saw my own future open up. I decided to go straight to my guidance counselor and ask for help to apply for a scholarship. "Mrs. Solis, my dream is to be an officer in the Army," I said excitedly. "I would like to find out how to apply for West Point and an ROTC scholarship."

But Mrs. Solis was not at all encouraging. "Ricardo, why do you want to go to West Point? You'll never be able to go there, because you are just a poor Mexican," she said. "Besides, you don't speak English well enough. Don't waste your time. What you should do is become a welder like your father."

I was stunned. It was clear to me that Mrs. Solis felt that the only kids who should go to college were Anglos and those from wealthy families. I knew it would do no good to plead with her to help me, so I simply said, "Okay, thank you, ma'am." But I was not going to stop there. I went immediately to the library and started doing some research. And the next day, I spoke to my two JROTC instructors, Major Marshall and Sergeant Grigsby, who then went out of their way to help me work toward getting a nomination to the U.S. Military Academy and applying for ROTC scholarships. Some of my teachers helped me write letters to our Texas senators and congressmen—and my father drove me all the way to Laredo and San Antonio so I could complete my physicals and aptitude tests.

In the end, despite a very high cumulative SAT result, my English score was not good enough to get me into West Point (I wound up a fourth alternate there, and a first alternate to the U.S. Naval Academy). The good news, however, was that I was awarded four-year ROTC scholarships by both the Army and the Air Force. In the spring of 1969, in order to decide between the two services, I went to Austin to visit the University of Texas, which was the nearest college that offered both Army and Air Force programs on campus. At that time, my heart was set on joining the Air Force. I wanted to fly like my big brother Mingo, and the good news was that I had qualified for a scholarship with a pilot track.

One building at UT Austin housed both the Air Force and Army ROTC programs, and when I arrived I went straight to the Air Force office, but was largely ignored. As soon as I walked through the Army's door, however, I was warmly greeted by Colonel (COL) Lawson Magruder. He spent quite a bit of time with me, and saw to it that I was given a guided tour of the Army's facilities at UT Austin. We went to the classrooms, the drill field, the shooting range, and even to the supply rooms. Where the Air Force ignored me, the Army made me feel valued. COL Magruder's willingness to spend a few minutes with a young person made all the difference in the world. And when I walked out of the ROTC building that afternoon, I had made my decision. I was going to be a soldier.

In June 1969, I graduated from Rio Grande City High School and went to the University of Texas on a full scholarship. My best friend, David Saenz (who had also received an Army ROTC scholarship), chose to attend Texas A&I University in Kingsville. Like many kids who leave home for the first time, I experienced some loneliness. I also felt out of place, because most of the students in my freshman class were Anglos. I was completely unprepared for the vastly different culture there, the massive size of the university, and the intense Vietnam War protests. For the first time in my life, I felt like a minority, and it was quite disconcerting.

I was able to make some good friends in ROTC, but there were also some significant problems with the program. A large part of the corps of cadets, for instance, were graduate students trying to avoid the draft. So their moral reasons for serving were not always the best. But the worst problem, by far, was the simple fact that we were in uniform on a major university campus during the worst antiwar demonstrations in the history of the United States. Every Tuesday and Thursday, when we were required to wear our uniforms, my buddies and I were harassed nearly everywhere we went—in our dorms, in our classrooms, on the quads. The worst moment for me, personally, came when the corps held its annual in-ranks inspection by the Inspector General. As we marched onto Freshman Field, the road was lined with thousands of antiwar protesters carrying signs and bullhorns. They yelled and just harassed the hell out of us. Then they followed us onto the field as we formed ranks. Our ROTC instructors had done a good job of counseling us to maintain discipline and to never lash out at the demonstrators. However, it was tough to keep my composure when one protester rushed up to me as I stood at attention, spit in my face, and called me a baby killer.

I went home for Thanksgiving during that first semester a disenchanted and unhappy young man. On my way back to Austin, however, I made the fortunate decision to stop in Kingsville to visit my buddy David at Texas A&I. He and I talked at length about the differences between the ROTC program on his small campus versus mine at UT. "I really want to embrace the military training," I told David, "but it's almost impossible in such a volatile environment."

"We don't have any of those problems down here," he replied. "Why don't you think about transferring? Our campus is a lot smaller and nobody ever harasses the cadets. There's also a big Hispanic population here and you'd feel more at home. Heck, we could even room together."

By the end of my freshman year at UT, I knew it was time to make a change. So in the summer of 1970, I transferred to

Texas A&I in Kingsville. I was now back in a familiar, calmer environment, and lots of my old friends were around. I was much happier, and rapidly integrated myself into the Texas A&I ROTC program. Everything I did focused on becoming an Army officer. I was also fortunate to have joined the Kings Rifles, a drill team of twelve to sixteen cadets that was perennially one of the best in the nation. We not only won numerous competitions (including a national championship); we also built a brotherhood that would last a lifetime.

The move to Kingsville, Texas, also changed my life in a way that I never expected. Maria Elena Garza was taking classes when I arrived on campus that summer and she lived in an apartment nearby. We had met during my Thanksgiving Day visit and we began to date almost immediately. Our relationship grew over the course of our college years and by the time we were seniors, I realized that she offered me the opportunity for true love and lasting happiness. Before I could propose, however, both of us had to fulfill some commitments. Maria Elena had promised her father that she would not get married for at least a year after college. She had also promised to get a teaching job and live with her grandmother in Escobares, Texas, a small community in the Valley. So after receiving her degree in education and obtaining a teaching certificate, she moved back to Hebbronville. I had to attend ROTC Summer Camp (boot camp for ROTC cadets), which I had postponed due to an injury during my first parachute jump at airborne school the previous summer. I had to successfully complete summer camp before I could receive my commission as a second lieutenant. So after graduating from Texas A&I with a double major in mathematics and history, I took off for Fort Riley, Kansas. After completing ROTC Summer Camp there, I was commissioned a second lieutenant in the U.S. Army.

I had already been notified that my initial assignment was going to be in the 82nd Airborne Division at Fort Bragg, North Carolina. That was a big surprise to me, because I had been told by the ROTC instructors that only West Pointers got to go to the

82nd Airborne. But I wrote down the 82nd Airborne as my first choice anyway and, in the end, I got lucky.

I was also lucky to be blessed in love. In August, while attending the Armor Officer Basic Course at Fort Knox, I called Maria Elena and proposed over the phone, and she immediately said yes. But she couldn't leave right away, because she had made a teaching commitment through the end of the year to Roma, Texas. She was also concerned about the promise she'd made to her father and the fact that she was going to have to leave the Rio Grande Valley. No one in her family had ever left Texas before, so it would be hard. In the end, though, her family supported her decision, her father gave us his blessing, and we were married on December 22, 1973. It was a simple, traditional Hispanic wedding ceremony in Hebbronville, Texas. After a honeymoon in Corpus Christi, Maria Elena and I were ready to begin our thirty-three-year journey in the service of our country.

My bride knew that I viewed this career as a calling, a sort of destiny. But there was a deeper, more profound calling that she and I shared—one that involved our faith and commitment to God. For me personally, the foundation of that commitment resulted, in part, from circumstances surrounding the death of my beloved Tio Raul back when I was still a senior in high school.

He had been suffering from diabetes and cancer for a while, but after he took a turn for the worse in April 1969, the doctors sent him for treatment in Galveston. After three or four weeks, my father received a call saying that Tio Raul was suffering from terminal cancer, and that we should come pick him up. "Besides," they said, "he's desperate to return home." Dad immediately pulled me out of school and, along with my older brother Ramon, we made the long drive to Galveston.

As soon as we walked into his room, Tio Raul greeted us in an urgent tone of voice. "I'm glad you're here," he said. "I want to go home. Right now!"

"Well, it's going to take a little while to process you out," said my father.

"No, no, no. I want to leave right now! Let's go!"

So while Dad took care of the paperwork, Ramon and I helped Tio Raul out to the car. We put him in the backseat on the driver's side so he could stretch out a little bit. And when Dad finished, he got into the backseat passenger side to comfort his brother during the ride home. Ramon did the driving and I sat next to him up front.

About two hours out of Galveston, my father turned to Tio Raul and said, "Go ahead and try to get some sleep. We'll be home soon." It seemed as though he was drifting in and out of sleep when, all of a sudden, Tio Raul pointed out the window in the direction of this wide-open field with a grove of trees at the far end, and said, *"Mira, mira, hay esta Mama y Poncho. Ya vienen a llevar me."* ["Look! Look! There are Mom and Poncho. They're coming to get me!"] Poncho was his older brother who had died in 1938 at a young age, and my grandmother had died in 1957 when I was six.

"No, brother," replied my father. "Settle down and rest. Everything will be okay."

"No, no! *Mira*!" said Tio Raul. *"Hay arriba de los arboles. Mira la luz! Hay viene Mama y Poncho para llevar me con ellos. Ya estoy listo."* ["Right over the trees. Look at the light! Mom and Poncho are coming for me. I am ready to go."]

I was looking into the backseat right at my uncle. I couldn't see anything above the trees in the field, but within a few seconds of speaking those last few words, Tio Raul reached out toward the field—and then closed his eyes and died.

The impact of that event changed my life forever.

CHAPTER **2**

Early Army Years

In mid-October 1973, I reported for duty to the armor battalion of the 82nd Airborne Division at Fort Bragg, North Carolina. Nobody seemed to notice me, however, because all hell was breaking loose. Equipment was moving back and forth; soldiers were scurrying around; tanks were being fueled, loaded up with combat gear, and moved to Pope Air Force Base to await aircraft and orders to deploy. As a brand-new second lieutenant, I literally had no idea what was going on. "We were put on alert this morning, Lieutenant," a sergeant told me. "We're preparing to go to war."

It turned out that I had walked smack dab into the Army's deployment preparations for the Yom Kippur War. About ten days earlier, on the Jewish holiday, Egypt and Syria had launched a surprise attack on Israel through the Sinai Peninsula and the Golan Heights. Because fighting continued through the Muslim holy month of Ramadan, the war came to be known as the Ramadan War by Arabs. Early on, Israel was caught completely off guard and had suffered serious losses. But the Nixon administration responded by launching a massive airlift (fifty-six combat aircraft, 815 sorties, and 28,000 tons of supplies and weapons), which quickly turned the tide of the conflict. When Israeli forces gained control of the Suez Canal, and advanced within forty-two miles of Cairo and forty miles of Damascus, the Soviet Union threatened direct military intervention.

It was at this point that U.S. forces were placed on alert. There was a very real possibility that elements of my division would be going to war. After all, the 82nd Airborne was America's "Guard of Honor," and *the* strategic reserve that could be deployed with eighteen hours notice to any contingency spot in the world. It was the Army's historic division from World War II, and filled with exceptionally dedicated young Americans.

After about a week of preparations and waiting, we were given the order to stand down. Fighting in the Middle East officially ended when the UN Security Council imposed a cease-fire and averted an escalation of the conflict into a possible global war. As it stood, 15,000 Egyptians, 3,500 Syrians, and 2,700 Israelis had already lost their lives. America's intervention, it turned out, had prevented the complete destruction of Israel.

During the alert, I tried to go with the flow and help out wherever I could. After things settled down, I was called in by the battalion commander and given my first assignment. Lieutenant Colonel (LTC) Albert Sidney Britt was a West Pointer—a tall man, very smart, with a good reputation in the division. "Rather than assigning you to a platoon in a tank company right now," he said, "we're going to assign you to be my headquarters tank section leader. You'll be in charge of two tanks and have general headquarters duties as we prepare for our gunnery deployment. You will be assigned as a platoon leader later."

Not realizing that he was testing me, I simply said, "Yes, sir. That's fine with me."

As I began my duties, I found conditions to be so bad in the Army that officers on night duty had to carry .45-caliber pistols to protect themselves from their own men. One platoon leader, I remember, was given a "blanket party" by his platoon. His soldiers wrapped him in a blanket, beat him up, put him in a wall locker, and threw him down the stairs. In another incident at

Fort Stewart, the mortar platoon took their jeeps and went off on a drunken rampage after partying at the NCO (noncommissioned officers') club. They led the military police on quite a vehicle chase, and put up a big fight before eventually giving up. We had discipline problems, leadership problems, racial problems, alcohol problems, and drug addiction problems. It was common for us to hold surprise health and welfare inspections in the barracks and find all kinds of illegal drugs. It was all very disturbing for a young, idealistic officer like me.

I eventually came to realize that I was seeing what we would later refer to as the "broken Army" in the wake of Vietnam. By this time, we had stopped reinforcing our troops in Southeast Asia, which is why I had not been deployed. But the long-term effect of that campaign proved absolutely catastrophic for the military. What caused it? For starters, civilian leaders in the White House micromanaged many aspects of the Vietnam War. They did not allow the U.S. armed forces to utilize the full extent of its resources to achieve victory. Instead, the military was forced to fight incremental battles that led to a never-ending conflict. And the Army itself descended into a dark cloud almost totally focused on Southeast Asia. That, in turn, resulted in it being overextended in virtually every area that one could imagine.

Back then, the tremendous demand for personnel could only be met by a full-scale draft. At least 400,000 to 500,000 soldiers were needed to fight the war in Vietnam. So eligibility requirements (physical, educational, and conduct) were dropped way down to reach the recruitment numbers necessary to sustain the Army. Kids were forced to join the Army for two years, and many of them just didn't want to be there. The never-ending individual rotation of troops into the war zone also created a major problem. Soldiers would come out of Vietnam, and then immediately receive orders to go back again. The psychological impact on them and their families was often devastating.

As the war continued, the Army also had to walk away from

its commitments to the broader capabilities of the institution. Training, professional development, maintenance and supply operations, garrison operations, and most key functions of a peacetime army all began to suffer. Rather than deploying as cohesive battalion or brigade units, soldiers were sent off individually or in small units, making unit integrity and efficiency on the battlefield difficult to achieve. New potential leaders were rushed through Officer Candidate School in three months, creating what came to be known as "90-Day Wonders." But worst of all, thousands of soldiers were sent into battle unprepared and improperly trained for the guerrilla warfighting conditions they would encounter in the jungles of Vietnam.

There was very low morale in the Army, and it was exacerbated by the fact that most of the American public did not support America's involvement in Southeast Asia. It would take more than a decade of determined, dedicated, and valiant efforts by some of the Army's best leaders to fix the organization and return it to the elite fighting force it formerly had been. In the meantime, though, I was just a young second lieutenant trying to navigate my way through the establishment and learn how to become an effective leader.

IMMEDIATELY UPON THE BATTALION'S return after a successful Fort Stewart gunnery deployment, I went home on leave. When I got back from my wedding, Lieutenant Colonel (LTC) Britt assigned me to Bravo Company as a platoon leader. Almost immediately, the platoon sergeant, Staff Sergeant (SSG) Pugh, pulled me aside for a heart-to-heart. "Lieutenant Sanchez, you have to understand how this works," he said. "Whenever we're in the field or in combat, you're in charge. But in garrison, when we're training, it's my responsibility to run things along with the other NCOs."

At that moment, I recalled that my brother Mingo had advised me to listen to my sergeants, because they were older and more experienced. So I acquiesced. "Okay, Sergeant," I replied. "Go ahead and take the lead."

"Thank you, sir," he replied. "Don't worry about a thing."

Within a week, I deployed my platoon on a training exercise at Fort Bragg. We were about fifteen to twenty kilometers into a daylong tactical road march when all of a sudden, two of our four tanks stopped dead in their tracks. "What's going on?" I asked SSG Pugh.

"Oh, we've got engine problems with the tanks, sir."

"Well, no kidding," I replied. "What kind of problems?"

After several minutes of watching our guys try to restart the engines, I asked them to check the fuel gauges. The two tank commanders (also both sergeants) then went up to SSG Pugh and whispered something to him. Finally, he came over to me and reported. "Uh, sir, we're out of gas."

"Okay, Sergeant Pugh," I said. "Get out the five gallon cans and send those two tank commanders back to get fuel."

"But, sir, that's a long way," he protested.

"It was their responsibility to make sure we were all fueled up and prepared to deploy. Send them."

"Yes, sir."

The two sergeants walked all the way to our motor pool, got some gas, and then caught a ride back. It was a big lesson for them, and an even bigger one for me. My platoon had not only wasted a full day, but even worse, we had completely lost 50 percent of our combat power, because we had failed to do our precombat checks. So in my very first training exercise on active duty, I had learned how crucial logistics are to a fighting army. I also realized that my big brother Mingo had forgotten to tell me that there were some NCOs who actually needed supervision and that, ultimately, the officer in charge was responsible for anything that went wrong in his unit. So I resolved to trust in my noncommissioned officers, but to make sure I checked up on everything in my area of responsibility. And I remember swearing to myself that I would never let something like that happen again.

When it came time for my first Officer Efficiency Report, LTC Britt called me into his office and said "Ric, I'm giving you a report that is fifteen points lower than I would normally give a West Point lieutenant," he said.

My initial thought was that his assessment had something to do with my overall performance as a platoon leader. But when he continued speaking, I knew it was much more than that. "Given where you come from, and the source of your commission," said Britt, "it's going to take you a long time to catch up to the West Pointers."

"Well, okay, sir," I replied. "What do I have to do?"

"Aw, it's just your experience, Ric. In seven or eight years, maybe you'll catch up. At this point, as an ROTC graduate, you just don't have the experience compared to a West Pointer." Of course, I was pretty disappointed with my first performance review and later realized that, to a certain degree, I had just been discriminated against.

MARIA ELENA AND I began our married life by renting a small two-bedroom house near Fort Bragg, where she secured a job teaching for the Department of Defense Dependent Schools, which serve the military's dependent children. Eventually, we saved up enough money to buy a town house, where we lived for most of our five years at Fort Bragg.

I recall discussing that first performance review with my wife, and being somewhat confused. Neither of us completely understood LTC Britt's reasoning, but we figured there must be some things they taught the cadets at West Point that I hadn't learned during four years of ROTC. So I resolved to try and do better, in part, by working to close any gaps that might exist between my background and that of a graduate of the U.S. Military Academy. And given what had just happened with my tank platoon, I decided to gain more knowledge about logistics in general, so I volunteered to be on the division's dining hall inspection team.

I knew that dining facility operations were critical to success as a company executive officer—and that was the next job I was hoping to get.

During one of my first inspections of Charlie Company's mess hall, I opened an oven and found a mousetrap with a dead mouse in it. Army procedures called for the mess hall to be shut down. But that action immediately shot up the chain of command and by the end of the day, LTC Britt had assigned me to be his new mess officer. "You found it! You fix it!" he said. So I left my platoon leader assignment after just four months and wound up trapped in a primary assignment as the dining facility officer responsible for all five mess halls in the battalion.

After I finished my job as a mess officer, I was moved into the executive officer position (XO) for Bravo Company. This billet (position), which involved responsibility for all logistics, was normally a first lieutenant's job; but I was a second lieutenant for several months before actually getting promoted. And that began a pattern over the course of my career, where I was consistently moved into a job before I'd actually been promoted to the rank appropriate for the position. It's a fairly common occurrence in the Army, though. Once an officer is selected for promotion and a position is available, he or she is likely to get assigned to that higher level of responsibility and authority while awaiting official promotion orders. Of course, in the interim, we don't get the salary commensurate with the job. This whole process is called "frocking." By mid-1975, I was promoted to first lieutenant.

After about a year as the Bravo Company executive officer, I was selected to fill a vacancy in the battalion headquarters S-3 (operations) shop. I reported to Captain (CPT) Scott Wallace who, rather unenthusiastically, gave me a desk in the back corner of the office and assigned me to be a training schedule review officer. That was okay, I thought, because reviewing training schedules was part of the S-3's responsibilities. The problem was that I received the training schedules once a week, but it only took me about an hour to finish reviewing them. So I looked around and wondered, "What do I do now?" That's when I began volunteering for everything that came along. I helped out with field exercises and motor pool safety inspections. I became the mortar safety officer, the gunnery safety officer, the range safety officer, and volunteered for jumpmaster duty at every opportunity. Overall, I did whatever I could to help out and keep myself busy.

Eventually, through CPT Wallace's recommendation, I was selected as aide-de-camp to the assistant division commander. Before I left the battalion, Wallace called me in for my Officer Efficiency Report counseling. He gave me an excellent review, and made an unusually honest admission. "Ric, when you were being assigned to my staff I didn't want you to come to the S-3 shop because you were a Hispanic officer. I'd had problems with Hispanics for most of my career, and I couldn't afford to have any more. But you came in and proved me wrong. Congratulations on your selection as aide-de-camp. Best of luck to you."

I liked Scott Wallace. He was a good and decent man who was simply reflecting the way it was in the U.S. Army at the time. The majority of Hispanics were enlisted personnel: noncommissioned officers and below. Very few had made it into the officer ranks, and every one who did make it was scrutinized under a magnifying glass. CPT Wallace may have come into the battalion somewhat biased, but that was not uncommon, and I certainly did not view it as deep-seated racism. However, his comments did make me realize that I was being more carefully watched than other officers my age. And I think having that knowledge helped me become a better leader, because I always felt I had to perform at a higher level than my peers who were not Hispanic.

Working in division headquarters opened up a whole new world to me. I was, for instance, able to observe how senior leaders made decisions, issued orders, and interacted with their subordinates. The officer I worked for, Brigadier General (BG) Richard Boyle, was a down-to-earth family man who took me under his wing to a certain extent. He kept me close to him, talked to me about his leadership style, and gave me advice that I would carry with me for the rest of my career. It was the first time any superior officer ever really took an interest in my development.

After spending about a year working as an aide-de-camp, I was given command of Charlie Company back in the armor battalion, which had been struggling to meet standards. In fact, it had just failed to complete a preliminary gunnery exercise. Charlie Company had about seventy soldiers and fourteen Sheridan tanks in it. These were small nineteen-ton armored reconnaissance airborne assault vehicles designed to be light enough so that they could be loaded in the back of a C-130 airplane and air-dropped into a combat zone by using a low-altitude parachute extraction system. At this point, I was still a first lieutenant, and it was rare for someone of that rank to assume command of a company, especially without having attended the Armor Officer Advanced Course. This was the first command where I was responsible for preparing a unit that could be sent into battle on a moment's notice. And I was definitely going to have my troops ready if that happened.

Charlie Company just had their commander removed because they had been unable to qualify all crews in the preliminary gunnery exercises. They were angry and their morale was low. But I gathered everybody together and challenged them right off the bat. "We have three tank companies and fifty-four tanks in this battalion," I said. "At the end of the gunnery training period, we're not only going to be ranked as top company, we're also going to have the top gun [best tank in the battalion] and the top platoon." One of the platoon leaders suggested that I had set the bar too high. But I told him that success was just a matter of confidence, fundamental gunnery, and leadership focus. In order to guarantee our readiness for war, I worked the company hard, especially on crew and individual gunnery skills. In the end, we had two of the top three platoons in the battalion, and we missed being rated the top company by only a couple of points. As a result, Charlie Company's morale skyrocketed.

. . . .

I was with the 82nd Airborne in Fort Bragg for nearly five years. It was a great place to begin my career in the Army, because it provided a tremendous variety of learning experiences, which I reflected on as my assignment started to wind down.

First of all, I looked back on what LTC Albert Sidney Britt said about me taking years to catch up with the West Pointers. Well, he was flat-out wrong. There was a difference in the training that I had received in ROTC, but nothing that put me at a gross disadvantage. And as for Britt personally, I realized that he was a vestige from an Army with glass ceilings. From his actions I learned that just because someone is senior to you doesn't necessarily mean that he knows what is best for you or what you will eventually be able to accomplish.

Second, I left my Fort Bragg assignment with a clear understanding that a leader must check up on everything in his area of responsibility. From my time as aide-de-camp at division headquarters, I observed several different and contrasting styles of senior leadership and chose one that best suited me personally. It became evident to me that people often *think* they know what the general wants, when in reality, they might be way off. Therefore, clarity in written and verbal communications is paramount in order to minimize misinterpretations.

From my command of Charlie Company, I learned that setting high expectations, establishing standards, and imposing discipline really does make a difference in a team's ultimate performance—and that every officer must have the moral courage to do what is right and to report, fix, and document any mistakes observed. And finally, I realized that I had the great good fortune to have a superior officer like BG Richard Boyle take me under his wing and provide some real mentoring and encouragement. I especially took note of his lifelong commitment to his family.

During our last couple of months at Fort Bragg, our daughter Lara Marissa was born, on April 3, 1978. I was in the room with

Maria Elena for the birth, and we cried out of pure joy and happiness. At that moment, we realized that our lives had changed forever—and we made a conscious decision that the family and our faith would always come first.

There would be only one caveat to that vow, and Maria Elena and I talked about it at length and came to an agreement. When called upon by the nation in crisis, I would put the Army ahead of the family. But short of war, unless I was deployed, I was going to do everything I possibly could to be at home and participate in family events.

DURING THE FIRST FEW months of 1979, events began to heat up in the Middle East. Iranian Islamic fundamentalists forcibly ousted the U.S.-backed shah of Iran and replaced him with the Ayatollah Khomeini, who had lived his previous fourteen years of exile in Iraq. Later that same year, in July, Saddam Hussein formally assumed power after engineering the resignation of his predecessor, President Hassan al-Bakr.

For years, Saddam had steadily consolidated his strength within the government and had been, for all intents and purposes, Iraq's de facto leader. But when al-Bakr began pursuing a treaty that would lead to the unification of Syria and Iraq, Saddam made his move. Immediately after seizing the presidency, he convened an assembly of all Baath Party leaders, read the names of sixty-eight members whom he labeled "disloyal," and had them arrested. Twenty-two were later sentenced to death. Within two months, Iraq and Iran were engaged in low-level conflict along various parts of their 900-mile border.

In Tehran, on November 4, 1979, a militant group of Iranian students stormed the American diplomatic mission by force. Supported by Iran's new regime under Khomeini, the militants held sixty-six U.S. citizens hostage for 444 days—and finally released them on January 20, 1981, the day Ronald Reagan was inaugurated President of the United States.

Under the impression that Iran was preoccupied with the American hostage crisis, Saddam Hussein escalated the ongoing border dispute by attacking several Iranian air bases in September 1980. But Iran immediately responded by bombing a number of military and economic targets inside Iraq. The two countries then engaged in a full-scale war that would last eight years and kill an estimated one million people. In the end, neither side would be able to claim victory.

After five years at Fort Bragg, I transferred to Fort Knox, Kentucky, to attend the Armor Officer Advanced Course, which was the next step in my officer professional development. The nine-month-long program was designed to broaden an officer's horizons as a warrior and to prepare him for service as a company commander. The advanced course also provided the fundamentals necessary to become an effective brigade or battalion staff officer. I learned to synchronize all systems (logistics, maneuver, intelligence, etc.) that are critical to success on the battlefield. This was important for me, because I had already commanded a company, and I was destined to be a staff officer for at least the next ten years.

Upon completion of the Armor Officer Advanced Course, I flew off by myself to my new duty assignment in Korea. Because there were no government family quarters immediately available, my first task was to find housing off base. Only then could orders be issued for Maria Elena and Lara to travel.

Yongsan Garrison in Seoul, where I was stationed, was a large, bustling military community that had been established after the Korean War. It hosted several major commands, including the United Nations Command, the Combined Forces Command, and the United States Forces Korea/Eighth United States Army, to which I was assigned. My job was in the Joint Staff's administrative office that controls all staff actions going into the four-star headquarters.

During the month I was by myself in Seoul, I tried to learn as much as possible about the history of the area, and I also studied the history of the Korean War. The three-year conflict had resulted in more than 700,000 fatalities (33,600 American, 58,000 South Korean, 215,000 North Korean, and 400,000 Chinese). The United States got involved after North Korea launched a surprise invasion across the 38th parallel in 1950. At that time, President Truman ordered the deployment of American forces in support of a UN mandate that the communist forces pull back immediately.

America's first deployment to the region had been a catastrophic failure. A unit named Task Force Smith had been hastily assembled and flown from Japan into the northern part of South Korea. Its mission was to delay the North Korean offensive in order to buy time for the deployment of additional U.S. forces. General (GEN) Douglas MacArthur (commander of all UN forces in Korea) later termed the operation an "arrogant display of strength." Hastily assembled, the task force was comprised of only 540 American soldiers, less than 20 percent of whom had combat experience. They were not only inexperienced, but improperly trained, and undersupplied for their mission. Worse yet, the task force leadership had little or no understanding of military operations in the Korean theater. Unfortunately, Task Force Smith was decimated in combat. In the intervening years, Army leaders would study the entire operation, and the Chief of Staff, Army GEN Gordon Sullivan, often used the catch phrase, "No more Task Force Smiths."

American-led forces eventually prevailed over the North Koreans, despite a massive infusion of troops by Communist China. In July 1953, a cease-fire was agreed upon and a three-mile-wide demilitarized zone was created along the 155-mile-long border between the two nations. But a formal peace treaty was never signed. Therefore, the United States and North Korea were technically still at war.

The Yongsan garrison was pretty nice compared to the rest of the military compounds in Korea. But as soon as you ventured outside the gates, it felt as if you'd crossed into a Third World city.

Conditions in Seoul were as filthy as could be imagined. In the evenings, I often saw monster rats running in and out of the sewer drains. When Maria Elena and Lara finally arrived, the three of us lived in a small 700-square-foot apartment in the Riverside Village complex, which was a couple of hundred meters from Yongsan. It had a tiny living room, a tiny kitchen, and only one bedroom. By western standards, it was abysmal. But by Korean standards, it wasn't bad. We didn't own a car, because we were within walking distance of the Yongsan compound. And when Maria Elena picked up a job teaching kindergarten to the dependent children there, our lives began to feel normal. We weren't living by the highest standards, but we were surviving, we were together, and we were fine.

Every morning, when I walked from Riverside Village to work, I could see the Han River, with its fast-flowing water, its rapids, and the lush vegetation that grew along its banks. Knowing the river was so close provided some psychological comfort. I guess it reminded me of the Rio Grande to some degree. The Han was much wider and had more water year-round. And in January and February, the river froze so solid that vehicles could drive across it. For both of the winters I lived in Seoul, the frozen river increased security concerns, because back in the early days of the Korean War, the North Koreans had driven across the ice to seize control of the city.

As far as my job went, it was a decent assignment, marking my first exposure to a combined joint command. The four-star commander was an Army general, and his three-star deputy was in the Air Force. So I was working with officers from all of the services—Air Force, Navy, and Marines, as well as other Army officers—and we were all working with Koreans at the coalition level. My specific job title was action control officer in the Office of the Secretary, Joint Staff, which controlled all staff actions going in and out of the command group. I was responsible for all combat service support functions—personnel and administration, logistics, public affairs, and legal matters. It was an office job, but I tried to make the most of it by learning all I could about theater-level, coalition, interagency, and joint command operations.

While in Korea, I observed the Army struggle to transform itself to a values-based organization. When I first arrived, the prevailing attitude of our military leaders and the soldiers was, "What happens in Korea, stays in Korea." And some fairly significant problems were routinely overlooked. In addition to severe difficulty with alcohol abuse, there were many stories about soldiers taking on local women and leaving kids behind.

Things began to change with the arrival of GEN John Wickham. As the four-star commander of combined forces in Korea, he immediately imposed new values on the U.S. garrison. He came right out and stated that we were no longer going to accept leaders exhibiting immoral behavior, and that heavy drinking would not be tolerated. GEN Wickham later became the Army's chief of staff and was instrumental in getting the institution back on track after Vietnam.

NEAR THE END OF my two years in Korea, I received an unexpected call. "Captain Sanchez," said the officer on the other end of the phone. "You've already commanded, you have good efficiency reports, and you're a math major. What do you think about going to the Naval Postgraduate School in Monterey, California?"

"I thought you had to go through a long application process to even be considered for graduate school," I responded.

"Normally you do. But in this case, the Army has anticipated a need, and identified you as a good candidate. It's a two-year program. You'll get your master's degree, but there's a utilization tour on the back end. You'll have to serve in that functional area for at least three more years. Why don't you think it over? If you'd like to go, you're in."

When I hung up the phone, I remember thinking that this

was exactly what I had hoped for with regard to my long-term plans. Having a master's degree would serve me well if, for some reason, I had to leave the Army. Besides, two years in Monterey, California, sounded pretty appealing. And I wasn't bothered at all by the utilization tour, because I knew I still had at least another six years as a staff officer. However, I did have a couple of other options, and Maria Elena and I talked about these at length.

I had been offered the opportunity to stay in Korea for another year. I could either move up to the 38th parallel and join the 2nd Infantry Division as the operations officer for a tank battalion there, or I could stay in Yongsan as the administrative officer to the deputy commander in chief. The possibility of getting that heavy armor assignment was very appealing. But it also would have meant leaving my wife and daughter behind in Seoul.

The biggest factor in the equation, however, was Maria Elena's pregnancy with our second child. Both of us were concerned about the quality of medical care in Korea. Lara, for instance, had suffered a severe skin rash and it had taken the doctors in the military hospital two months to finally diagnose it correctly. In the end, we decided that the heavy armor assignment would have to wait. Family came first. So we decided to accept the opportunity at the Naval Postgraduate School and have the baby in California.

I was also motivated to move back to the States because of my father's declining health. He had suffered a couple of alcohol-induced strokes shortly after I arrived in Korea. After receiving a call that he was probably not going to make it, I flew back home and stayed at his bedside for about ten days. Eventually, Dad recovered and came out of the hospital. The doctors warned him to stay off the booze and he did—for about two months. But then he started up again, suffered another stroke, and this time, he didn't come out of it. Eventually, he wound up in a nursing home.

Before my dad's strokes, I drank a little bit. But after seeing what it did to him, I swore off alcohol forever. And I never touched another drop. His situation also cemented my perspective on how to deal with soldiers who abused alcohol. I would tell them, very clearly, that I did not drink. But I'd also make certain not to impose my personal standards on them. I wholeheartedly supported the Army's regulations and policies against alcohol abuse. So I informed my men that if they chose to drink and got themselves in trouble, they would not receive any leniency from me.

When Maria Elena, Lara, and I arrived at the Naval Postgraduate School in California during the summer of 1981, I was both hoping to see my father healthy again and looking forward to the birth of our second child. While Dad remained stable in South Texas, Maria Elena gave birth to our first son on December 11, 1981, at Fort Ord in Monterey. We baptized him in the Catholic faith, named him Marco Ricardo, and called him Marquito (Spanish for "Little Marco"). We were so joyful to be blessed with a son and a daughter, which is what we had always thought of as the perfect family.

At the Naval Postgraduate School, I enrolled in the two-year operations research–systems analysis engineering program, which was known to be one of the best in the country. During my time in Monterey, Maria Elena and I forged a number of lifelong relationships, including two officers in the Colombian Navy, Alfonso Calero and Guillermo Barrera. Both men would later go on to become admirals and our paths would cross many years later when I worked as director of operations for U.S. Southern Command in Miami. Unfortunately, my time at the Naval Postgraduate School was also marred by a personal tragedy that would change my life forever.

During the fourth quarter of our program, in September 1982, my brother-in-law Jorge came to visit us and invited Maria Elena to accompany him back to Texas for a couple of weeks. Her parents wanted to see four-year-old Lara and meet their new grandson, Marquito, who was by now nine months old. Of course, they loved him at first sight—especially my mother-in-law, who raved that he was a gorgeous child.

I stayed back in Monterey to attend class, but Maria Elena

and I spoke every day by telephone. Near the end of those two weeks, though, I started to get antsy. "You know, I think you really need to come home," I told her. "I can't explain why, but you just must come back as soon as possible."

"Well, we have a couple of things to do this weekend," she replied. "Everything will be all right. It'll probably be Wednesday or Thursday before we fly home."

On Sunday afternoon, Maria Elena and her parents planned to take the kids down to Zapata, a small town close to Laredo, spend the night at her sister Maricela's house, and then continue to Rio Grande City to celebrate our nephew's first birthday. As they loaded up the car for the one-hour drive from Zapata, Maria Elena's sister suggested she take along the car seat for Marquito. We had always used safety seats for our children, but in this case, her father's small Chevy had bucket seats in the front and no seatbelts in the back. So there was really no way to secure the car seat. "We're just going to be on the road for an hour," Maria Elena told her sister. "I'll hold him in my lap. It'll be fine." So they all got in the car and took off. My father-in-law drove, my mother-in-law sat in the passenger-side front seat, Maria Elena sat behind her mother holding Marquito. Sitting beside her were Lara and her fifteen-year-old sister Bellita.

About halfway into the drive, with her mother dozing in the front seat and Lara reading *Green Eggs and Ham* to Marquito, my wife suddenly felt an urge to pray. She silently thanked God for bringing Lara and Marquito into her life. And at that instant, she felt a deep sense of loss she has never been able to explain. About five minutes later, at 1:19 p.m., a pickup truck loaded with kids made a quick left-hand turn right in front of my father-in-law's car.

"Dad, watch out!" shouted Maria Elena from the backseat.

Her father, going about fifty miles per hour, hit the brakes and turned to the left in an attempt to avoid a collision. But it was too late. The little Chevy smashed almost broadside into the pickup truck. Maria Elena's father was knocked unconscious, and her mother suffered the full force of the impact and was killed instantly. In the backseat, both girls were driven to the floor and knocked out. Maria Elena reached down and pulled Bellita back onto the seat to keep her from suffocating Lara. Then she looked down at Marquito, whom she was still holding tightly. My son was unconscious and had obviously suffered a massive head injury. His breathing was labored, his pulse weak. About a minute later, there was no breathing at all, and his face turned ashen.

At that very moment, I was in church back in Monterey. The priest had just begun his homily, which this day was about children. Suddenly, I felt some tugging on the silver chain I wore around my neck. Whenever I held Marquito in my arms, he would grab that chain and tug at it. It was as if he was right there with me.

After church I went straight home to call my wife and tell her that I wanted her to come home. As soon as I walked up to the door, I could hear the phone ringing inside. I answered it. It was Maria Elena calling from the hospital. "Dear, we've been in an accident," she said, "and Mom and Marquito were killed."

I flew back home that night. My four-year-old daughter had a broken collarbone and was in the hospital with a concussion. My wife had a broken arm and was in a cast. My father-in-law had suffered some chest injuries and an impacted hip, but was going to be okay. My sister-in-law had facial injuries and a concussion, but would fully recover. No one in the pickup truck was seriously injured.

Maria Elena and I were torn between taking care of Lara in the hospital and attending to Marquito's funeral plans. We talked. We cried. We prayed together. And that night, we made a decision that we had to go on. Yes, we would mourn our son. But we had to live for our daughter. We desperately wanted Lara to have a normal childhood. It would take a Herculean effort to not get consumed by this tragedy. If we allowed that to happen, though, it could destroy our marriage and our family.

A few days after the accident, we buried Marquito and my mother-in-law, Chavela, in Escobares, Texas. During the Catholic rosary, Marquito's little body rested in an open coffin. It is an image I will keep with me always. At that moment, I knew I had missed out on too much of my son's short life, and there was no way I could ever make up for it. I had spent too much time studying, too much time thinking about advancing my military career, too much time away from him. It was something I would regret for the rest of my life.

Me quede con las ganas de abrazarlo y decircle que lo amo! [I still long to hold him in my arms and tell him I love him!]

CHAPTER 3

The End of the Cold War

Upon our return to California, representatives of the Naval Postgraduate School offered to let me skip the semester's last two weeks of classes. "If it's all right with you," they said, "we're going to take whatever your scores are right now, and give you those grades so you don't have to take any final exams. And if there's anything else you need, anything at all, please don't hesitate to let us know." I appreciated their offer and immediately accepted it. I needed a little extra time before returning to class since Maria Elena was still undergoing medical treatment for her injuries. However, when the Army called and gave us the opportunity to leave graduate school and transfer to a new environment, we politely declined. We had a strong support network of friends right there in Monterey, and we felt it was best for us to remain with them.

In grappling with the tragedy, Maria Elena and I began attending regular Bible study. My wife's faith had always been strong, and she was certain that her mother and our son were safe together in heaven. While I knew that was true, I was also looking for faith as a more concrete source of strength across the board. I began reading the psalms at night before going to bed and took comfort in a number of them, especially several lines from Psalm 36:

The children of men take refuge under the
shadow of your wings.
You will make them drink of the river of your pleasures.
For with you is the spring of life.

I also had an immediate connection to Psalm 144, which caused me to think deeply about my career as a warrior fighting for his country.

Praise be to the Lord, my rock,
who trains my hands for war,
my fingers for battle.
He is my Loving God and my Fortress.

The death of my son strengthened my belief in God immeasurably. And that faith has driven me in ways that I cannot fully explain. I have, for example, never been afraid in battle. If I was going to be taken in the service of my country, I reasoned, then that was my destiny. Besides, as I considered the consequences of death, I would now have my Tio Raul and Marquito to help me make that transition, just as my grandmother and my Uncle Poncho had helped Tio Raul. From this point on, whenever anyone said to me, "*Si Dios quiere* [If God wills]"—it had a new, even deeper spiritual meaning.

As my time at the Naval Postgraduate School wound down, I began lobbying for an assignment back to the armor force. I had worked with the lighter Sheridan tanks, but ten years into my career, I felt I needed to get well grounded in heavy armor. Fortunately, the Army agreed and sent me to Fort Knox, Kentucky, the Home of Armor. Now I was finally going to be involved with the sixty- and seventy-ton M60 and M1 tanks. This was the best heavy armor in the world, with armor protection, powerful engine capacity, and incredible firepower.

When I arrived at Fort Knox in July 1983, Major General (MG) Frederic (Ric) Brown, commanding general of Fort Knox and the Army's chief of armor, selected me to be the chief analyst for a special study group, the Future Armored Combat System Task Force. Our mission was to design the future tank to be fielded in the year 2000, including every possible technology that might contribute to a revolutionary leap-ahead system. For the next seven months, we looked at all kinds of leading-edge technology applications, not only to design the future leap-ahead tank for our Army, but to find a replacement for the lightweight Sheridan tank. Eventually, we came up with multiple tank designs that employed advanced computer technology throughout all systems, including fire control, target acquisition, navigation, and maintenance. Our study group also developed the concept known as "battlefield management system," which envisioned a flat-panel display inside the tank to provide the vehicle commander with the ability to see the positions of both the enemy and our own troops.

There were also several spin-off initiatives that made a significant impact on the armor force. The development of an improved kinetic energy bullet was justified, funded, and initiated, as was the M1A2, which today is the world's most advanced tank. We also came up with a new armored gun system concept to replace the Sheridan tank, and initiated a new task force to study the entire family of armored vehicles. This all signaled the beginning of an Army transformation effort known as the Future Combat System.

MG Ric Brown subsequently appointed me his special assistant and placed me in charge of a small group in the office of the director of combat developments (which managed all aspects of matériel, force structure, and equipment issues for the armor force, including future systems and capabilities). As a result, I spent quite a bit of time interacting with senior leadership in Washington and throughout the Army. In the end, the Army not only made a huge monetary investment in our newly designed programs, but significantly pushed forward the timetables for research, development, and deployment of enhanced warfighting capabilities.

MG Brown's confidence also thrust me into a position where I could directly observe operations at the highest levels of the Army. I saw how the different departments interacted with each other and how they dealt with Congress. Even though it was a staff job, I wound up right in the middle of discussions regarding every major decision that related to the future capabilities of the armor force, because MG Brown dragged me to almost every meeting he had with the Army's key decision makers.

I remember one specific instance at Fort Knox when the Army Vice Chief of Staff, GEN Max Thurman, was being briefed on the armor force that MG Brown envisioned for the year 2000. As the briefer discussed leader development by the armor school in the 1995-to-2000 time frame, GEN Thurman turned to MG Brown and said, "Ric, do you know who your armor force generals are going to be in the year 2000?"

With a puzzled look, MG Brown replied, "Sir, I don't know."

"Well, you ought to start thinking about it. Let me tell you, there is this young captain who's going to be one of the Army's future leaders." The Vice Chief of Staff then proceeded to discuss this captain's past record, described in detail the planned career path, and stated, matter-of-factly, that he was going to be made a general unless he screwed up somewhere along the line.

"His name is John Abizaid," said GEN Thurman. "Watch out for him. Oh, and Ric, the armor force needs to start identifying and developing its future leaders now."

I was extremely surprised that the Vice Chief of Staff would be so open about the early identification and management of the careers of young officers who seemed destined to be future senior leaders in the Army. "This guy Abizaid is my rank and he's already been identified as a future general officer and put on a fast track," I thought to myself. "I wonder if he knows he's going to be a general?" Of course, I had no idea that John Abizaid and I were destined to cross paths again in both Kosovo and Iraq.

Overall, I spent three wonderful years at Fort Knox. During that time, Maria Elena and I were blessed with two more children: Rebekah Karina on Christmas Day, 1983, and Daniel Ricardo on December 11, 1985. With my family growing, the never-ending staff assignments, and the lure of a higher salary in civilian jobs, I seriously considered leaving the service for the one and only time during my career. I had been working a great deal with General Dynamics Corporation and, after we'd finished the preliminary study of the Army's future tank designs, they offered me a job as an operations research systems analyst at $50,000 per year. At that time, I was a captain making only $600 a month, so it was very tempting. But one evening at the dinner table, Maria Elena looked at me and said, "You know, ever since I've known you, the only thing you have ever dreamed of being was a soldier—and your goal has always been to be a battalion commander. If you get out now, you will never know whether you could have made it. And you'll walk away from everything you've always wanted to do."

I thought about that for a second, and then said, "You're absolutely right. I don't think we need to talk about this job offer anymore. We're going to stay in the Army." I often think about my wife's unselfishness at that moment, and of the things she's gone through as an Army wife. I am a very lucky man to have her in my life. And I tell her so, often.

MEANWHILE, THE SITUATION IN the Middle East had by no means calmed down. On October 23, 1983, a terrorist suicide bomber detonated a truck loaded with explosives at the U.S. Marine Corps barracks in Beirut, Lebanon. A total of 241 American servicemen were killed (220 Marine, eighteen Navy, three Army) and sixty were wounded. An almost simultaneous attack occurred at the nearby French barracks, killing fifty-eight and wounding fifteen. The servicemen injured were part of a United Nations multinational force sent into Lebanon to maintain peace amid a burgeoning civil war. This event, one of the first major terrorist attacks against the American military, was believed to have been perpetrated by members of the Lebanese Shia militia (an early Hezbollah group) and backed by Iran and Syria.

Two days later, on October 25, 1983, 9,600 U.S. troops were involved in the invasion of Grenada, a small island in the Caribbean Sea. Their mission was to rescue some 600 American medical students who had become hostages of Cuban and Grenadian forces in the wake of a bloodless coup (believed to have been backed by the Soviet Union). Nineteen Americans were killed and 116 wounded in the first deployment of U.S. troops into direct combat since the Vietnam War. The mission ended with the successful evacuation of all the medical students, but a postinvasion analysis determined that the operation was marred by inadequate intelligence, unacceptable interservice rivalry, disorganization, and poor leadership. As a result, the Army revised its training procedures and got serious about achieving joint warfighting interoperability and effectiveness.

Two months after Grenada, in December 1983, President Reagan appointed Donald Rumsfeld a presidential envoy and dispatched him to Baghdad to meet with Saddam Hussein. His message was straightforward: Washington was willing to resume diplomatic relations with Iraq, because its potential defeat in the now three-year-long Iran-Iraq war was contrary to the best interests of the United States. Rumsfeld returned for another personal meeting with Saddam in March 1984 to continue his diplomatic efforts. Despite the fact that the United States had openly condemned Iraq's use of lethal chemical weapons against Iran, diplomatic relations between the two countries were fully restored in November 1984. (They had not existed since the 1967 Arab-Israeli war.)

On April 15, 1986, right about the time I was notified of my selection to attend the Command and General Staff College (CGSC) in Fort Leavenworth, Kansas, President Reagan ordered a series of air strikes on Libya. U.S. intelligence indicated that strongman

Moammar Kadafi had sponsored a terrorist attack on a German nightclub popular with off-duty American servicemen, in which one soldier had been killed and more than sixty wounded. The United States responded with 200 aircraft dropping approximately sixty tons of bombs. The President's message to terrorists around the world couldn't have been clearer. Such attacks on American citizens were going to be met with harsh retaliation.

Six months later, on October 1, 1986, Reagan signed into law the Goldwater-Nichols Act, which fundamentally changed the command relationship structure between the military and the civilian executive branch of the government. It was an attempt to solve problems related to interservice rivalries that were especially evident during the Vietnam War, the failed Iranian hostage rescue mission in 1980, and the invasion of Grenada. Unfortunately, senior military leaders in the Pentagon were bitterly opposed to the changes.

Essentially, the new legislation had the chairman of the Joint Chiefs of Staff reporting directly to the President as well as the Secretary of Defense. Prior to this, the Chairman's position on the Joint Chiefs rotated regularly among the Army, Navy, Air Force, and Marines, which tended to result in favoritism toward the military service in power at the time. The new law also removed the Joint Chiefs of Staff from the formal chain of command for regional combatant commanders. Henceforth, combatant commanders would report directly to the secretary of defense. And finally, the new legislation required the President of the United States to annually outline a national security strategy and present it to Congress.

Goldwater-Nichols took effect after I enrolled at the Command and General Staff College in Fort Leavenworth. It turned out that exposure to the highest levels of Army leadership, combined with the fact that I had performed well, had been a real plus for my career. I was promoted to major and selected for resident study at Fort Leavenworth as the next step in my officer professional development program. At CGSC I focused on learning to become an effective division and corps-level staff officer.

After three years in the extraordinarily fast-paced work environment at Fort Knox, I was determined to refocus on my family. Maria Elena had simply been shouldering too much of the responsibility, and I wanted make up for it. Of course, I did not abandon my studies. Rather, I dug in extra hard on the warfighting parts of the curriculum, such as the division and corps tactics and joint task force operations blocks of instruction.

As graduation approached in the spring of 1987, I received an opportunity that was too good to pass up. LTC Mike Jones, commander of the 3rd Battalion, 8th Cavalry of the 3rd Armored Division in Germany, called and asked me to become his S-3 (battalion operations officer). Mike and I had served together on the task forces at Fort Knox as we worked the future tank options, and had become fast friends. By now, he had been commanding a tank battalion in Germany for a year and was headed into a major training exercise at the Hohenfels training area. This exercise would provide an assessment of his unit's tactical proficiency and he really needed me to take over his S-3 shop. After thirteen years in the Army, I was finally close to an assignment that would allow me to serve in a hands-on leadership position in a heavy armor battalion. There was no way I was going to walk away from that opportunity. So we packed up and headed to Gelnhausen, Germany.

From 1987 to 1990, I served with the Army's 3rd Armored Division, which had played a storied pivotal role in the U.S. victory in Europe during World War II. At this time, and during the entire Cold War, 3rd Armored was the North Atlantic Treaty Organization's (NATO) primary point guard for the defense of Europe. Our mission was to be prepared to stop Soviet forces should they attempt to advance along the most obvious route of attack: the Fulda Gap, a break in the Vogelsberg Mountains located between Frankfurt and the East German border. During those three years, I served as battalion operations officer and executive officer for the 3rd Battalion, 8th Cavalry (Mustangs) in Gelnhausen, and as deputy operations officer at division headquarters in Frankfurt. Now, I felt that I had faced every possible challenge expected in leading an armor battalion—short of war. We had conducted major maneuver exercises across the German countryside. On multiple occasions, we had been tested in the Combat Maneuver Training Center against a professional opposing force and the 3-8 Cavalry had excelled in supply, maintenance, and gunnery competitions at the Army level.

During my second year in Germany, MG George Joulwan assumed command of the 3rd Armored Division, and immediately began to implement leading-edge training concepts developed by the Army at the National Training Center, Fort Irwin, California. He totally restructured training at Hohenfels to embrace the professional standards of the National Training Center in the United States. Joulwan's leadership resulted in a vast improvement of our preparational exercises.

MG Joulwan also led the division during a REFORGER (Return of Forces to Germany) exercise, one of the most massive and inspiring training exercises I had ever participated in. We literally had battalions, divisions, and corps—with thousands and thousands of men and equipment—maneuvering all across the European countryside. Joulwan was right there in the field with us—commanding, observing, directing, mentoring, praising, and critiquing as we went along.

Another significant part of our professional development while in Germany involved conducting staff visits to the more famous European battlefields. The battlefield tour that may have impacted me the most was our study of the Battle of the Bulge. During World War II, on December 6, 1944, with 200,000 men and 1,000 tanks, Germany launched a surprise attack through the Ardennes Forest in Belgium. Their goal was to drive all the way to the English Channel, split the Allied armies, and reverse the course of the war. The main battlefield was a seventy-five-mile snow-covered stretch of dense woods.

Our group visited the battlefield in December, under the same conditions, at the same time of year. It was snowing, raining, and intensely cold as the historians walked us through the dark woods and narrow trails. We moved along the enemy's path of attack, and stood at encampments where resting American forces were taken by surprise. The real story of this battle, we learned, involved the heroic efforts of individual soldiers. Often isolated and unaware of the broad picture, they did whatever they could to impede the Nazi advance. American soldiers organized themselves into small groups to fight the enemy. They burned gasoline supply depots so that German tanks couldn't be refueled. They waded through waist-deep snowdrifts to attack the flanks of advancing forces. And in the end, they bought enough time for GEN Eisenhower's reinforcements to arrive and GEN Patton's 3rd Army counterattack to be successful. On Christmas Day, the 2nd U.S. Armored Division stopped the Germans cold in their tracks, and by the close of January, the enemy had retreated out of Belgium. Barely three months later, the war in Europe was over.

This walk through the Battle of the Bulge convinced me that small-unit leadership was of the utmost importance in a battlefield situation. In the end, it all comes down to that young soldier, sergeant, or officer who has the courage to stand his ground and continue the fight when the going gets tough and all seems lost. One person, no matter what level of command, can turn the tide of a battle, and impact the entire theater of operations. That was a lesson I would never forget.

In the late 1980s, as my time with the 3rd Armored Division in Germany wound down, a number of social issues seemed to take center stage in the American military. First and foremost among them was a focus on racial diversity. On a national level, President Reagan created a National Hispanic Heritage Month to recognize the many contributions of Hispanic Americans throughout U.S. history. Each uniformed service implemented some sort of formalized effort designed to throw light on the issue of diversity. The Army, for instance, sent teams around the world to highlight the difficulty it was having in creating and retaining a diverse officer corps. When they came to Germany, I was surprised to hear them refer to us as a "lily-white armored force."

During this time, however, the Army's armored forces made concerted efforts to attract, identify, retain, and promote minority individuals. To some degree, I benefited personally from these programs. Clearly, the armor force did not want to lose me to another branch in the Army, which was great for me, because I wanted to stay in armor. The bad news was that, ten years down the road, we were probably going to be short on midlevel minority officers. I saw that as an unfortunate probability, because the world was changing. It was clear to me that diversity was going to play a much more important role in the affairs and strength of America. Just consider what was happening in the late 1980s. Personal computers and the World Wide Web had just been invented. Businesses were aggressively expanding abroad and international borders were beginning to blur. Perhaps most amazingly, in November and December of 1989, the Berlin Wall came down and East and West Germany were reunited. Just two years later, the Soviet Union itself would collapse and dissolve. These global changes were unimaginable just a few years earlier.

Back in 1987, upon first arriving in Europe, Maria Elena and I took the duty train from Frankfurt to West Berlin to visit some old friends. As we traveled across East Germany en route to Berlin, it reminded me of crossing the river into Mexico as a kid. We had, quite literally, entered a different world. On the other side of the Berlin Wall, the buildings were stark and gray, and the people were desperately poor. Because they didn't have electricity, they had to burn fires for both light and warmth. But that was all going to change now. It would take time for development to take hold. But there was no doubt that new industries, new companies, and new money were going to flood into Eastern Europe. Accordingly, the American military was destined to change, as well. It would no longer be necessary to guard the Fulda Gap, for example, because the Soviet Union would no longer be there to drive tanks through and wage a war. As a matter of fact, the 3rd Armored Division was deactivated and dissolved in 1992. The end of the Cold War was clearly going to change everything.

AS MY TIME IN the 3rd Armored Division was coming to an end, the armor branch assignments officer asked me for my assignment preference, and I responded, "Any M1 tank battalion in the Army." After a couple of months, I received a letter in the mail. "Welcome to Fort Benning," it read, "Home of Infantry, the 197th Infantry Brigade (Separate)(Mechanized) and the 2nd Battalion, 69th Armor."

I had no idea there was a tank battalion at Fort Benning. I quickly learned that the 2-69 Armor was a training support battalion for the infantry school. Their focus was school support with a primary mission of training infantry soldiers how to interact with and employ armor forces. "Well, I wasn't specific in my request," I thought to myself. "I got what I asked for—command of an M1 tank battalion. So I'll just have to go to Georgia."

I had mixed feelings about leaving Germany. Maria Elena and I were going to miss Europe. In addition, our family had also grown with the birth of my son Michael Xavier on August 28, 1988. Maria Elena's due date coincided with a gunnery training exercise and I was just wrapping it up when I received a call that her water had broken and she'd gone to the hospital. After a rushed five-hour drive, I was surprised to find that the baby had not yet been born. As soon as I walked into the room she went into contractions. So I was able to keep my record intact. I had been in the room for the birth of every one of our children.

Even though somewhat nostalgic about leaving Germany, I was also excited about the future. I had been promoted below-the-zone to lieutenant colonel and my assignment to command a tank battalion constituted the realization of a childhood dream. We would be back in the United States, a bit closer to home—even though Fort Benning (located in south-central Georgia near the Alabama border) was still a thousand miles away from the Rio Grande Valley.

Immediately after my change of command, after the receptions were over and all the dignitaries were gone, I called the battalion chaplain, CPT John Betlyon, into my office. This was the first time I ever had a chaplain reporting to me, and I wanted to take advantage of it. I asked CPT Betlyon to give me a thought for the day—every day of my command, wherever I might be located, his mission was to provide that prayer. Most often, he'd drop by the office every morning and leave me an inspirational quote. Sometimes it would be four or five lines, sometimes a couple of paragraphs. My staff began reading these passages and they grew in popularity to the point that every company wanted a copy. I never pushed the idea, but I certainly was glad to share some faith-based thoughts with my soldiers. I also made it a point to have the chaplain offer us a prayer at the conclusion of every meeting.

It soon became apparent that my faith provided something of a bond among my entire battalion. I found it interesting that when they wanted to know more about me personally, they went to my wife. "Ma'am, can you tell us a little bit about your husband?" they would ask.

"Well, for one thing, he's a good father," replied Maria Elena. "He'll tell our children that it is far worse for them to lie to him than anything they might do. So my advice to you is to own up to it if you make a mistake. But don't ever lie to him, because then you'll be in a lot of trouble."

One of the most important people to a battalion commander is his executive officer. In my case, that turned out to be a very bright and dedicated major named Fred D. (Doug) Robinson, who had come in two weeks before my arrival. Our individual skills complemented each other and we very quickly became a good team. Both of us realized that the battalion had become too comfortable with the school-support mission. Training and deployment readiness had stagnated and tank maintenance was not up to par. There was a lot of work to be done.

At this point in my career, I had a firm view of my responsibilities as a commander. All of my instincts and all of my focus had always been to train properly and be prepared to go to war if our nation called. *That* had to be our top priority. *That* was what we were supposed to do. After all, even though the Berlin Wall had come down, anything could still happen. In the previous two years alone, Saddam Hussein had used chemical weapons to kill 5,000 Kurds in northern Iraq, and 20,000 U.S. troops had invaded Panama to restore democracy and oust the corrupt dictator, Manuel Noriega. No one really knew when or where U.S. troops would be needed next.

So in June 1990, Doug Robinson and I changed the entire concept and focus of training and readiness in the 2nd Battalion 69th Armor. First of all, whenever any element of the battalion deployed for training, we would take everything we owned to the field. This included all our maintenance, supply, and other warfighting equipment. We had to ensure that we could identify what we needed to fight, and then we could transport and maneuver with those materials. Second, when we went into the field for training, we *lived* in the field. I was going to force our troops to survive and operate under expeditionary conditions even if they were just out on the gunnery range.

This message sent a lightning strike across the battalion. Previously, while training at Fort Benning, most units would sustain themselves from their permanent facilities. Often, they would go out to train and come back home at night to sleep. If they had a maintenance problem, they'd bring the tanks back to the motor pool for service.

"What the hell is this guy doing?" some of our soldiers asked, referring to my orders. "We have never been alerted for deployment! Besides, we have always been a school-support battalion.

We're just a training aid for the infantry school! They'll never send us to war! The Infantry School would never allow it."

Despite the early resistance from most senior noncommissioned officers, I trusted my instincts and stayed with the plan. "We will do everything possible to ensure that our battalion is trained and ready to deploy should we be called," I told my soldiers. "It is my responsibility to train you for war. That is the highest priority for this battalion and that's what we are going to do!"

So for the months of June and July, our entire battalion trained every day. We hauled everything we had into the field. We performed our maintenance out there. We became proficient at maneuvers, logistics, and gunnery. We strengthened team unity and instilled the warrior spirit. It was painful for us to go through. But by July 31, I knew the Panthers of the 2nd Battalion, 69th Armor were close to combat readiness.

Two days later, on August 2, 1990, Saddam Hussein invaded Kuwait.

CHAPTER **4**

Desert Storm

When the Iran-Iraq war ended in 1988, Saddam Hussein was in command of the fourth largest army in the world. The Iraq economy, however, was struggling badly. Unable to pay back the $14 billion he had borrowed from Kuwait, Saddam attempted to raise the price of oil through OPEC (the Organization of Petroleum Exporting Countries), but Kuwait responded by increasing the flow from its own massive oil fields, which kept world prices low. Outraged by that action, plus the fact that Kuwait refused to waive Iraq's debt, Saddam amassed his military forces along the Kuwait border in late July 1990, and then summoned American Ambassador April Glaspie for a private meeting. What happened next has been the subject of much debate.

It may have been that Saddam Hussein misinterpreted Glaspie's comments that the United States had no opinion on inter-Arab conflicts, and inferred that America would not intervene should Iraq invade Kuwait. After all, the Reagan and Bush administrations had repeatedly blocked efforts by Congress to impose economic sanctions on Iraq while simultaneously allowing American companies to sell billions of dollars in arms to Iraq. Clearly, Saddam became convinced that any U.S. military response to an invasion would be inconsequential.

Therefore, eager to pull Kuwait back into Iraq (the two had been separated by Great Britain in 1913), desiring of its vast oil reserves, and perhaps itching to demonstrate his own immense military power, Saddam Hussein launched the invasion at 2:00 a.m. on August 2, 1990. More than 120,000 Iraqi forces participated in the operation, including armored, mechanized, and motorized infantry divisions, special commando forces, squadrons of helicopters and fighter-bombers, and members of the elite Republican Guard. Kuwait's meager 16,000-man army was easily beaten and, after only two days of fighting, its monarchy was overthrown, its government dissolved, and Saddam Hussein announced that Kuwait was now Iraq's nineteenth province.

The United Nations responded swiftly by passing resolutions condemning the invasion, demanding a withdrawal of Iraqi troops, and instituting economic sanctions against Iraq. On August 8, 1990, President George H. W. Bush ordered American troops to the region, ostensibly to prevent an invasion of Saudi Arabia. Operation Desert Shield was now under way. Eventually, that movement resulted in the buildup of approximately 500,000 U.S. forces over the next five months.

Shortly after President Bush gave the order to deploy the American military, MG Barry McCaffrey showed up at Fort Benning to inspect my parent brigade, the 197th Infantry. McCaffrey—a graduate of West Point and veteran of Vietnam (where he had been seriously wounded)—was commander of the 24th Infantry Division (Mechanized). My brigade had been attached to McCaffrey's division after the 48th Brigade of the National Guard had encountered significant preparation challenges during its predeployment training. He visited the three battalions in the brigade in order to determine our overall combat readiness. In order to gauge how long it would take us to complete our training and get ready for deployment to Saudi Arabia, McCaffrey's first question to me was, "What is your training plan?"

"Sir, we are trained and ready to deploy right now," I responded. "We just completed gunnery training. Every element of the battalion has deployed in its entirety and our battalion recently completed some tough maneuver training. We have a training plan that will continue to hone our warfighting skills as we prepare for deployment. The Panthers will be ready to fight."

Impressed, McCaffrey gave us some guidance for the next two weeks while we were getting organized for combat operations. As 2-69 Armor got close to the deployment date, our personnel numbers increased. We built combined arms teams by detaching two companies, one to each of the infantry battalions in the brigade. We had to build the team rapidly as we honed our gunnery skills and did limited maneuver training in the Georgia countryside. By the third week of August, my task force was set for war. But it was only when the eighteen-wheelers arrived in our motor pool to load up tanks with live ammunition that some of our noncommissioned officers accepted the fact that their school-support unit was really being deployed.

Our brigade departure ceremony from Fort Benning reminded me of the ones I used to attend as a kid when my brother Ramon went off for his yearly Reserve training. This one was much larger, of course, but the ceremony was essentially the same—with the speeches, the families, and the flag waving. This time, it was the real deal. As the buses pulled away from Kelly Hill and I waved goodbye to my family, I thought, "This may be the last time some of us will ever see our loved ones." I prayed to God to give us strength. We were going to war, and everybody knew it.

After two weeks of training and integrating into the 24th Infantry Division, the Panthers arrived in the Port of Dhahran, Saudi Arabia, at noon on September 1, 1990, and waited two weeks for the arrival of our tanks and equipment. During that time, we kept the troops busy with mission rehearsals, Scud missile drills, individual training, and reconnaissance of defensive desert positions. But we were all cramped up in a stuffy, miserably hot tent city near the port and couldn't wait to get to the forward assembly area where we could spread out a little bit. The danger of Scud missile attacks was very real. So when our equipment finally arrived, we wasted no time in downloading it all, conducting combat checks, forming our units, and moving out to the desert.

Our assembly area was ninety miles to the northwest, but still forty or fifty miles from the Kuwait border. As part of the first mechanized division on the ground, our armor battalion task force was comprised of over 1,000 men, two tank and two infantry companies, and an abundance of other supporting equipment, including Humvees, cargo trucks, fuelers, command and control vehicles, and ambulances. As soon as we arrived in our assembly areas, we launched reconnaissance teams to scout out and determine the best routes to our defensive positions. Then our priority was to figure out how to survive in the harsh desert environment—to assess the effects of heat on our ability to shoot, and how to limit activity during peak temperature periods of the day, for example. For the first sixty days, we survived on MREs (meals ready to eat) and literally lived off our vehicles and what we carried with us. Officers, sergeants, soldiers—all of us were living under very austere expeditionary conditions. Actually, if the 24th Infantry Division had been ordered into battle during those initial two months, we would not have been able to sustain ourselves as a fighting force. We had limited ammunition, and fuel supplies were so low that movement of combat vehicles, especially tanks, was prohibited. So I had to shut down our tanks, and ordered that they only move under very strict controls.

During this limited movement period, we focused on platoon tactics; nuclear, biological, and chemical defense training; medical evacuation procedures, and individual work on all critical combat skills. We conducted forty- to fifty-kilometer treks on our Humvees that included day and night navigation with compasses. (We had no global positioning systems at that time, just some old, outdated maps.) We also set up a small-arms range to practice shooting in the desert heat. On the first day, the troops showed up without flak jackets or load-bearing equipment (LBE). When I told the NCOs that we would be shooting in full combat gear, the order was received with skepticism. "Sir, we've never done it that way before," complained one of the senior NCOs. "Most of the soldiers will have difficulty qualifying."

"Are you going to take off your flak jackets and LBE in battle so you can fire your weapons?"

"Well, no sir."

"Then don't you think you should practice fighting that way so that our soldiers can be prepared when we send them into battle?"

That ended the conversation, and sure enough, we started out with an 80 percent failure rate in our qualifications. After some remedial training in full combat gear and in the heat of the desert, the troops gradually got used to the different environment and became proficient.

As we had prepared for deployment, Doug Robinson and I agreed that it was critical for our tank battalion to sustain itself in the field for at least two weeks with limited maintenance help from anybody. So before we left Fort Benning, our maintenance section was told to bring everything we would need—including spare parts for engines, hulls, turrets, communications systems, wheel vehicles, fire control systems. When our maintenance chief warrant officer came up to me and said, "Sir, that's a lot of stuff!" I simply responded, "We may be out there for a long time, and our survival will depend on our equipment. So we're going to take it all, Chief." That decision paid off when we got out to the desert and consistently reported high operational readiness rates.

I thought I might be in trouble, however, when the assistant division commander paid us a visit in mid-October. "How can you have an 80 percent operational rate?" he asked. "What are you doing?"

I thought he was complaining, because we were too low. "Sir, I don't know," I replied. "We're doing what we're supposed to do. We're servicing our tanks and maintaining our equipment."

"Well, how in the world are you doing that?"

"Sir, we brought our maintenance tents and service kits

and materials with us, and we have been servicing all of our vehicles."

"You have? This I've got to see."

So I took the general out to our maintenance area, where our mechanics were working. He saw engines under service, gun turrets being worked on, and soldiers buzzing like bees all over the place. "Who thought of this?" he asked.

"Sir, this is what armored units are supposed to be doing," I responded.

"Well, this is great," he finally said. "I haven't seen services being done anywhere else in the division. No wonder you guys have the highest operational readiness rates. Good job, Ric."

"Thank you, sir," I replied. "But the real credit goes to my executive officer, my maintenance warrant officer, and especially our mechanics. The work they've done is unbelievable." I later learned that the 24th Infantry was experiencing tank operational readiness rates below 60 percent.

The longer we stayed in the forward assembly area, the better we were able to adapt to the harsh environment. For the most part, we helped ourselves. But once in a while, we received aid from some rather unexpected sources. Within two weeks after we left the Port of Dhahran, I received a report one afternoon that one of the companies had gone on alert because a civilian SUV was approaching one of our perimeter positions. I immediately contacted the company command post to get an update and I dashed out to the location thinking the worst. Upon arriving, I saw the SUV very close to one of our tanks, and became quite concerned. But then the lieutenant came up and said, "Colonel Sanchez, you need to come see this. These guys have cold Cokes."

"What? Cold Cokes? Who are they?"

"Sir, they're Americans. They work for the oil companies in Dhahran. They've been driving around looking for American soldiers, and there are four or five other four-wheelers en route. They want to give us Cokes and ice."

It turned out that there were about two or three dozen American expatriates who adopted our battalion and took it upon themselves to bring us soft drinks, snacks, and ice nearly every weekend that we were stationed in our forward assembly area. They all became good friends to us (especially the two leaders, James Steve Cothern and Lee Ingalls), and participated in our Thanksgiving meal and our Christmas celebrations and services.

Food was always a topic of conversation among the troops. We all missed cold drinks, hamburgers, pizza, and the usual staple of fast foods that we had back home. I remember one night in the battalion command center, a group of us were sitting around talking about home when one of the guys brought up the goat-grabs for which Saudi Arabia was so famous. "Hey, we used to do that all the time when I was growing up in South Texas," I said. "Goat is a delicacy and a staple back home. I loved it while growing up."

"Really?" asked Doug Robinson, our executive officer.

"Oh, yeah. *Cabrito*! That's good food. The best!"

Well, wouldn't you know it, within a few days Doug dragged me over to the field behind the kitchen tent where somebody had parked a tank. "What is your tank doing over here?" I asked.

"Oh, you've got to come see this, sir," Doug replied.

When we walked around back, I could see that the gun tube was elevated, there was a dead goat hanging from it, and the mess sergeant was dressing it.

"Doug! What the heck is this?" I asked.

"Well, sir, we were out training and—*I'll be damned, sir*— this goat just ran into my Hummer. And I just couldn't leave it out there, sir."

After everybody stopped laughing, I looked at Doug and said, "You know you're going to get us into trouble, don't you? Did you find the Bedouin to pay him for his goat?"

"No, sir. It was just a herd out there."

"Hmm. Well, do you guys know how to cook it?" I asked.

"Yes, sir. I think so," replied the sergeant. "I was planning on making a stew."

"Well, I guess it's okay," I finally relented. "Serve it up to anybody who wants it." Of course, most of the guys wouldn't touch it. But for me, it was just like having a holiday meal back home. Only this time, I had more than my fill.

Shortly before Christmas that year, I received another present that turned out to be better than the goat. We got a call from brigade headquarters that the new M1A1 tanks were available and they could be delivered to us if we felt we had time to make the transition. At that point, it looked like we weren't going to displace from our current assembly areas until at least mid-January, so our division commander figured we had just enough time to transition and train our troops on the new vehicles. Our old tanks, the M1s, had 105-millimeter guns. But these new M1A1 tanks sported 120-millimeter guns—and that was a significant increase in firepower. Doug and I were intimately familiar with the M1A1 and there was no doubt we could integrate the new tanks in the time available. But I was really gratified when the new ammunition started arriving as we were shooting our M1A1 qualifications.

"Look, sir!" said my master gunner. "These are those new sabot rounds. You know, the ones with greater kinetic penetration capabilities."

"Right, I know exactly what they are," I replied. "Fantastic!"

What we were receiving at that moment was the ammunition that had been developed through the future task force that I had served on back at Fort Knox. We had been looking fifteen to twenty years down the road, and proposed the expedited development of a new kinetic energy round by 1988 (KE-88) in order to stay abreast of Soviet armor protection enhancements. Even though the Army had put a rush on development and manufacture, KE-88 was just starting production. For me, it was a very satisfying moment, because I had personally participated in this improvement. For my soldiers, the new tanks and ammo represented both upgraded firepower and protection. So we were all feeling fairly confident as we approached the moment of combat.

From September through November our theater logistics

system started to flow and gradually our fuel, ammo, and spare parts stockpiles were built up. We received tents, cots, mail, and better food. To this day, I still remember that first hot meal in late October or early November: steak and eggs. And when the division lifted the restrictions on combat vehicle movement, I immediately started combat maneuver training of our entire team. These exercises were crucial because many of our troops had never experienced training under desert conditions. We began by training companies and platoons. When I felt the battalion was ready, I conducted a training exercise that involved deploying every single person and every piece of equipment we would take into battle. If a tank or Hummer wasn't running, we would hook it up and tow it. Once we had uncoiled from our assembly area, we deployed every element into its combat formation, in march order, and spaced according to the correct doctrinal distances. Every leader would move in the vehicle they expected to use in battle. Then we moved a distance of approximately thirty to forty kilometers.

About halfway through the exercise, I asked Doug Robinson to take over command of the formation. Then I left my tank, jumped in a Hummer, sped ahead of the formation and drove up on top of a nearby hill, perhaps 150 feet in elevation, so I could have a broader view of the formation. I was unprepared for the emotion I felt as I looked out through my binoculars. Every vehicle and all 1,200 soldiers under my command were headed directly toward me—and it hit me that the orders I would issue would have an impact on all their lives. So I lowered my head and said a short prayer: "Please God, give me the wisdom and the courage to make the right decisions, that I may keep these brave soldiers alive in battle. Amen."

IN LATE DECEMBER, MG McCaffrey gave us the order to break camp and move 200 kilometers to the northwest—past Kuwait up near Saudi Arabia's border with southern Iraq. "Be prepared to fight immediately upon arrival," said McCaffrey.

By now the United Nations had given Iraq a deadline of January 15, 1991, to withdraw peaceably from Kuwait. The Bush administration had also built a strong coalition of thirty-five nations that had agreed to participate in removing Iraqi forces by military force, if necessary. Approximately 660,000 troops had been assembled (74 percent, or 500,000, were American) and $56 billion had been pledged to finance the war (most coming from Saudi Arabia, Kuwait, and other Persian Gulf countries).

As we prepared for the move, I issued a directive that every soldier be allowed only one duffel bag on their combat vehicles. Every other bit of space was for our combat needs (ammunition, food, tools, and spare parts). Of course, after nearly four months of deployment, many of our soldiers had received all kinds of "stuff" from back home. But we simply could not haul it all with us into battle. So everybody got fair warning that there were going to be inspections, and anything found in excess of that one duffel bag would be thrown out. In the end, we wound up digging large holes in the desert and dumping in all kinds of personal equipment.

Upon arrival at our attack position, the living conditions became expeditionary once again, just like they had been in September. As we waited for the order to attack, our standard operating procedure was that in order to minimize our exposure to any enemy artillery fire, nobody was to sleep above ground in vehicles or tents. We dug foxholes and slept in them. My personal foxhole was five feet deep and three to four feet across, with a little tarp thrown over the top. Every night, as I went in, I searched to see if there were any scorpions or spiders inside. We were back to eating MREs, doing laundry out of buckets, and using trench latrines and makeshift showers or having sponge baths. And even though it is a soldier's right to complain, I heard very few grumblings from the time we had left our assembly areas.

It wasn't long before all of the division leadership was summoned to MG McCaffrey's headquarters for a final briefing on the division attack plan. Fighting would begin with massive air strikes to soften up Iraqi defensive positions. The main attack would come out of Saudi Arabia directly into Kuwait with the full force of U.S. VII Corps (1st Armored Division, 3rd Armored Division, 1st Infantry, 1st Cavalry, and 11th Aviation Group). The 24th Infantry would be part of the U.S. XVIII Airborne Corps's sweeping "left-hook" through the desert of southern Iraq. We would be the reserve battalion of the 197th Infantry Brigade, which would be positioned farthest to the left during the division attack. The French 6th Light Armored would be on our left protecting our flank. The 24th Infantry was the main effort in the XVIII Airborne Corps mission to execute what came to be known as the "Hail Mary" concept. We would slide to the north undetected, attack into Iraq and cut off Highway 8 (the main road between Kuwait, Basra, and Baghdad). This road formed one of two critical lines of communication crucial to the survival of the Iraqi Army in Kuwait. Our brigade objective was an area astride Highway 8 south of Nasiriyah along the Euphrates River, not far from Tallil Air Base.

On January 17, 1991, Operation Desert Storm officially began with massive air strikes on Baghdad and other key military positions across Iraq. More than 1,000 sorties per day were flown, and thousands of Tomahawk cruise missiles were launched from Navy ships in the Persian Gulf. The destruction that rained down on Iraq was astonishing—and it led Saddam Hussein to declare that "the mother of all wars" had begun.

In 2-69 Armor, we all knew the order to attack would come any day now. The anticipation was high, and the tension barely tolerable. So I was completely unprepared when word reached me on February 7 that my father had died back in Rio Grande City. MG McCaffrey contacted me, personally, and asked if I wanted to go home. If I did, he said, it would be all right with him.

I had to drive three hours to get to a telephone so I could call home. My brother told me that Dad had passed away quietly on February 2, my mother's birthday. They had tried to contact me, but couldn't get through, so they went ahead with the funeral and had buried him the day before. I called Maria Elena and talked things over. "There's really nothing for me to go home for," I said. "They've already had the funeral."

"Then do what you feel in your heart is right," she said.

I knew what was right.

We were on the verge of launching this attack.

I belonged with my men.

The 2nd Battalion, 69th Armor (designated Task Force [TF] 2-69) launched our attack on the afternoon of February 24, 1991, and crossed the Iraq border headed northeast. Our movement was relentless with continuous operations day and night. If we slept, we slept in our tanks while they were moving. We only stopped to refuel and change drivers. "We will not stop unless absolutely necessary," I had said to our officers and NCOs. "You will continue to attack. If your electronic fire control systems fail, you will continue to fight with iron sights. The only reason for you to stop is if you experience catastrophic mechanical failure. If that happens, we'll have mechanics fixing vehicles on the move, if possible."

We encountered only minimal resistance that first night. One of our company commanders reported that he was taking fire on the left flank, but was not quite sure what it was.

"Return fire," I responded.

"But, sir, we don't know who's firing at us."

"Return fire," I said again.

There was a long pause on the other end. Finally, the young captain said, "Sir, what if there are women and children there?"

"Are you taking fire, Captain?"

"Yes, sir. We're taking machine gun fire."

"Then return fire. That's what you have to do. We'll figure out the rest later."

He returned fire and we continued pressing toward our objective. We never did learn whether it was a lone Bedouin or an enemy position.

All that night and through the next day, we pressed forward as the reserve unit trailed our two lead infantry battalions (Task Forces [TF] 1-18 [1st Battalion, 18th Infantry] and 2-18 [2nd Battalion, 18th Infantry]). By evening on February 25, we had reached our initial objective about 235 kilometers inside Iraq, and were moving through the Shamaya Desert. As darkness fell, we slowed down considerably due to a rainstorm that created near zero visibility. As we moved north, we encountered an enormous *wadi* (a desert canyon or gulch).

At that point, our scouts had to go forward and map out a route so that we could get our task force and the trailing elements of the brigade (the brigade tactical command post, field artillery, medical, fuel) to our objective on time. So I called in the scout platoon leader and gave him his mission. "Find a way through the wadi," I said, "and if you encounter the enemy, avoid contact, but mark their positions so we can later take them out."

I distinctly remember thinking back to my tour of the Battle of the Bulge, and the importance of small-unit leadership, when I asked this young platoon leader if he had any questions. He was about three years out of West Point, of average height and a slight build, but a great leader and the best lieutenant in the battalion. Turning to me with a look of concern, he said, "Sir, we might get into a fight."

"Yes, that's right," I replied.

"We might lose people."

"That's exactly right, Lieutenant."

"Well, what am I going to do, sir?"

"Son, you are going to accomplish your mission."

When I looked into that young man's eyes, it suddenly occurred to me that this was the first time I had ever issued an order sending my soldiers into a situation where the possibility of death was very real. For the rest of my career, to my very last order in uniform, I always thought back to that young lieutenant, and then asked myself the question, "Have I done everything possible to ensure that my soldiers are properly trained to win in battle and to ensure that these young soldiers will come back safely?"

We continued the attack and followed our infantry battalions into the Euphrates River Valley, where our attack slowed considerably due to the swampy terrain (called *sebkha*). Still trailing in the reserve role, we were on the left flank behind TF 2-18, while TF 1-18 was on a separate route on our right flank. As the sebkha steadily worsened, TF 2-18 was forced to move single file at a snail's pace to try to get safely through the area. Eventually, the entire attack stalled.

Shortly before midnight, I received a call from our brigade commander. "Ric, can you redirect your battalion to link up with Task Force 1-18 on the right flank, execute a forward passage of lines and assume their mission?" he asked. "They have been in contact with the enemy and now their tanks are running out of fuel. We have to cut off Highway 8 by dawn."

"Yes, I can," I responded. "We have plenty of fuel in our tanks and my fuel trucks are full. I'll link up with TF 1-18, execute the forward passage of lines and leave behind my fuelers to get them going again. We can be at the objective by dawn." I immediately directed all combat systems—tanks and infantry carriers—to move south and out of the swamp. We pivot-steered to extract ourselves from the narrow mud road, maneuvered to the south, linked up with our fuel trucks, and then moved westward to rendezvous with our sister battalion. Major (MAJ) Doug Robinson stayed behind with orders to follow along with all remaining elements as fast as he could.

After a thirty-mile movement, we reached the other task force at 2:30 a.m., completed the battle handoff, and passed into enemy territory in the middle of the night on our way to Highway 8. By morning light, Doug caught up with us, and we all arrived at our objective between Basra and Nasiriyah. No sooner had we set up our east and west defensive positions to cut off traffic than we encountered Iraqi troops coming from both directions. The enemy had no idea we were there. We quickly destroyed their vehicles, and the survivors surrendered. With similar engagements continuing throughout the day, we rapidly piled up hundreds of Iraqi prisoners, whom we moved to a detainment area away from the road and out of harm's way. We placed them under guard, gave them blankets and food, and provided medical treatment to those who needed it.

During a lull in the action, I went over to our collection point to make sure our prisoners of war were being treated properly. When I spoke to one of the English-speaking Iraqis, he said that their officers had deserted them days earlier—and that they were shocked to have been treated so well. "We were told that if captured by the Americans, we would be executed immediately," he said.

"No, that is not what we do," I told him. "You're safe now."

That conversation proved to me the need to be ruthless in battle, but benevolent in victory. Once committed to battle, it is our responsibility to bring to bear every possible element of combat power to accomplish our mission and preserve the lives of our soldiers. But once victory is achieved, we must take care of our prisoners and treat them with dignity and respect. Sometimes that's a very difficult balance to maintain, because being benevolent in victory comes immediately after the most violent phase of a battle. But the leadership and discipline of the U.S. Army allows us to do so, and it is what makes us the best combat force in the world.

At about ten thirty that morning, the brigade commander called me in to his tactical operations center. He had received orders from MG McCaffrey, directing us to conduct a raid on Tallil Air Base, which was about seventy kilometers away. "Major General McCaffrey wants to send a very clear message to the Iraqis that we have combat forces operating in their rear areas," said the brigade commander. "Your mission is to go into the air base and create havoc. Destroy everything you can—but don't get decisively engaged. You have to be back by dusk, refuel, and rejoin our attack. Can you launch the attack by noon?"

"Yes, sir, that's a very short timeline but I believe we can do it," I replied.

"Good. You must execute as soon as possible. You must also maintain your defensive positions. We can't give up the highway."

As soon as I left the brigade command post, I called my company commanders and the task force operations center to give them a warning order. The company commanders were waiting for me by the time I arrived. I placed a map on the front slope of the tank and we developed our maneuver plan. One infantry and one armor company team, our artillery battery, and the battalion scouts would conduct the raid on Tallil. I would personally lead a force of over 200 soldiers, while Doug assumed command of the remainder of the battalion and hold our defensive position on Highway 8.

Our maps and satellite images were outdated, so when we arrived at Tallil, we found the entire perimeter of the air base surrounded by thirty- to forty-foot berms. Our original plan was to enter from the south. Instead, we had to maneuver around the berms until we arrived at the main entrance. As we circled to the north, our formation passed a group of Bedouins and their children. Standing there with herds of camels, sheep, and goats, they began pointing and screaming at us. We also passed an extensive array of fighting positions in which the Iraqis had placed their Soviet-built MiG fighters for protection.

The berms kept us from being seen and muffled the sound of our tanks, so when we attacked the main gate, the Iraqis were startled to see us. Our tanks streamed into the compound shooting anything that moved. The enemy defenses were easily overwhelmed, although we continued to take small-arms fire. With the constant pinging of the bullets against our armor, we made a big sweep through the base and shot up buildings, hangars, trucks, and aircraft (including some helicopters and C-130s). On the way out, one of our tanks got hit and its engine started to burn. I quickly gave the order for that crew to get the tank as far outside of Tallil as possible and then evacuate. After the crew got out safely, we fired two rounds into the M1A1 so it could not be used against us later. As we exited the compound, I gave the order to destroy as many Soviet MiGs as possible while we withdrew.

We arrived back at our defensive position on Highway 8 at 6:00 p.m., and Doug Robinson was waiting with multiple tankers to refuel us. We had practiced this refueling process endlessly so that we could refuel the entire battalion in only eight minutes. Move 'em in, fuel 'em up, send 'em out, and back into battle. Doug had the rest of the battalion ready to continue the attack, so when our tanks were filled, we all headed south along Highway 8. There was no break for the Tallil raiding party. At this point, we were 370 kilometers into Iraq.

During the middle of the night, while the battalion stopped for a short rest, one of our soldiers reported hearing the groans of a wounded soldier. When we searched the area, we found an abandoned Iraqi who had lost both legs and an arm. We picked him up and kept him alive for a day and a half until we could evacuate him by helicopter. During that same rest break, I got out of my tank to stretch and take a look around. It was a dark night, but about five meters away, I observed what looked to be the shape of a man lying in the middle of the road. But when I walked up there, I was shocked to see nothing more than a spot. This was an Iraqi soldier who had been killed and then his body run over by everything that was traveling up and down the highway. It was an awful reminder of the horror of war that will stay with me forever.

Shortly after we continued our attack down the road, word arrived (at 2:30 a.m.) that a cease-fire was going to take effect at daylight. Meetings had been held the day before in Kuwait, and coalition forces had agreed to allow the Iraqis to withdraw as long as hostilities ceased. So we stopped right where we were and immediately set up temporary defensive positions. In the morning, we expanded our area of operations and, on orders, began searching for stockpiles of chemical or other weapons of mass destruction. While we found no WMDs, we did uncover several logistics sites and a number of very large conventional weapons stashes. Most of the weapons were so new that they had not even been unwrapped. During our afternoon situation update, one of the NCOs brought a beautiful Kalashnikov pistol and gave it to me. "Here, sir," he said. "We found this in the weapons stash and wanted you to have it."

"Well, thanks, Sergeant," I said with a smile. "But you guys are going to get me in trouble for having this. I don't believe we will be able to keep weapons as war trophies."

I kept that pistol while we were in combat, but turned it in once we got back to Saudi Arabia. As for the other weapons found, we soon received an order from division headquarters to destroy all logistics and weapons sites. And that resulted in some unbelievably huge explosions when the ammunition in the sites ignited.

Within twenty-four hours of the cease-fire, elements of the division became involved in a number of engagements with the enemy as they withdrew north along Highway 8. As a result, MG McCaffrey ordered an offensive. My orders were to immediately move all combat elements of my battalion southeast along Highway 8 to an attack position, assume the role of division tank reserve, and be prepared for commitment to battle. The sense of urgency in that order left no room for delay, and we left so fast we didn't even bother to take down our tarps. Major fighting took place on part of the road between Basra and the Kuwait border, which later became known as the "Highway of Death." After the war, there was a great deal of controversy associated with this battle, in part because it was such a lopsided victory for the coalition. But from my perspective, having monitored the battle, I believe MG McCaffrey made the correct decision to attack.

After holding in place for several days, we received the final cease-fire order on March 3. "Enemy soldiers or their vehicles are not to be fired upon," read the order. "Any enemy unit now has the authority to move across our friendly lines. We are anticipating no further engagements."

Task Force 2-69 returned to battle positions somewhere between Basra and Nasiriyah, and then moved five to ten kilometers south of the highway. This area had seen heavy fighting, and was littered with bodies, equipment, and minefields. Our assignment since the cease-fire declaration was to police the battlefield. "Collect, bury, and mark," we called it. Collect the dead, put them in temporary burial sites, and mark the coordinates of those locations so they could be provided to the Iraqis later. As we did so, heavy plumes of black smoke drifted over us and choked the air. It was from all the oil wells that Saddam Hussein had set on fire in Kuwait.

The next day we received orders to prepare for redeployment. That meant finding a suitable position for the battalion to consolidate and reorganize, which wasn't as simple as it sounded, because there were minefields all over the place. So we sent out our advance parties to scout an area farther south. Their task was to identify a suitable location for us, determine a route to get there, and then mark it for daytime occupation. The scouts did their jobs well, but unfortunately, we were delayed and darkness set in. As we proceeded along, following our advance party, it became more and more apparent to me that we were not really aware of our location. From my perspective, it appeared almost as if we were stumbling around in the dark. So I finally said, "Stop! Everybody just stop! This doesn't feel right."

A few minutes later, we heard an explosion. But we didn't know if it was an attack or just a wild animal tripping a mine. After we made sure that none of our men were hurt and that we were not under attack, I issued another order. "Okay, nobody move. Just stay in your vehicles. We'll figure out what's going on in the morning."

We all slept in our combat systems that night and, sure enough, at first light it became apparent that the lead element of the battalion, about 150 people and twelve tanks, had driven right into a minefield. So we backtracked every vehicle out of there on the exact same path we had made going in. It took us hours to do it, but we didn't lose a man or sustain any damage.

Later that day, as we prepared for redeployment, our battalion leaders began to speculate: Where were we going to go next? We were planning on heading to Baghdad. It made sense. We were already in the country, there was no significant opposition, and we were kicking ass. Why wouldn't they let us continue the attack into Baghdad and force a complete surrender. Of course, that was our thinking at the battalion level. In Washington, our political leadership had different ideas. Within a couple of days, we were given the order to withdraw from Iraq back to Saudi Arabia, along virtually the same route we came in on. Task Force 2-69 never set foot in Kuwait.

The Persian Gulf War officially ended on March 3, 1991 (on the day we received word of the official cease-fire), when Iraq formally agreed to accept all the terms laid out by the United Nations—including a pledge to dismantle and not pursue weapons of mass destruction. Kuwait had been liberated and the United Nations, the United States, and the coalition had realized all goals. A year later, on March 30, 1992, President George H. W. Bush answered the question as to why we did not go on to Baghdad to take out Saddam Hussein:

> We certainly had the military capability to go on to Baghdad. But once we had prevailed and had toppled Saddam Hussein's government, we presumably would have had to stay there and put another government in place. And what would that have been: a Sunni government, a Shia government, a Kurdish government, or another Baathist regime? How long would U.S. forces have been required to stay in to prop the government up? And how effective could it have been if the government we put in had been perceived as a puppet of the U.S. military? To involve American forces in a civil war inside Iraq would have been a quagmire, because we would have gone in there with

> no clear-cut military objective. It's just as important to know when not to use force as it is to know when to use it.

By the second week of March, Task Force 2-69 was back in Saudi Arabia preparing for a rapid redeployment of all our people and equipment. Fortunately, everything went right on schedule, and I arrived back at Fort Benning just in time for my daughter Bekah's first communion—on Easter Sunday.

MY RETURN HOME ON Easter was a particularly symbolic and important moment for me—not only to celebrate the resurrection, but also to give thanks that all the men under my command had returned safely to their families. It also gave me pause to reflect on my time in Iraq. I had spent a week waging war against an enemy in an area of the world known as the "Cradle of Civilization." Five thousand years ago, people living there used the fertile valleys of the Tigris and Euphrates rivers as a supply of water for irrigation. They created the very first society, the very first big city, and the earliest known public code of laws and values. The ruins of that big city, Babylon, are today located barely ninety kilometers (fifty-six miles) south of Baghdad on the eastern side of the Euphrates.

The Tigris and Euphrates river valleys were quite similar in nature. Their banks were four to six feet high, the flowing water was dirty, and there was quite a bit of vegetation on the sandbars. And that provided quite a stark contrast to the dry, brown, austere desert landscape. The landscape reminded me of the Rio Grande River Valley in South Texas. How ironic, I thought. The Cradle of Civilization looked just like home. And that's where the enemy was. But I wondered: Did they have to be the enemy?

CHAPTER 5

In the Box

Prior to Operation Desert Storm, our parent unit (the 197th Infantry Brigade) had been a separate mechanized unit, with long-term plans for the Army to roll us into the 24th Infantry Division. But in the wake of the war, a decision was made to expedite the process, so the 197th was renamed the 3rd Brigade and became part of the 24th Infantry, and we went through all the mechanics of restructuring logistics, maintenance, supply, and command and control operations.

In addition, our soldiers went through all the phases associated with an army's return from battle. It began with the euphoria of returning home victorious. There were parades and welcome-home ceremonies that included speeches, family gatherings, and flag waving. And then we had to help our soldiers refocus and adjust both professionally and personally to a peacetime environment. That turned out to be a fairly significant challenge because, while in Iraq, they had been entrusted with life-and-death decisions and had, indeed, performed in an exemplary manner. But it was a tremendous letdown to be forced back into a restricted garrison environment where peacetime training procedures were dictated to them. Our battle-proven soldiers and leaders did not react well, for instance, to having a firing range safety NCO tell them how to clear their weapons after conducting small-arms training.

It became obvious to me that we needed to make some changes to figure out a way to treat these warriors more appropriately once they returned home. Our leadership struggled with this issue for the first six to eight months back at Fort Benning while we made some desperately needed changes. We also came to realize that the most important thing we could do was to be patient. We simply needed to give our men and women the time required to readjust, both physically and mentally.

Within a few months of our return from Iraq, there was a noticeable spike in family and marital issues. So the Army provided counseling and support. Overall, the reintegration of returning warriors was haphazard, because the Army did not yet fully recognize the importance of the family unit. We also experienced a couple of cases of Gulf War Syndrome, an unusual illness specific to veterans of Operation Desert Storm. One officer had a brain problem, another a blood infection. The medical community tried to determine if and where they might have been exposed to chemical agents while in Iraq. The only thing they determined was that it might have happened after the cease-fire when we were blowing up those large ammunition dumps. There was a possibility that chemicals might have gone unnoticed and were subsequently released into the air during the large explosions. Other than that, there were no indications that chemical weapons were used against our troops.

For a period of time after our redeployment, 2-69 Armor also experienced some discipline problems. Not showing up on time and some general nonchalance when performing duties were things that occurred sporadically while our soldiers struggled to reintegrate into the peacetime garrison environment. I did not expect, however, to be so deeply impacted when one of my men was arrested for drunken driving. There is a special bond that forms with people who've been in combat together. And this one captain, who had been in charge of our tactical command post, had not only done a terrific job in Iraq, but was someone who had great potential. Unit policies required mandatory actions in all DWI cases—and I was really torn about how to handle it. Do I follow regulations to a tee? Do I take care of my battle captain, and let it slide? What do I do?

I ended up going to my brigade commander to ask for advice. "Sir, because of my personal relationship with this young man," I said, "I may not be objective."

"How can I help you, Ric?" he asked.

"Well, I want you to either take this case from me, or monitor my execution of it and give me your best advice on whether I'm doing the right thing for the Army."

"Okay. Do what you think is right, and I'll monitor."

After a lot of soul-searching, I decided to put a letter of reprimand in the captain's official file. He was devastated when I told him about it, in part, because he had believed I was going to cut him some slack. As a result of my action, he could stay in the Army for another couple of years, but he was going to have to leave after that. I hated to do it, and it tore me apart as a person. But I really felt it was the right thing to do, not only for the Army, but for that young man's personal well-being. I had seen what alcohol abuse had done to my father, and those memories played no small part in formulating my decision.

When my command tour at Fort Benning came to a close in June 1992, I had already been selected to attend the U.S. Army War College in Carlisle, Pennsylvania. Three hundred of the top officers in the Army (along with some from the other services) were chosen to participate in this next level of professional education. It was a one-year program that focused on preparing us to begin operating in a national strategic or theater strategic environment as staff officers. While the course curriculum was interesting and stimulating, one of the key insights I took away from my Carlisle experience was a clear picture of the egotism among officers who believe they are among the chosen few anointed to become generals.

There were two major events that occurred during an officer's time at the War College. In the early part of the year, selections for promotion to full colonel were announced. At seven thirty in the morning, they brought us all into one big auditorium and kept us there for about five minutes while they posted the selection list on the bulletin board outside. Then they released everybody all at once to look at the list.

The levels of expectation varied among the group. Some, like me, expected nothing. Others believed, with their whole heart and being, that they would be on the list and, if they weren't, it would be the end of the world. I remember the reaction of one lieutenant colonel, who was absolutely incredulous that he wasn't on the list. "This is a mistake—a big mistake," he said. "I'm going to make some calls. I was assured that I would be on the list."

After the crowd dispersed a little bit, I went up to take a look, and was elated to see my name posted. I had been selected for a "below-the-zone" promotion to colonel, which meant I had been chosen early—and that was rare. I felt blessed.

The next big cut came a couple of months later, in the spring, when the list of those selected to command brigades was released. The reactions were pretty similar, only this time there was a colonel who was so outraged that he threatened to leave the Army. "This is not the brigade I want," he said. "If they don't change it, I'm going to quit!" Well, they didn't change it and he quit. At first, many of us felt that the Army should have forced him to serve out his obligation. But then we came to the conclusion that it was probably better to get rid of such a self-serving officer.

Most of us who were selected for brigade command were very grateful and very happy to go wherever they wanted to send us. Of the four below-the-zone armor officers eligible for command, only one was selected—and that was me. Personally, I couldn't believe it. This was 1993, and both my promotion and my transfer to Fort Riley, Kansas, were to take place in mid-1994. For the intervening year, the Army decided to send me to Washington, D.C., to work at the Inspector General's (IG) office.

I left Carlisle with a concern about how much true dedication to the Army and to our nation really existed in the officer ranks, and how much it was just about personal ambition. And that caused me to look back and reflect. I remembered the conversation Maria Elena and I had about my dream to command a battalion. I had now done that in combat. I had achieved my dream, and now the Army wanted me to move up and command a brigade. So I decided to continue to serve in the Army as long as the nation wanted me.

Once in the Washington, D.C., area, Maria Elena and I rented a house in Burke, Virginia, about thirty minutes from the Pentagon, where I worked. During my early years as an officer, I had learned to report violations as I saw them, and I knew if one was against a general officer, the Inspector General in Washington would investigate it. Now I was one of those investigators.

Every single allegation, whether anonymous or not, was thoroughly investigated. Our team of eight to ten colonels and lieutenant colonels had access to virtually every database you could imagine—hotels, airlines, credit cards, telephones, you name it. The process was disciplined and thorough. We began with a preliminary inquiry. If there was no substance to the allegation, we would simply close the case and the subject investigated would never even know about it. However, if we found substance, we launched a formal inquiry—with witnesses, experts, subpoenas—the whole ball of wax. Usually, the last person we questioned was the general under investigation. And by then, we pretty much knew what he had or had not done, and what the result was going to be. In fact, I don't recall a single situation when the subject ever gave us any new information that changed the course of the inquiry.

The objectivity of the process, which had an explicit goal of protecting both the individual and the Army, convinced me of one very important thing. If an allegation was ever made against me, the worst thing I could do was lie, because even the worst IG investigator would find the truth. There were simply too many people, too many cross-investigations, and too many ways to figure out what a person did or did not say and do. It was the same lesson my parents had taught me when I was a kid. "The worst thing you can do is lie to me," they said. "And if you lie, I'll find out and you'll get whipped twice!"

During my year at the Inspector General's office, I gained a pretty good understanding of why generals screw things up. It was usually for one of two reasons. They were either ignorant of a regulation or policy or they were arrogant and figured they could get away with anything. Those generals who made a mistake were fairly easy to deal with. We would find that their intent was right, but they simply did something wrong. So they would get a hard slap and it would be all over. However, those generals who believed they were God's gift to the Army could find themselves in real trouble. I remember one general officer who sat in front of me and sneered his contempt for the entire process. "Look, if this person was still in political office," he said, "I wouldn't be sitting here. Because he'd be protecting me and you wouldn't be able to touch me."

I never forgot that ugly side of generalship. It taught me how *not* to be a general, if I was ever fortunate enough to rise to that level of leadership.

After my year was up in Washington, D.C., we went home to South Texas for a couple of weeks before making the drive to Fort Riley. While I was visiting my mother, she casually mentioned that Benito Gonzales, the father of the big family that lived across the street, wanted to talk to me. "Sure, Mom. I'll go over and see him," I said.

"Ricardo!" said Mr. Gonzalez, who did not speak English. "*Pasale y sientate aqui, dejame ensenarte algo!*" ["Come in! Sit here, I want to show you something."] He took me over to his kitchen table and pulled out a box that had medals and military records in it. "Look, I was in the Army," he said. "I fought in World War Two for the United States and I am very proud of my service."

Mr. Gonzalez then started telling me about his experiences. It turned out that he had served in a Mexican-American unit that had been put together from the Rio Grande Valley. "Our officers were all gringos," he said. "Some of our sergeants could speak English so when the gringos gave us orders, our sergeants would translate. In Europe, when they wanted us to attack, they would point in the direction they wanted us to go, say '*Ataquen*' ['Attack!'], and we would attack."

I never realized what the Hispanic elders in our community had gone through during World War II. Benito Gonzalez had deployed to Europe and participated in the campaigns in France, Germany, and Belgium, including the Battle of the Bulge. There was a World War II combat veteran living across the street from me the entire time I was growing up and I never knew it. He had proudly served our country and then returned to the poverty he had left behind.

I spent several hours with Mr. Gonzalez, asked him questions and listened to his stories. When it was time to leave, I thanked him for his service to our nation and told him that all Americans owed him a debt of gratitude. With tears in his eyes, he saluted me and told me how proud he was of the man I had grown up to be. I had served in Desert Storm three years before, but I knew who the real hero in the room was.

On the drive up to Kansas, Maria Elena, the children, and I stopped for a weekend to see MG Richard Boyle and his wife Fran, in Kilgore, Texas, where he had retired several years before. We had stayed in contact ever since we first met back at Fort Bragg when he was assistant division commander and I was his young aide-de-camp.

MG Boyle and I reminisced about the old days, and he was particularly interested in my experiences during Desert Shield and Desert Storm. Near the end of the weekend, I asked him what retirement from the Army had been like—and he surprised me with his answer.

"Well, you know, Ric, it's been very hard," he said. "All the things you do, all the wonderful blessings you receive in the uniformed service of our country . . . it's all great that you've done it. But the Army, the system, doesn't really care."

"Really, sir?"

"Yeah, really. Once I was retired, the Army didn't care that I had given all those years. It was almost as though I had never been in the service. In the end—and you must never forget this, Ric—in the end, the only thing you'll have left is your family, your friends, and your faith."

I thought a lot about that piece of advice during the 600-mile drive from Kilgore to Fort Riley. MG Boyle had been the first senior officer who really took an interest in me. I liked him very much, and was disturbed that he did not seem happy in retirement. The main impact that had on me was to reaffirm the pact that Maria Elena and I had made many years before. Short of war, our family came first.

It turned out that Fort Riley would be one of the best family environments we would experience during my military career. Driving onto the post, we all marveled at the magnificent limestone buildings that lined the streets. But it was when we drove around Schofield Circle with its beautiful green parade field that Maria Elena and the kids really got excited. On Schofield Circle, there stood huge limestone houses occupied by colonel and lieutenant colonel commanders. One of those houses would be our home for the next two years. It contained 6,400 square feet of living space, including a huge living room and kitchen—and each of the kids would get their own bedroom. When I walked in for the first time, I couldn't help but think back to Korea and the tiny little apartment that my wife and daughter had endured. We had come a long way.

Twenty-one years earlier, I had been commissioned a second lieutenant at Fort Riley. Now I was returning to command 3,200 soldiers in the 2nd Brigade (Dagger) of the 1st Infantry Division. From the very first day, two things occupied my mind. First, the history of Fort Riley, which began in 1855, and the 1st Infantry Division history were evident everywhere you turned. The Civil War, Custer at the Little Bighorn, Pancho Villa and patrolling the Mexican border, World War II, Korea, and Vietnam were all part of the heritage there at Fort Riley. And I was determined to uphold that legacy at all costs. Second, I never forgot that when, as a young lieutenant I had walked into Fort Bragg the first time, it was during an alert for the Yom Kippur War—and sixty days after I showed up at Fort Benning, my battalion was informed we were going to war in the Persian Gulf. I didn't know if something similar was going to happen this time, but I was darn sure going to have the Dagger Brigade prepared if we were ordered to deploy.

During our training, I received a call one afternoon from the MPs (military police) that there had been gunfire in the barracks behind our headquarters. I immediately ran across the parade field to check out the situation and learned that one soldier had killed another with a shotgun and was holding the MPs at bay. As soon as I had that basic information, I got to a phone and called the division commander, MG Randy House, and informed him of the situation.

"Okay, Ric," he responded calmly. "What help do you need from me?"

"Well, sir, at this point, I don't know that I need any help from you."

"Okay, sounds like you've all got control of the situation, and that everything is being handled correctly. When you have time, call and give me an update."

Back at the barracks, I conferred with the battalion commander and the MPs and we came up with a plan to end the standoff. No sooner had they started to move in when the young man with the shotgun turned it on himself and committed suicide.

That was a very disturbing experience for all of us. But I must admit that my boss made it easier for me. Knowing that it was a life-and-death situation, and that there was tremendous pressure, MG House did not interfere and add to it. Rather, he offered support, and then showed his confidence and trust by letting us handle the situation ourselves. I really appreciated what he did and told him so later.

Randy House was a Texan. He had graduated from Texas A&M University, and was a true Aggie. I discovered very quickly that he was an innovative and brilliant leader who focused on training soldiers to make sure they were prepared for war. His entire philosophy was in lockstep with my own, and I watched in awe as he single-handedly changed the conventional wisdom about training at Fort Riley. House's expertise was honed when, as a colonel, he had been involved in developing innovative training procedures at the National Training Center (NTC).

MG George Joulwan had exposed me to the methods of the NTC with his massive exercises at Hohenfels when I was in Germany with the 3rd Armored Division. And now Randy House would not only educate me on NTC history and methodologies, he would personally be involved in mentoring me as my brigade prepared to go through the rigorous two-week NTC program.

Located in Fort Irwin, California (about 100 miles northeast of Los Angeles), the NTC was formed in the early 1970s after the Vietnam War when Army leaders decided they had to fix training procedures. Participating troops find themselves in a simulated fighting environment, going through both offensive and defensive scenarios against a professional enemy. The live-fire exercise is exciting and challenging. Units and individual leaders are monitored and observed twenty-four hours a day, seven days a week. At the end of the experience, evaluators conduct an after-action review to discuss all aspects of unit and leader performance. It is designed, in part, to provide senior leaders insight into their natural tendencies under the pressures associated with the battlefield.

MG House's expertise and experience would prove invaluable

in helping to prepare both brigades of the 1st Infantry Division to participate in the NTC's exceptional training program. The 1st Brigade would go in January to February 1995 and we (2nd Brigade) would go eighteen months later near the end of my tour.

In December, the colonel in charge of the 1st Brigade came up to me and asked us to support him. "Ric, we could use some help," he said. "Can you spare some people and equipment?"

"Of course," I responded. "We'll give you whatever you need." And we did. We supported them totally and without hesitation, because we were part of the same team.

In the meantime, the Dagger Brigade began our own training to get prepared for war—just as we had done when I was a battalion commander back at Fort Benning. Interestingly enough, on this issue, my reputation had preceded me. Some of the officers under my command had heard that I was unyielding when it came to physical training (PT). "Hey, when Sanchez comes in," they were told, "you are never going to cancel PT because of the weather. And you better plan to go to the field with everything you own."

When a couple of my battalion commanders mentioned what they had heard, I explained my basic philosophy to them. "Look, if we're fighting," I said, "we have to focus on training the way we will fight. We're not going to call the war off just because it's hot or cold outside. We better figure out how to survive and fight under those conditions. In other words, gentlemen, my mantra is: 'We are going to train as we will fight.'"

Sure enough, during that first summer, I stirred up some controversy with my methods. Because of what I had learned in Desert Storm, I directed that the entire brigade dress in full battle gear when we went to the field. Of course, we were in Kansas in August, and it was hot—but not as hot as it was in Saudi Arabia or Iraq during the Gulf War. Not long after we began training, we started to have a series of minor heat-related injuries. MG House, who was ever-present during our training exercises, came up and asked what was going on. "Ric, I've heard you're having some heat injury problems," he said. "Do you want to keep training with all of your combat gear? Maybe you ought to consider scaling back a bit."

"Well, sir, we're making sure everyone is retrained. We are getting our NCOs completely engaged in monitoring our soldiers. I don't like having these heat injuries and I will never place my soldiers in a life-threatening situation. But due to your concern I'll review my standards, sir."

"Is there any way to minimize the exposure?"

"Well, maybe I can get them out of their gear for short periods of time during the hottest part of the day. But we have to continue training this way."

"Okay, Ric, I trust your judgment. But you need to be careful there are no serious injuries."

We completed our summer training and the Dagger Brigade performed extremely well. I was confident that we were much closer to being ready for war if the nation called. Fortunately, that didn't happen, but we maintained that readiness status through the next year until it was time for the train-up leading to our visit to the National Training Center in December 1995.

I could survive the hottest days in Iraq without any problem. I guess that came from growing up in South Texas. But the chill on the plains of Kansas that winter was something else altogether. Some of my battalion commanders kidded me because I had dressed up with five or six layers of clothing and looked like Nanook of the North. I was definitely prepared. But when the wind-chill factor hit minus-52 degrees Fahrenheit, we stopped training. There was just no way we could continue with those Kansas winds blowing the way they were. Since we were deployed to the field for a minimum of two weeks, we simply tried to survive in our vehicles until the winds subsided and the temperatures rose a bit.

As the time approached for deployment to the National Training Center, our brigade had equipment shortages that needed to be fixed. So we sent a note over to the 1st Brigade asking for a little help, which I thought would be no problem after we had helped them the year before. But my guys came back and informed me that we were not going to get the equipment we needed.

"Why not?" I asked.

"Well, sir, the brigade commander issued an order that no equipment was going to be lent to us."

"Issued an order?"

"Yes, sir. When we asked one of the 1st Brigade's battalion commanders to confirm, he replied, 'That is accurate. I'd be violating orders if I helped you.'"

In my mind, it was not worth the trouble to make a big deal out of this issue. So I simply worked through it. We pushed hard to fix our broken equipment, and borrowed equipment where we could from other division units before deploying to Fort Irwin, California.

MG House was present every day for the entire two weeks we were training "in the box," so to speak, at the National Training Center. For me, his presence was worth every bit as much as any exercise the Dagger Brigade went through. After one of our initial key battles, in which we were soundly thrashed by the professional opposing force, House drove over to our position at the end of the battle, got out of his Humvee, and handed me an ice-cold Dr Pepper. I had been up for thirty-six hours straight and I was dying of thirst.

"Here, Ric, take this and sit down," he said.

I sat down and took a swig of the Dr Pepper. "Thank you, sir," I said. "That really tastes great."

"Okay, let's talk," he said. "Where were you on the battlefield?"

"Well, sir, I was behind the lead battalion."

"Did you think you had good situational awareness at the critical point in the battle?"

Clearly, from the performance of the brigade, I had not been at the right place to make critical decisions. "No sir, I'm still trying to figure out where I belong during a brigade fight."

"Ric, you've just got to move to the sound of the guns," said House. "You've got to understand how far forward you can go. Usually, you don't want to be with the company and battalion commanders. But if it's absolutely critical, you might have to be. So don't be afraid to move forward. Sometimes, you might even have to be at the decisive point with a young platoon leader. If that happens, just make sure that your second-in-command understands the situation."

During the next battle, as the brigade fought through a mountain pass, I was way up front on the battlefield. And sure enough, House came up to me immediately afterward, handed me another Dr Pepper, and put his arm around me. "Ric, you were a little too far forward this time," he said. "You can't be in the breach while it's being fought. With a little experience, you will learn."

Overall, I came out of the NTC feeling pretty confident. Of course, just like most brigades, the Dagger Brigade did not fare very well against the professional opposing force. But I had learned a great deal about myself, my subordinate leaders, and my soldiers. We had all grown tremendously as a unit. I also was impressed by MG House's ice-cold Dr Pepper mentoring. He had a love for soldiering, not to mention an approach that helped me develop as a leader. His methodology was in sync with my own natural tendencies. But now, because I saw the same leadership style coming from a general whom I truly admired, I felt my own approach had been validated. A leader does not have to be abusive, obscene, or denigrating. But in order to lead soldiers effectively, he must be tough, must be demanding, and must lead from the front.

During my last few months in command at Fort Riley, I received a personal letter from GEN Barry McCaffrey. "Ric, I'd like you to come to SOUTHCOM and be the deputy chief of staff here in Panama. I look forward to working with you again." At the time, GEN McCaffrey was in charge of Southern Command, and had responsibility for the entire South American and Central American area of operations.

MG House had been making some phone calls on my behalf in order to get me a good follow-on assignment. When I saw him the next day I said, "Sir, I need to let you know that I just got a note from General McCaffrey. He's asked me to go to SOUTHCOM to be the deputy chief of staff. Maybe I should tell him no."

"Ric, I need to stop making calls for you," responded House. "You never say 'no' to a four-star."

"Okay, sir," I responded. "I guess I'll tell Maria Elena that we're going to Panama."

PART II

SENIOR LEADERSHIP IN A POST–COLD WAR WORLD

★ ★ ★

CHAPTER 6

Joint Interagency Operations and SOUTHCOM

I spent the next three years, 1996 to 1999, working as a principal staff officer in U.S. Southern Command (SOUTHCOM), one of the major joint combatant commands in the Department of Defense. This phase of my career would prepare me for my later experience in Iraq in a number of areas. First, it exposed me to the decision-making processes of national-level leadership, specifically with the Joint Chiefs of Staff and the executive branch of the national government. I regularly interfaced with the Congress and various other elements of the legislative branch. Second, SOUTHCOM helped me to understand the different cultures among the various departments of the Interagency (Department of State, National Security Council, Joint Chiefs of Staff, and Department of Defense). My dealings with the Department of State gave me an understanding of the world of diplomacy and the foreign service officer mind-set that became invaluable when I later had to deal with Ambassador L. Paul Bremer's Coalition Provisional Authority, and the subsequent establishment of the new U.S. embassy. Third, dealing with multiple South American nations and building regional coalitions provided insights into the complexities of national rules of engagement and, most important, the need to understand the impact of individual national interests on coalition operations. This experience was instructive when as commander of ground forces in Iraq, I had to forge relationships among the different Iraqi factions and coalition partners.

One of the most important experiences during my SOUTHCOM assignment was learning how to deal with unconventional leaders. The complex coalition, joint, interagency operational environment of SOUTHCOM often led to untraditional command relationships. SOUTHCOM could not issue direct orders when dealing with national leaders of the counterdrug coalitions and the U.S. Interagency. As a result, our leadership strategies depended heavily on building consensus and advancing common interests to achieve unity of effort. Appreciating the different interagency service cultures and operating constraints was crucial to getting their cooperation and commitment.

My time at SOUTHCOM took place right in the middle of a historic period that would see a seismic shift in the U.S. military's approach to fighting wars. Post–Cold War events, before, during, and after my time at SOUTHCOM in Panama and Miami set future strategies for the handling of low-level conflicts around the world. During this period of global change, an expectation was built into the nation that U.S. troops could intervene only in regional conflicts with minimal casualties for six months to a year. American leadership grew comfortable in believing that our long-term commitments in the wake of conflict would require only a peacekeeping presence. That expectation, in turn, resulted in a front-end political expediency to enter troubled areas of the world with willingness and ease, but often without a carefully crafted exit plan.

We can trace the beginning of this paradigm shift to the breakup of Yugoslavia in 1991 to 1992. The resulting ethnic violence reached a crescendo with brutal Serbian attacks on Bosnia's Muslims. While the United Nations swiftly passed a series of resolutions aimed at ending the conflict, a public debate took place in the United States over calls for international intervention. The argument seemed to have been won by Joint Chiefs Chairman Colin Powell when he argued cogently for intervening only after certain conditions were met. GEN Powell's specifications, now known as the "Powell Doctrine," included: a risk to the national security of the United States; positive international support; clearly defined political objectives; full consideration of risk to U.S. troops and their length of stay; a well-defined exit strategy; and clear approval from the American people. GEN Powell cited the Persian Gulf War and the U.S. invasion of Panama as examples of successful interventions. Throughout the decade of the 1990s, his doctrine would be tested with crises in Somalia, Haiti, Bosnia, and Iraq.

Before leaving office, President George H. W. Bush deployed 25,000 U.S. troops (on December 9, 1992) to Somalia to assist the United Nations in humanitarian aid for famine victims. By June 1993, President Bill Clinton had reduced U.S. presence to 4,200 troops for peacekeeping operations only. However, on October 3, 1993, two Black Hawk helicopters were shot down in the capital city of Mogadishu, provoking a firefight that ended with eighteen U.S. soldiers dead and eighty-four wounded. In the battle's aftermath, the body of one of the helicopter pilots was dragged through the streets by enemy combatants—an incident caught on film. Those macabre television images outraged the American public and caused the Clinton administration to send in reinforcements to stabilize the situation. By March 31, 1994, all American forces were withdrawn from Somalia.

Six months later, in September 1994, violence erupted in Haiti. This was several years after its democratically elected president, Jean-Bertrand Aristide, was overthrown during a military coup. When U.S. troops tried to enter the country to help with UN efforts, they were refused admittance by General Raoul Cédras, the country's dictator. Rather than provoking a military confrontation at that time, President Clinton withdrew the ship carrying the troops, and then put together an invasion plan involving 25,000 U.S. forces. A last-minute deal brokered by the triumvirate of former President Jimmy Carter, GEN Colin Powell, and Senator Sam Nunn, pushed Cédras to leave the country. Aristide was restored to power, and U.S. troops landed in Haiti without opposition. The peacekeeping force withdrew in less than a year, by March 1995.

During August and September 1995, in response to the deadly Serbian bombings of a marketplace in Sarajevo, NATO (North Atlantic Treaty Organization) launched two weeks of air strikes over Bosnia using 400 planes and 5,000 troops from fifteen different nations. This action finally brought the Bosnian Serbs to the negotiating table with Croat Muslims and resulted in a peace agreement forged in Dayton, Ohio.

In Iraq, Saddam Hussein was a problem throughout the 1990s as he sparred with the United States and the United Nations time and again. After the Gulf War, no-fly zones were established over the northern and southern portions of the country to protect Kurds and Shiites, respectively, from Saddam's air attacks. On June 27, 1993, the United States launched an attack against the Iraqi intelligence service in response to a failed plan to assassinate former president George H. W. Bush. After an "oil-for-food" program began in 1995, Saddam threatened to withdraw cooperation from UN weapons inspectors unless economic sanctions and the oil embargo were lifted. Over the next several years, UN inspectors were expelled and let back in several times; ultimately, the United States and Great Britain launched four days of air strikes in December 1998 against suspected Iraqi WMD (weapons of mass destruction) sites. Despite those attacks, Saddam Hussein continued to spar with the United Nations about allowing weapons inspectors access to Iraq.

Throughout the 1990s, both the U.S. military and the Clinton administration were adjusting to the post–Cold War turmoil. In 1994, the Army announced a modernization initiative that would integrate new technologies across the board. At the same time, a fresh plan was put in place to reduce the number of active-duty forces by 40 percent. Five years later, in October 1999, the Army Chief of Staff, GEN Eric Shinseki revealed the development of a new concept for a medium-weight force capable of deploying anywhere in the world in only four days. It would be a new high-tech Army designed specifically for twenty-first-century wars. GEN Shinseki believed that the Army had to "build irreversible momentum" in order for this transformation to last beyond his time as Army Chief of Staff.

In the meantime, the Clinton administration developed a series of strategic policies regarding when to employ military force, conditions for U.S. involvement, and how the government should manage such operations. In May 1994, for instance, Presidential Decision Directive 25 mandated getting involved in peacekeeping operations only when goals were linked to concrete political solutions—when a strategy existed that included integrated political, economic, and military solutions fit within a firm budget estimate.

Two years later, in March 1996, the administration made it clear that a well-defined exit strategy had to be developed before entering another country by force. In any case, military intervention could take place only when justified by some combination of the following seven reasons: (1) to defend against direct attacks on the United States, U.S. citizens, and allies; (2) to counter aggression; (3) to defend key economic interests; (4) to preserve, promote, and defend democracy; (5) to prevent the spread of weapons of mass destruction, terrorism, international crime, and drug trafficking; (6) to maintain reliability with the international community; and (7) for humanitarian purposes.

Finally, in May 1997, the Clinton administration issued Presidential Decision Directive 56 (PDD 56), which was designed to force interagency cooperation and synchronization among government agencies. Essentially, it ordered the Pentagon, the Department of State, the Central Intelligence Agency (CIA), and other key government agencies to work together to create a new program for educating and training personnel for peacekeeping missions, in order to reduce or even eliminate the government's bureaucratic "stovepipe" method of operations. Meanwhile, the executive branch was attempting to synchronize the work of intergovernmental agencies (the Interagency) in attaining national security objectives. Unfortunately, there were no enforcement mechanisms put in place to hold the Interagency accountable—and this shortcoming would haunt us in Iraq.

During all of these swirling changes, I found myself as the Director of Operations at SOUTHCOM, one of the few commands in the Department of Defense where joint and Interagency operations were a daily reality. All five military services (Army, Navy, Air Force, Marines, Coast Guard) and most major civilian departments (State, Defense, Justice, CIA, Drug Enforcement Administration, and others) worked together on our national counterdrug mission. SOUTHCOM's operations encompassed the entire Western Hemisphere and revolved around synchronizing, integrating, and optimizing all participating elements of the government to interdict the flow of drugs. Our mission included working with most of the countries in the Caribbean and South and Central America—their governments, their armies, and their civilian humanitarian-related agencies—to build credibility and capability. It was a challenging place to work given the lack of national focus and resources committed to Latin America.

Before going to Panama as deputy chief of staff, I had really been looking forward to working for GEN Barry McCaffrey again. On the battlefield, there was no better warrior or leader, and there was no doubt I could learn a lot by working in his command group. But that opportunity never materialized, because President Clinton appointed him to be the director of the Office of National Drug Control Policy a few months before I made it to Panama.

Upon my arrival, the Deputy Commander in Chief (DCINC) had moved up a notch to temporarily fill McCaffrey's position, which made me the de facto chief of staff, at least temporarily.

My position proved problematic, because two of the key staff officers, the director of operations and the director of intelligence, were generals.

For me, the most difficult aspect of my new position involved dealing with the DCINC, who was a Navy admiral. I found out very quickly that the staff feared him because he did not handle bad news very well. Even the generals on staff would come up to me and say things like, "Ric, the admiral isn't going to like this. Can you take it in and prep the battlefield for us?" In instances like that, I'd go in and be the first to convey the bad news and, often, I'd have a bunch of papers thrown at me. The first time that happened, the admiral cooled down and apologized. "Ric, please understand that this was not directed at you personally," he said. My willingness to buffer the staff earned their respect and helped tremendously when the new combatant commander arrived and I transitioned back to my primary job as deputy chief of staff.

GEN Wesley K. Clark showed up in the late summer of 1996 to assume command of SOUTHCOM. He was a 1966 graduate of West Point, and a Vietnam veteran who had been wounded in combat and been awarded both the Silver and the Bronze Stars. I soon found out that he had an insatiable desire for detailed information, and had the unbelievable ability to retain and use it all. GEN Clark was also an uncommonly good strategic thinker whose visions for the future were almost always compelling and inspiring. He kept us all moving with multiple approaches to our problems and, as a result, a variety of creative strategies rapidly evolved. GEN Clark's unparalleled ability to build personal relationships with the Panamanians and most Central and South American leaders made all of our jobs easier.

Maria Elena and I had hoped that our tour in Panama would be for only one year. So as spring approached, I began inquiring about the possibility of being transferred back to heavy armor in Germany. And it wasn't long before GEN Clark called me in to discuss my assignment options.

"Ric, I want you to stay another year and be the operations officer [J-3] for the command," he said. "Will you accept it?"

Remembering my earlier lesson that you never say "no" to a four-star, I immediately responded, "Sir, I had hoped to go back to Germany, but short of that I will definitely stay and be the operations officer."

Clark then smiled and shook my hand. "I can't really tell you the other news that goes along with that," he said, "but I'm sure you know." GEN Clark was telling me that after twenty-five years in the Army, I was going to be promoted to brigadier general. I was humbled, happy, and couldn't wait to get home and share the news with Maria Elena and our children.

Almost immediately, I began a transitioning process with the incumbent J-3, and forty-five days later I assumed the full responsibilities of the job. Because I would be interacting with various international military leaders, I was immediately frocked—which gave me all the authority and responsibility of a brigadier general without the pay. My actual promotion didn't take place until later in the year.

My first major task as operations officer was to move U.S. military assets out of Panama to the new headquarters in Miami. Based on a 1974 treaty between the United States and Panama, the Canal Zone would be turned over to Panama by the turn of the century and the United States would withdraw. The move of SOUTHCOM's headquarters constituted the first step in formally ending the U.S. military presence there. The official change actually took place on an airplane over the Caribbean Sea as we stood up communications in Miami and shut them down in Panama.

No sooner had we moved the command element than some members of Congress began questioning our overall budget, which was a very large amount of money. So they dispatched a delegation of staffers to be briefed on our plans, but they sent them first to Panama instead of Miami. The Army element of SOUTHCOM was still waiting for their turn in the cycle to be transferred out, and they sent a young major over to the airport to meet and update the congressional staffers. Feeling slighted, the staffers put a hold on all of our funding. When word reached me, I made plans to have breakfast with the delegation when they stopped in Miami on their way back to Washington.

"We're very sorry you weren't treated properly in Panama," I said. "We should have sent a higher ranking officer."

"Oh, it's fine," one of them responded. "Just come see us in Washington."

"But I can answer any questions you have."

"No, it's too late. You have to come to Washington and justify it to us."

So the next week, I went to Washington, D.C., and spent an inordinate amount of time justifying a budget. Eventually, we won the day and our budget did not get cut. The entire incident taught me an important lesson about the egos and power of Congress—down to the staffer level, no less.

By the time I was fully entrenched in my job as operations officer, we had yet another new combatant commander. GEN Clark had been transferred to Germany as the Supreme Allied Commander, Europe, and would lead the U.S. European Command. Charles E. Wilhelm—a Marine and a Vietnam veteran who had also won both the Silver and the Bronze stars—replaced him. GEN Wilhelm immediately began holding short morning updates in the conference room next to his office with key members of his staff. During that first week, I had given the general a very detailed weekly update report. But in the next morning meeting, he picked up the report and turned to me. "Look, Ric, don't feel bad," he said, "but do you all see this report from Sanchez? What the hell do I need to know all this stuff for?"

Of course, at first, I wanted to crawl under the conference table. But I soon realized that he was just trying to establish how he wanted to interact with his staff. "This is twenty pages of material that you all need to be handling," he said. "As my staff, you all need to be asking yourselves the question, 'Where do I need Wilhelm to engage?' If there is an issue with the Secretary of Defense, or the Chairman of the Joint Chiefs, then bring it to me. Other than that, it is your responsibility to use every bit of authority you have as colonels or generals to deal with the Interagency, the Joint staff, the Department of Defense, and everybody else out there. Just keep me informed and do not allow me to be surprised."

I was glad to stop that report, because it was very detailed and took a long time to prepare. But the more I thought about what GEN Wilhelm had just said, the more I realized that he was giving us a huge lesson in empowerment and delegation of authority. All of a sudden, with this guidance I was a brand-new one-star general with the duty to interact all the way up to the four-star level—across all departments and elements of power in a joint command. And if that speech wasn't enough to get his staff to take on as much responsibility as possible, Wilhelm found other ways to do it. One day, for instance, a couple of us handed him what we thought was an important white paper. But when he called us back into his office a few hours later, there was a big "DOA" (dead on arrival) written on the front of the document.

"Oh, boy," I thought to myself. "Here it comes."

"Gentlemen, it is very apparent that the people on your staffs wrote this paper," he said. "Now, they are very capable, but they do not have your level of experience. Sometimes the generals have to get involved—and that's what needs to happen in this case. I want you to go back, rewrite this, and include your own perspectives."

It was not an unreasonable request. Wilhelm was essentially saying to us, "Hey, you're generals now. I expect you to know when your judgment, experience, and abilities are required." Over the course of the next two years, GEN Wilhelm showed great patience and trust in his staff. I learned invaluable lessons from him about how to be a general.

As the 1997 summer assignment cycle started, I was hopeful for an Army billet in Germany. But once again, another four-star would ask me to stay on for an additional year. GEN Wilhelm wanted to restructure his staff by making Brigadier General (BG) John Goodman (the current director of strategy, policy, and plans) his new chief of staff. There were six main offices on the joint command's staff: Personnel (J-1); Intelligence (J-2); Operations (J-3); Logistics (J-4); Strategy, Policy, and Plans (J-5); and Communications (J-6). GEN Wilhelm asked me to retain my responsibilities as the operations officer (J-3) and assume additional duties as the director of strategy, policy, and plans (J-5). After accepting the move, John Goodman and I worked closely for the next year as I performed both jobs. And it proved to be a whirlwind of activity and learning.

Everything I did involved coalitions and combined joint operations. In forging new post–Cold War strategies, we held regional conferences that included all nations in the Southern Hemisphere, Central America, and the Caribbean. We brought in ambassadors, military leaders, and key officers in all U.S. interagencies (State, Justice, CIA, etc.) to meet with their international counterparts. We sought to achieve mutual cooperation in the areas of security, counterdrug activity, and humanitarian assistance. As a result, I learned to set up joint task forces, craft joint manning documents, and negotiate bilateral activities with host nations.

As it became clear that our staff had to be prepared to handle natural disasters and all missions simultaneously, GEN Wilhelm asked us to adjust our procedures accordingly. John Goodman and I, for example, started a major effort to restructure the staff to establish crisis action teams that would allow us to perform multiple operations at the same time. When one of the deadliest storms on record, Hurricane Mitch, struck Central America in October 1998, the SOUTHCOM staff was well prepared. Catastrophic flooding in Nicaragua and Honduras left nearly 20,000 people either dead or missing. SOUTHCOM's Task Force Aguila, commanded by Colonel (COL) Duz Packett, organized and deployed within a matter of days. Duz performed magnificently as he orchestrated humanitarian assistance operations in those countries for nearly six months. SOUTHCOM's swift actions not only saved lives, they also opened doors to help forge better political and military relations with Nicaragua and other Central American countries. And it wouldn't have happened without GEN Wilhelm's foresight to create and exercise our response capability in the first place.

Wilhelm was also the first combatant commander to bring in the Joint Warfighting Center out of Norfolk, Virginia, to conduct exercises designed to facilitate joint interagency operations (just as PDD 56 had ordered). A team of about sixty people came in to evaluate everything we were doing at headquarters. They built scenarios based on events that were actually taking place in our theater of operations, they monitored and observed us, and at the end of the forty-five-day focus period, they formally evaluated us. All the while, we were still performing our regular work. It turned out to be both an incredible grind as well as an essential education.

The Joint Warfighting Center placed heavy emphasis on Phase IV planning in association with coalition operations. Phase IV refers to the part of a campaign that occurs after decisive combat—when operations are focused on stabilization and reconstruction. In the past, Phase IV was often overlooked, because the mind-set of U.S. military leaders was transfixed on winning wars rather than the peacekeeping that came afterward. However, our experiences after the Cold War repeatedly taught us that thorough Phase IV planning was required prior to commencing combat operations. Otherwise, a war that was won through tactical success could turn into a strategic defeat. The nation had struggled with this issue in Bosnia, Haiti, Somalia, and Panama.

The Joint Warfighting Center also pointed out that most military leaders would rather fight the war, win it, and then pass off the Phase IV operations to other U.S. government agencies. History, however, has proven that when properly resourced, the U.S. military is more effective in rebuilding a nation's capacity to govern and secure itself. There is no other government agency that has the communications, logistics, and strategic planning capacity that are crucial in the immediate aftermath of a conventional conflict. On the other hand, various agencies within the U.S. government are virtually incapable of performing this mission, because they often function in a stovepipe fashion, have limited deployment capacity, and hardly ever train for such missions. These points were extremely important for me to remember—and they would come back to haunt me later in both Kosovo and Iraq.

The majority of our time in SOUTHCOM was spent in helping strengthen existing counterdrug interagency and coalition relationships, and in crafting the future footprint of America's counterdrug strategy in the Western Hemisphere. In this effort, I traveled to every country in South America except Uruguay and Argentina. I visited outposts in the jungles of Colombia, took a boat up the Amazon River, and flew in a Puma helicopter for two days along the border between Venezuela and Colombia. My missions were to form personal relationships and build strategies that would take the United States beyond bilateral relationships to regional coalition solutions.

The fact that I spoke fluent Spanish certainly came in handy. I would walk in for the first meeting, we'd shake hands, they would greet me in English, and I would respond in Spanish. My ability to do so created a tremendous rapport, which invariably resulted in smiles all around and the dismissal of translators.

During one of my visits to Colombia, I was able to visit with my old friends Alfonso Calero and Guillermo Barrera, whom I had met at the Naval Postgraduate School in Monterey, California. Both were still in the Colombian Navy, and Alfonso was now an admiral. When I arrived in Bogota, they arranged for a big family dinner. Alfonso and Guillermo were instrumental in establishing relationships with key leaders from the Colombian government, the armed forces, and the police. As soon as Alfonso told them that we had been in graduate school together, and that we were great friends, I was accepted as someone they could trust.

The culture in Central and South America had many similarities to the one that I had grown up with in the Rio Grande Valley. Personal relationships were paramount, and written agreements were viewed with far less importance than the commitment and honor of a "handshake." They valued the fact that they could talk to someone frankly, make an agreement, and then trust that the agreement would be carried out. In other words, your word was your bond. That was exactly what my father and uncle had drilled into me when I was a little boy. "If you say you're going to do something, and you shake on it, then that's it! You don't back away from that commitment." In forming coalitions with these South American and Central American countries, we certainly ratified them with formal written documents. I learned, though, that to build a successful coalition operation, one-on-one personal relationships had to form the foundation.

My experience at U.S. Southern Command proved to be the best possible developmental job for a new brigadier general. I gained a thorough knowledge of interagency, counterdrug, and coalition warfare operations. And the challenges of conducting multiple crisis operations across an entire continent prepared me for my role in future conflicts.

I was, however, completely unprepared for the politics and parochialism of the branches of the armed services. As we crafted new strategies for coalition and joint interagency task force building in a post–Cold War era, the military services invariably fought us tooth and nail. "We can't give you people for this stuff," they said. "We'd have to take them out of hide, and we're not willing to do that."

In the end, I realized that such negativity was about much more than committing forces to participate in joint operations. It illuminated a narrow outlook and a greed for power. In essence, the old and legendary interservice rivalries reared their ugly heads at a time when we could least afford them.

CHAPTER 7

Kosovo and Coalition Warfare

In the summer of 1999, after three years at SOUTHCOM, I was finally on my way back to Germany. My new position was to be assistant division commander for support in the Army's 1st Infantry Division. Maria Elena and I had loved our initial assignment to Germany and we were excited about rejoining the Big Red One. But once again, I was headed to an organization on the verge of going to war—this time it was in Kosovo, a region in southern Serbia that borders Albania, Macedonia, and Montenegro.

In my mind, Kosovo exists as a microcosm of what I would face in Iraq three years later. The United States led a military coalition of eight countries, each of which had their own national caveats, geographical restrictions, rules of engagement, and tactical requirements for the conduct of operations. Accordingly, I would learn the difficulties of coalition command while at war. Very quickly it would become obvious that I had to consider national interests, internal national politics, cultural sensitivities, and international political implications when developing the plan of action. From a political standpoint, the United Nations Mission in Kosovo would be faced with all of the security, political, and economic challenges with which the Coalition Provisional

Authority (CPA) would later struggle in Iraq. These included: lack of personnel and financial resources; nonexistent national and international guidance; and a business-as-usual approach. All my experiences in Kosovo would help me later as I led the military operations of the thirty-six-nation coalition and assisted the CPA in development of a cohesive political-military strategy and execution of an overall campaign plan for Iraq.

The ethnic hatred in this area of Eastern Europe had existed for centuries, at least back to 1389 and the battle at Kosovo Polje (also known as Field of the Blackbirds), in which Serbia fought to a standstill with Muslim Turks of the Ottoman Empire. To this day, Serbs celebrate it as a great victory, because they believe the battle stopped the spread of Islam to the West. In the aftermath of World War I, Serbia was incorporated into Yugoslavia and then, in the wake of World War II, the country became a satellite of the Soviet Union. Animosities between the Serbs (as Eastern Orthodox Christians) and the Albanians (as Muslims) were kept under control through the iron-fisted rule of Josip Broz Tito. But when Yugoslavia split apart in 1991 to 1992, Kosovo wanted to break away from Serbia and become an independent nation. The Serbs, however, were adamantly opposed to letting that happen, because they felt a strong historical linkage to the region, and did not want to hand over the region to Albanian Muslims. As a result, Serbia imposed a harsh rule over Kosovo.

In 1996, a year after the Dayton Peace Accords were signed (formally ending the war in Bosnia), the Kosovo Liberation Army started a three-year guerrilla war against Serbian security forces—and the Serbs retaliated by murdering thousands of Albanians. In essence, Yugoslav President Slobodan Milosevic (aka "the butcher of the Balkans") continued the ethnic-cleansing campaign he had started against Muslims in Bosnia years before. In an effort to prevent more genocide, NATO launched a limited bombing campaign against Yugoslavian targets on March 24, 1999. Over the course of the next seventy-eight days, 1,000 aircraft operating from bases in Italy and carriers in the Adriatic

Sea conducted 38,000 sorties against specific military targets in Yugoslavia. Rather than back down, however, Milosevic initially stepped up the ethnic cleansing in Kosovo, which pushed hundreds of thousands of Albanians to flee into neighboring Macedonia and Albania. Finally, on June 10, 1999, after NATO and the United Nations threatened ground intervention, Milosevic agreed to withdraw Serbian troops from Kosovo and allow the region to become a UN protectorate.

When I arrived in Schweinfurt, Germany, in late June, some elements of the 1st Infantry Division were already deployed to the region. As the assistant division commander for support, my primary responsibility was to oversee the adequate manning, equipping, and sustainment of our forces. Most of the time, I would have to operate from Camp Able Sentry, the American post in the Former Yugoslav Republic of Macedonia near the Kosovo border. So I was immediately thrust back into a coalition warfighting environment. The United Nations Kosovo Protection Force (UNKFOR) was led by German Army General Klaus Reinhardt, who reported to GEN Wesley Clark (Supreme Allied Commander, Europe).

I made my first trip into Kosovo in early July 1999. While the war had stopped, there was still some fighting going on as ethnic Albanians poured back into the country and started displacing the Serbs. In turn, several hundred thousand Serbs fled Kosovo in a mass exodus to Serbia. After arriving at Camp Able Sentry, I immediately flew into Kosovo to link up with our deployed forces and reconnoiter the only route available for transporting supplies and equipment into the province—the Kačanik River Valley. We flew below the mountaintops so I could get a good view of the route: a two-lane road cut on the side of the mountains. This path became a strategic priority for protection and maintenance, especially given the many old bridges along the way. We flew for about twenty minutes to the northwest, up the river valley, and finally out onto the central plains of Kosovo near the city of Ferizaj. As soon as we rose out of the valley, we observed fires burning in villages as far as the eye could see.

The helicopter landed on a hilltop where BG Bantz J. (John) Craddock (our assistant division commander for maneuver, and the U.S. ground force commander in Kosovo) had set up his tactical command post. (John Craddock and I had known each other since 1978 when we attended the Armor Officer Advanced Course in Fort Knox, Kentucky.) After a short meeting with John, and an operations update from his staff, I got back in the Black Hawk and flew around the entire U.S. sector to get an understanding of the environment.

The major cities of Ferizaj, Vitina, Kamenica, Strepce, and Gnjilane contained neither modern buildings nor sophisticated infrastructure. In the villages spread around the countryside, all you could see was livestock, outdated farm equipment, dirty canals, and austere housing. The villages lacked hotels, service stations, and major shopping centers. It was Mexico all over again—a Third World–type country where the people were pitifully poor. At that moment, I knew we were doing the right thing. The United Nations, the United States, and the coalition forces had stopped a genocide and could now make a difference in the lives of these people by setting them on the road to progress and prosperity.

Over the next five months, I spent most of my time at Camp Able Sentry, which was located adjacent to the airport at Skopje. The airport served as the sole port of debarkation for forces, equipment, and supplies headed into Kosovo. Given the vulnerability of the Kačanik Pass, we had our hands full ensuring that supply lines were in place and protected. Eventually, UNKFOR established other supply routes into Kosovo, which alleviated the stress on the Kačanik route and ensured a timely delivery of supplies to coalition forces.

In mid-August, MG John Abizaid took over command of the 1st Infantry Division in Germany. When his appointment was announced, I immediately remembered that he had been the young captain mentioned years earlier by the Vice Chief of Staff of the

Army during our discussions at Fort Knox. John had been tagged as a future general and sure enough, he had made it, and now he was my boss. Shortly after Abizaid arrived, John Craddock was reassigned and replaced by BG Craig Peterson, who had prior experience in Bosnia. After Peterson's four-month tour was up, Abizaid placed me in that position, commander of Multi-National Brigade East, also known as Task Force Falcon, which I assumed on December 10, 1999. Task Force Falcon was a UN coalition force comprised of elements and soldiers from eight nations—the United States, Poland, United Arab Emirates, Greece, Russia, Lithuania, Jordan, and Ukraine. I was now the U.S. ground force commander in Kosovo.

The mission of Task Force Falcon was three-fold: (1) conduct military operations and eliminate resistance; (2) prepare conditions for transfer of regional power to civilian authorities; and (3) assist in reestablishing civil and governmental structures.

Two-thirds of our mission resembled that politically taboo phrase "nation-building." But in fact, that's exactly what we had to do in Kosovo, because there was no government, no police force, and no army. After all, the Serbs, who had been running everything, left as UN forces occupied the province and ethnic Albanians returned. There were, of course, many Serbs still in the province, but they were located primarily in isolated villages and in protected neighborhoods within major cities. And each of those villages had its own loosely organized, small militias for protection, which destabilized our coalition forces. To top it off, there were low levels of violence associated with a growing insurgency, perpetrated by foreign elements (smugglers and criminal elements) moving in and out of Serbia.

The UN Mission in Kosovo was responsible for establishing security forces and forming a provincial government in Kosovo. But they didn't have the resources or the expertise to achieve that mission, so the military had to do it. Unfortunately, we didn't have the resources or expertise, either—at least, not at first.

My commanding role proved to be extremely complex: From

one minute to the next, I could be in a video teleconference engaged in a geopolitical discussion (strategic), coordinating or ordering action in support of the United Nations or other national sectors (operational), or out on the ground running a patrol with a platoon at the battalion level (tactical). Task Force Falcon consisted of a reinforced brigade with a staff that had no experience with this kind of mission. As a matter of fact, other than me, there was no one within the headquarters who had ever served in a joint interagency or coalition environment. During the course of our training exercises, MG Abizaid and I discussed the deficiencies of the staff, and he committed himself to fixing the problem. "I'm going to help in every way I can," he said. "If we don't have what you need, we'll go to the theater level for help."

True to his word, Abizaid quickly reinforced the headquarters by sending staff officers from Germany who had the right skill sets, grade levels, and experience to get the job done. So many officers were sent to Kosovo that, at one point, Abizaid kidded me that he was struggling back in Germany, because I had everybody on his staff. There was no doubt that Kosovo was the commanding general's top priority and the division's main effort. Because of Abizaid's commitment, our underresourced brigade headquarters expanded to a point where I became confident that we could achieve our mission.

Within a week of assuming command in Kosovo, we suffered our first casualty when Army SSG Joseph E. Suponcic was killed by a land mine along a trail near the village of Kamenica. There were a lot of Serbs who were angry that coalition forces had come in and disrupted their lives. But it was extremely difficult to ferret out these small pockets of resistance, because virtually all communications were person to person—within tribes and other small sectarian groups. We didn't understand the people, the culture, or the source of conflict well enough. Nor did we have much success in penetrating the Serb organizations or in linking up with Albanian Muslims to help us. Moreover, we learned very quickly that in this environment, our high-tech intelligence capabilities and sophisticated sensors were of little or no help. After all, the enemy did not have sophisticated command and control systems or computers.

In Kosovo, I gained a reputation inside the intelligence community as both a hands-on and extraordinarily demanding individual. I dedicated a period of time every day to interface with our intelligence staff. I also worked with my superiors to establish and fill the position of director of intelligence. However, the Intelligence requirements were so complex and intense that a couple of people came and went in that job before I was satisfied that we had the leadership and expertise necessary for the mission. We were dealing with a different type of enemy in Kosovo. And I had to be very clear with our intelligence personnel that we needed to find new ways of gathering information. The lives of our soldiers depended on it.

We were gradually able to improve our capacity and capability by taking advantage of the diversity of our joint interagency coalition, which included the CIA, Special Forces, and foreign intelligence services. We pulled everybody together to formulate strategies, synchronize operations, and come up with innovative solutions. For example, we set up an organization designed specifically for human intelligence operations. We called it G-2X. The "G-2" represented the standard intelligence function of any headquarters staff, and the "X" represented the human aspect. With that organization, we became very effective in preventing enemy attacks on the lives of our soldiers.

As we became more proficient with our intelligence capabilities, we also began capturing more insurgents. So we were forced to set up and continually improve our own detention facilities and interrogation center at Camp Bondsteel. In Kosovo, there was no question that the Geneva Conventions applied to all prisoners. No one from the Department of Defense came in to help us, and no one ever questioned the morality of our approach.

Task Force Falcon did experience one notable case of prisoner abuse, however. In the spring of 2000, we received information from locals about an incident involving members of the 82nd Airborne Division deployed in Vitina. Our preliminary investigation determined that a sergeant and four of his squad members had captured a couple of Serbs while they were planting land mines. The prisoners were taken into an abandoned house, beaten, and threatened with death unless they talked. The investigative report stated that the soldiers were focused on gaining intelligence necessary to save lives and, because the prisoners had not been seriously harmed, no charges should be filed. I did not agree, however, and subsequently recommended a court-martial. It was clear to me that the squad could have conducted tactical questioning, but instead chose to violate both the Geneva Conventions and our own standards regarding interrogation and prisoner treatment. There was no doubt in my mind that, if we had let them get away with it, other soldiers on the ground would have thought it was okay to do similar things. In the end, we referred the sergeant and his men for court-martial, where they were convicted and sent to prison.

Around this same time, I also witnessed one of our officers become afraid to take action for fear of an investigation—and that caused me serious concern. The incident involved a young battalion commander in the military police (MPs). One of his platoons had gone on a normal patrol into the depth of a valley that dead-ended into the mountains near Strepce. When a group of Serbs cut off their route back, the young lieutenant called for reinforcements, which was the right thing to do. However, the Serb mob organized itself and, as reinforcements entered the valley, they began attacking the soldiers.

From the sides of the surrounding mountains, the mob, which included women and children, started rolling down boulders and tree trunks, and throwing all kinds of objects. The reports we were monitoring at Camp Bondsteel indicated that soldiers were dropping and the toll was mounting. The MP commander was conducting an organized withdrawal through the valley, but it was extremely slow going. I deployed attack helicopters to provide support, but the proximity of the mob to our soldiers did not even allow warning shots to be fired. Fearing the worst, I decided to fly up there in my command helicopter to assess the situation. As we circled the area, I could clearly see the Serbs attacking the soldiers. So I got on the radio and spoke directly with the MP commander.

"Sir, we're taking all kinds of injuries," he said. "I have about fifteen wounded and some appear to be serious. Some of my soldiers went down when hit by the boulders."

"You've got to fight back," I said.

"But, sir, I can't fire into the mob. There are some women and children up front."

"Well, you're going to have to make a decision on the ground, Commander. I can't make the decision for you, but if you are taking this level of casualties, I recommend that you shoot to drive them back."

After a couple of hours, the MPs finally withdrew from the valley. Fortunately, there were no fatalities. During the after-action review, I asked the commander why he had not opened fire.

"Sir, I just didn't know if the rules of engagement allowed me to do that," he said. "I was afraid there might be investigations if we had killed someone."

"First of all, the rules of engagement do allow you to open fire in a situation like that," I said. "Second, you should never, ever allow the fear of an investigation to hinder your decision-making process when engaged with the enemy. You will *never* make the right decision if you do that. Always use your best judgment based on what you know at the time. You won't have perfect knowledge, but if you are anywhere near right, I am going to support you." Overall, this young officer performed magnificently in Kosovo and later commanded an MP brigade with distinction in Iraq. But what happened in Kosovo was a significant learning experience for both of us. I realized that training for rules of engagement had to be conducted by warfighters and not by lawyers.

There were no security forces in the province (either army or police), so the military had to provide stability while such forces were being built. The United Nations, however, did not have any kind of resourced plan to reconstitute either the police or a new Kosovo army. And even though doing so was not the responsibility of coalition forces, MG Abizaid and I decided to take on the challenge.

We reasoned that the now-defunct Kosovo Liberation Army (KLA) was the best place to start. Nearly all of its leaders and soldiers were still in the country, and we knew who their battalion commanders were. So, with General Reinhardt's approval, we decided to create a new organization to bring them back into the fold. The United Nations was opposed to the establishment of any group that had a security role, so our challenge was to identify structures, roles, and functions for a new civil-oriented organization. We called this force the Kosovo Protection Corps, and limited its missions initially to disaster relief and humanitarian assistance. As soon as we put out the word, thousands of former members of the KLA across Kosovo enthusiastically showed up to enlist. We gave them military-style uniforms, organized them into units across the sector, and began training them for humanitarian operations and disaster relief. The initiative was so successful that the Kosovo Protection Corps soon spread beyond our sector, and before long, it was operating across the entire province. As the Kosovo Protection Corps grew, it relieved the pressure on our military units and made a positive contribution to the security and stability of the region.

In addition to building our intelligence capabilities and forming the Kosovo Protection Corps, we also interfaced with civic and religious leaders. Ethnic Albanians viewed the United States as the saviors of Kosovo, and were especially appreciative of President Clinton and GEN Clark, whom they viewed as heroes. Of course, that made engagement and cooperation much easier in their villages. Serbs, on the other hand, were angry with Americans for "destroying our way of life," as they phrased it. "Everything was fine in our country before the Americans came," they said. "We had peace and we could go anywhere." Predictably, they resisted cooperating with us.

It took a lot of time to gain the confidence of Serbs. For instance, when some of their leaders expressed concern that they could not safely travel to and from Serbia to visit relatives and purchase supplies, we began providing military escorts for thousands of Serbs to do so. And as we implemented projects in their villages to improve conditions, they gradually came to the conclusion that we did, in fact, care about helping them.

Eventually, we started bringing people together to give them a voice in the political, economic, and security issues facing the sector. For example, we hosted a weekly dinner for all involved leaders, ranking members of the Kosovo Protection Corps, UN regional administrators, and senior coalition military leaders. After starting out with a structured meeting to discuss the previous weeks events and issues, we'd have a social dinner, which promoted strong relationships. These dinners became so popular that everyone wanted an invitation.

At one dinner, the UN regional administrator praised our soldiers. "General, your troops look magnificent," he said. "Your patrolling of our sector reminds me of the days when my mother was holding me by the hand as the American Army came through Rome to liberate us during World War Two. I can still remember it to this day. American soldiers and their tanks rolling through the streets. We were waving American flags. For the first time in many years, we felt safe."

That comment made me proud that our soldiers were continuing the great legacy of the U.S. Army. But the administrator's comment also made me realize that the images of war imprinted on the minds of Albanian and Serb children in Kosovo would last a lifetime. The hatred they witnessed on a daily basis would take decades to overcome.

In traveling around the province, I also received comments from Kosovars that gave me pause to think about the coalition's overall mission. For instance, at a social function in Pristina, I was taken completely off guard when one of Kosovo's leading citizens approached me with a question.

"General, what are you?" he asked.

"Sir, I'm an American general officer and commander of U.S. forces in Kosovo," I responded.

"No, no, General. What are you?"

"Well, if you mean my heritage, I'm Hispanic, a Mexican American. My grandparents were from Mexico."

The gentleman paused a moment, and then with a look of utter dismay said, "But how can that be?"

"Well, they migrated to the United States, where my parents were born," I said. "I was born in Texas."

"No, General. What I mean is, how can a minority be the senior commander of U.S. forces here in our country?"

This gentleman's comment completely floored me. He found it inconceivable that a minority could be the overall leader of an integrated force that included members of the majority. He had no understanding of the progress in equal opportunity and non-discrimination that had been achieved in America. This was a profound moment for me; as an American, I did not understand the mind-set of his culture—or that he, in turn, did not understand who we were, what America represented.

I probably shouldn't have been as surprised as I was that a leader in Kosovo didn't understand what was going on in the United States, because, at times, it seemed that our leaders didn't have the faintest idea what was going on in Kosovo. I received several impracticable directives from Washington that, most of the time, made me wonder who could be dreaming them up. Much of the time, the orders displayed an ignorance or complete disregard for Kosovo's vague culture. For example, a July 1999 directive from Washington ordered U.S. forces to make sure that, within sixty days, all the schools were integrated when we opened them for the next school year. Integrated! The people in Kosovo had been fighting a civil war against each other. The Serbs had no desire to interact with the Albanians, or vice versa. To think they would send their children to integrated schools was completely foolish. It took the United States more than a hundred years after its own Civil War to integrate schools. You would think that people in the State Department would have considered that before issuing such guidance.

In addition to Washington, we also had a major problem with the United Nations, which was responsible for the entire administrative effort in Kosovo. Within a month of assuming my command, it became painfully obvious to me that the United Nations was still struggling to develop a feasible strategy for the province, had few executable plans, and had no way of coordinating operations across the province. The regional UN representatives were not on the ground in Kosovo for the initial six or seven months of operations. And while UN leaders were willing to transfer the responsibility for certain missions to the military, they refused to give up any of their authority, or hold anyone in their organization accountable for results (or lack thereof). The situation was so bad that I started out my briefings to visiting delegations (congressional, Army, etc.) with a slide that read: "United Nations: UNcoordinated, UNplanned, and UNresourced." It was important to reveal the situation on the ground to higher authorities, because every single bit of progress in Kosovo during the first eight to twelve months of the operation was being engineered by the military, and the military alone.

Of course, our multinational military coalition also had its problems. It had been rapidly formed through the NATO structures at the national level, and little to no information had been communicated regarding the limitations that each nation had on how their troops could be used. So we were forced to learn the various caveats, restrictions, and constraints as we executed operations on the ground. When the 82nd Airborne was deployed to quell violence in Mitrovice, for instance, the French commander and his forces did nothing but watch as an American unit was pummeled by Serb resistance. Working through these formidable complexities required negotiations between the French commander, the KFOR commander, and GEN Clark.

We also experienced some international incidents with the

Russians, who were huge supporters of the Serbs. In one case, I ordered the Russian battalion commander to be prepared to stop a convoy of Kosovar Serbs that, according to our intelligence, was planning to instigate violence in a nearby ethnic Albanian community. But the Russian battalion commander stated that he could not comply with that order. I immediately appealed to the KFOR commander who, in turn, engaged the commander of all Russian forces in the province. But he, too, refused to comply. Given the time sensitivities, I took the matter directly to GEN Clark, who escalated the dispute all the way to Moscow. Finally, and barely in time, an order came down from the Kremlin for the Russians on the ground to execute my order.

The Russian military was also struggling in the wake of the demise of the Soviet Union. At times, I believe they still had one foot planted in the Cold War. When I met with the Russian lieutenant colonel who reported to me, for example, he was accompanied by a political commissar, whose job was to monitor everything the young Russian officer said and did while in my presence. And if I issued an order, I never really knew if it was going to be carried out. The guy obviously calling the shots was the political commissar, not the lieutenant colonel.

In the long run, I learned to deal with the reality that most coalition nations had to clear their orders with their national military and political chains of command. My own chain of command was also muddled. Because I was serving under the UN Kosovo Force, my operational chain went straight to the Kosovo Force commander, who was General Klaus Reinhardt. I also reported to MG John Abizaid, who was my division commander for all training, manning, equipping, and logistics issues. In addition, there was a separate operational chain of command for the United States, which went directly up to Wesley Clark, who made all the final decisions where disputes were involved between NATO and U.S. interests.

Despite the complexity of the chains of command in Kosovo,

and the ongoing problems with Washington and the United Nations, we were still able to build significant stability in the province. By the middle of 2000, when I rotated back to Germany, most of the cities and villages were stabilizing. The Kosovo Protection Corps was functioning well, and anxious to become the Army of Kosovo. Local governing councils were operational. Schools were open and the economy became vibrant as resources began flowing back into the province and new construction began. Unquestionably, the Multi-National Brigade East and the U.S. military had made a huge and positive impact in Kosovo. Still, there were many lessons to be learned from U.S. involvement there.

First, the UN coalition was totally unprepared to deal with a nation-building situation that involved a regime change, the complete dissolution of political structures, a void in security structures, ethnic and religious hatreds, poverty and economic strife, individual village militias, and people who did not recognize formal borders.

Second, there was a lack of national priority, focus, and competence on the part of the U.S. government. Washington was quick to demand action without a clear understanding of culture, the situation on the ground, or potential collateral damage. Many of their directives were simply unrealistic.

Third, the coalition completely depended on the U.S. military for a successful outcome in Kosovo. The military, however, had not initially resourced post–major combat operations (Phase IV), and was sorely lacking the organizational structures, funding, and personnel necessary for the majority of tasks that had to be accomplished.

Fourth, the U.S. military was forced to take the lead in the following areas: fighting a low-level insurgency; building police and security force capacity; restoring basic utility services; creating an effective intelligence capability in a Third World–type environment; building detention facilities; and reestablishing the social, economic, and political base of the Multi-National Brigade East sector.

Accordingly, I made several key recommendations in my After Action Report:

1. For the first twelve to eighteen months after combat operations, or until a civilian organization has its capacity deployed, the U.S. military must have both responsibility and authority for the mission. It is the only organization that has the strategic and operational planning capacity, the command, control, and communications, and the logistics capacity necessary for success in such an environment.
2. A grand strategy must be in place before hostilities commence. Operations in Kosovo were being conducted in a strategic and operational vacuum due to the lack of a clearly defined vision for the province. Due to a lack of consensus over Kosovo's future, the governmental, nongovernmental, national, and international organizations constantly struggled with competing objectives.
3. The headquarters staff must have the manpower, processes, and expertise to tackle the daily complex strategic, operational, and tactical challenges of the mission.
4. Units deployed on such missions need national-level support and assistance on a "push" basis in order to succeed.

I filed my Kosovo After Action Report according to procedure. But it went into a Pentagon file cabinet and died there.

MY MOST DIFFICULT EXPERIENCE during Kosovo revolved around the tragic rape and murder of a twelve-year-old ethnic Albanian girl named Maurita. An American sergeant had committed the crime. As soon as I heard about it, I went straight to our morgue, where the young girl's body had been taken. When I saw her laying out on the table, it immediately took me back to visions of my young son's death, when I had seen him resting in the open coffin before burial. I became very emotional and empathized deeply with the community's pain.

After writing a letter to the little girl's family, I set up a meeting with local leaders where I personally expressed my condolences and briefed them on our legal process and the status of our investigation. When I concluded my remarks by saying, once again, how deeply sorry I was for Maurita's death, the gentleman on my left said, "Well, that's okay, General. That was God's will. Now let's talk about you releasing those of our people you have in custody."

"Sir, I'm not here to discuss your people in custody," I replied. "I came here because of my concern for the family, for this village, and for the impact it might have on our relationship."

"Oh, don't worry about that. God willed the little girl's death. Let's get on to something important."

I remember later telling my aide that I didn't know how to react to what that gentleman had said to me. As a Catholic, I understood the idea of God's will, but not to the point of completely dismissing what had happened, or any possible impact it might have on the community. It was a huge lesson about the cultural differences between my faith and that of Islam.

He was a Muslim and I was a Catholic. The words were virtually the same, but the meaning seemed so far apart.

It was God's will.

Si Dios quiere.

CHAPTER 8

Unleashing the Hounds of Hell

In December 2000, I was back in Belgium walking the Battle of the Bulge with several historians. I had done this more than a decade earlier with my battalion commander, Lieutenant Colonel (LTC) Mike Jones. Back then, I was a young major, and we focused on the tactical actions of the battle. This time, I was with my new boss, GEN Montgomery Meigs and his principal staff, and we were studying the battle from the perspective of operational and strategic leadership.

While the Battle of the Bulge was being fought, GEN Dwight Eisenhower (Supreme Commander of Allied Forces in Europe) had to contend with British Prime Minister Winston Churchill, U.S. President Franklin Roosevelt, the Russian leadership, and rifts between British General Bernard Montgomery and some of the American generals on the ground in Belgium. Ike was deeply involved in every aspect of the battle from the tactical actions to the strategic political issues. The big lesson we took away from this trip was that a general officer must understand political realities and how they impact his military objectives. In order to make the right decisions, a commander must stay in close, almost daily, contact with the battlefield.

GEN Montgomery Meigs, the four-star commander of U.S.

Army Europe, was a knowledgeable historian who, twice a year, took his staff of young generals around Europe to study the battles of World War II. In these exercises, we particularly focused on Eisenhower's military leadership. What was Ike thinking? What was he struggling with? How were politics impacting the fight, and how was the fight impacting the politics? Meigs also brought in renowned military historians who illuminated the perspective of both the Allied and German leadership, their decision-making processes and the challenges they faced.

During the Battle of the Bulge, the German generals executed their orders (against their better judgment) and ended up losing not only the battle but, a few months later, the war itself. History has frequently demonstrated that politicians have made decisions that were against the best judgment of the military. When that happens, a general has only two options: to execute or to retire. In the case of the Battle of the Bulge, the German generals had to obey, or else the Nazis would have shot them. Most of the time, a general who is confronted with legal orders that are in direct conflict with his best military judgment will execute those orders to the best of his ability.

GEN Meigs, a West Point graduate and Vietnam veteran, felt it was his sacred responsibility to develop his subordinates to be future leaders of our Army. As soon as I became director of operations for U.S. Army Europe in June 2000, he took me under his wing. GEN Meigs had the human touch—something a lot of senior generals lack. He constantly encouraged me to follow my natural instincts, which were to stay close to soldiers and let them know I cared about them. And he intuitively understood the thought processes of minority officers. Without in any way being condescending, he made certain that subordinates like me received all the professional development and support necessary to positively contribute to the nation.

GEN Meigs was also an exception to the rule in that he was willing to change with the times. Most of the Army four-star generals had served their entire careers believing that "the Army fights and wins the nation's wars." The Army was still struggling with, and didn't necessarily embrace, the joint warfighting vision expressed in the Goldwater-Nichols Act of 1986. Although Meigs was looking to embrace joint warfighting, there was tremendous pressure from his peers to keep things the way they were. Clearly, the Army lacked senior leaders with joint service experience.

After my three years at SOUTHCOM, I showed up as director of operations at U.S. Army Europe and found myself to be the only person on GEN Meigs's staff who had worked extensively in a joint environment. And Meigs went out of his way to encourage me to lead the command into a post–Cold War Joint operating environment.

One of the first issues I handled was the Joint Integrated Priority List, which was an important document on funding within the Department of Defense. When it first hit my desk, I asked members of my staff how they had previously prepared it for the commanding general. "Well, sir, we just fill out some basic stuff and send it back to higher headquarters," came the response. "We don't bother him with this."

"This should not be considered a bother," I responded. "This is the combatant commander's mechanism for prioritizing his requirements. It focuses on the budgets of all the services, including the Army, Navy, Air Force, and Marines, and can have a huge impact on our funding."

I immediately went in to GEN Meigs and asked him to get personally involved. "Sir, this list is the mechanism where each joint combatant commander provides priorities to DoD," I explained. "Look at your quality of life item, for instance. You've been having problems getting it funded out of the Army's budget. But if you place it high on the Integrated Priority List, then the combatant commander will embrace it and the Army will have to fund it."

After GEN Meigs got involved with our submission, U.S. Army Europe succeeded in raising the priority and funding of quality of life requirements for our soldiers and their families throughout Europe. Because of the Cold War bureaucracy, we had only been working with Army budgets. The Joint Integrated Priority List opened up a new avenue for improving funding.

In addressing future training requirements, we designed and aggressively pursued a training program that would incorporate joint warfighting. When GEN Meigs presented the concept to GEN Eric Shinseki, the Army Chief of Staff, at a briefing, many of his four-star peers balked. "No, we've got to first stay focused on sustaining our core competencies as an Army," they said. "If we don't make sure our soldiers and units have mastered their warfighting skills, we'll never be able to fight in a joint environment." This statement demonstrated a deficient understanding of the complexities of joint warfighting. GEN Shinseki liked our idea, but stressed that all four-stars had to be on board. It took a while, but Meigs and Shinseki succeeded in convincing the other senior leaders that training in isolation was an outdated methodology that needed to evolve. So U.S. Army Europe moved forward with the integration of joint training into our division and corps exercises. Training in isolation was no longer an option.

Another initiative undertaken by U.S. Army Europe was to consolidate Army bases, primarily in Germany and Italy. This idea, called "efficient basing," was a continuation of the U.S. effort to reduce our presence across Europe in the wake of the Cold War. Our plan would result in the consolidation of forty or fifty installations into four or five major bases with new housing, improved training, and other support facilities. It would also achieve efficiency—hence the name—across the entire realm of sustainment, providing more modern amenities for soldiers and their families and, in the long run, recouping our investment and saving the nation hundreds of millions of dollars.

At a briefing by our higher headquarters, a congressional staffer was introduced to the idea and the next thing I knew, I was on a plane to Washington, D.C., with Bill Chesarik, the architect of the concept. Our task was to brief the efficient-basing concept to Congress for possible funding. Having dealt with congressional delegations on these types of issues before, both on future tank designs and on eliminating funding in Panama, I knew that our success would depend on our ability to prove the cost benefit of the proposal. After spending a couple of endless days and nights to develop our presentation beyond our original two diagrams, we briefed senior leaders in the Army, the Department of Defense, and finally Congress.

Shortly after we arrived back in Germany, GEN Meigs was informed that Congress had embraced efficient basing, and that an expedited funding timetable was to be provided. Bill Chesarik and I had been in the right place at the right time and, by the time I left Germany in 2006, efficient basing would become a reality.

Several months later, in the spring of 2001, I was relaxing at home on a Saturday night when I received a surprise phone call from GEN Shinseki. "Ric, I've decided that you're going to command 1st Armored Division," he said. "Congratulations. We have great confidence in your leadership. Best of luck to you and Maria Elena. You're going to do a great job as division commander."

When I hung up the phone, I must have had a look of utter astonishment on my face. "Who was that?" Maria Elena exclaimed. "What happened?"

When I told Maria Elena the news, we sat down and discussed our good fortune. Both of us felt blessed, and we talked about an old phrase from the Gospel of Luke: "From everyone to whom much is given, much will be required."

We knew the future was going to be very demanding for our family. But at that moment, we had no idea just how demanding it would become.

ASSUMING COMMAND OF THE 1st Armored Division in June 2001 was not the smoothest transfer I'd ever experienced. GEN George Casey had just brought the division back from Kosovo, was trying to reset it, and saw nearly half of his personnel rotate out after the combat deployment. By the time I took over, every single brigade commander was new, 60 percent of the battalion commanders had moved out, and, like me, every department head on my headquarters staff had just arrived. Predictably, our first few meetings were completely disorganized. Nobody could answer any questions, because everyone was still learning the nature of their jobs. And all of our office equipment was still in boxes, because the division had just relocated to Wiesbaden as part of the force drawdown initiative. Finally, I just told everybody that we weren't going to have any full staff meetings for thirty days. "I'm not going to ask you any questions during that time," I said. "I'll go around the division and handle issues as they come up. Meanwhile, I want you all to get unpacked and organized. When we meet again, you all need to know what is going on in your sections."

By the summer of 2001, we were all taking note of the changes occurring in Washington, D.C. President George W. Bush and the Republicans had taken over the White House after eight years of the Clinton administration, and we were all expecting some kind of a seismic shift. The new secretary of defense was Donald Rumsfeld, who had formerly held the same position during the Gerald Ford administration when the Vietnam War was winding down. Almost immediately, I started hearing grumblings from the Pentagon that Rumsfeld was a micromanager. There was also quite a bit of debate emerging about the military philosophy of conservative ideologues in the Bush administration.

As a new national security strategy came into being, among the first things we heard about was the notion of a "first strike" policy. Most military officers were fairly conservative, and our initial reactions to such an idea were not negative. Maybe we did need to be a bit more aggressive in taking out our enemies, we thought. But the ensuing discussions that advocated unilateralism—going it alone—did not seem to consider the realities of the last ten years or so. Nearly every crisis or contingency deployment faced by the nation had been handled in the context of a coalition environment.

The greatest concern among Army leaders came when there was serious talk about drawing down the size of the Army from ten divisions to eight. If that happened, the nation would be accepting significant strategic risk in the event of a major conflict. We simply would not be able to sustain an extended deployment of combat forces. This basic fact seemed to have been ignored by advisors to the Bush administration. Moreover, in association with an Army drawdown, Defense Secretary Rumsfeld was aggressively advocating a major shift toward air power and a resurrection of the missile defense system. The Army, he said, was too cumbersome, because it was still locked in the Cold War. To a degree, he was absolutely correct. There was nobody more locked into old Cold War thinking than some of our generals. Our doctrine was not evolving, we were still training to defeat the Soviet Army in central Europe, and our combat formations were not rapidly deployable. Overall, the tension between Rumsfeld and the Army leadership had reached a contentious and counterproductive point.

Within the 1st Armored Division in Germany, I was monitoring all of the strategy and philosophical changes coming out of Washington, but my main priority was to ensure that Old Ironsides (as the division was known) was trained and ready for war. Because most of our soldiers had rotated out after the division returned from Kosovo, we had to reestablish our training readiness. As I had done before, every unit was sent to the field with all their gear to train under expeditionary conditions. Usually, field exercises were handled by the brigade commanders. But I was determined to be "in the box" for every battalion- and brigade-level event, a commitment I had learned from MG Randy House.

In the lead-up to our training exercises, I heard a lot of complaints. We've never taken everything to the field like this before. We've never stayed out there for weeks at a time. What does the boss want? Why is he doing this?

I understood what our soldiers were thinking. They were going to experience some tough new conditions and learn some difficult lessons. But I had experienced Desert Storm. And I had been to Kosovo. I knew what my soldiers were going to encounter if deployed, and I wanted them to be prepared for the worst of conditions. So I heard their complaints, but pressed on anyway.

ON SEPTEMBER 11, 2001, I was in a V Corps commanders' conference at Patrick Henry Village in Heidelberg, about an hour's drive from Wiesbaden. About midafternoon, in the middle of a briefing, an aide walked in and handed Lieutenant General (LTG) Scott Wallace a note informing him that a plane had flown into one of the towers at the World Trade Center in New York City. During a short break, a number of us watched the CNN live report, and we saw a commercial airliner fly directly into the second tower. At that point, everybody knew we were most likely dealing with a terrorist attack. The question on our minds was whether it was limited to New York or was part of a much larger assault against the United States.

LTG Wallace immediately canceled the rest of the meeting and directed us to place all forces on a heightened force protection posture, and to ramp up security at all installations so that our facilities, equipment, and people were properly protected. We had no idea where the enemy might strike next, and some of our high-value facilities in Europe were exposed and vulnerable. After receiving Wallace's guidance, I huddled with my operations and intel officers (who were also present for the meeting) to discuss our next moves. "I don't know who the enemy is yet," I said, "but we're going to war."

On the drive back to Wiesbaden, we heard the terrible news that the Twin Towers had collapsed, that a plane had flown into the Pentagon, and that another had crashed near Shanksville, Pennsylvania. The North Atlantic Treaty Organization immediately invoked Article 5, which meant that all nations that had signed onto the NATO agreement were now joined together in united military operations. It wasn't long before the direct involvement of the terrorist group, al-Qaeda, and its leader Osama bin Laden became apparent. Knowing that al-Qaeda was based in Afghanistan and protected by the Taliban regime, the United States, through NATO, demanded that it be allowed into the country to search for bin Laden. When Taliban leaders refused, there was an immediate and mad scramble by the Department of Defense to prepare for war. On September 20, 2001, President Bush delivered an ultimatum to Afghanistan that unless bin Laden and his associates were turned over to the American government and al-Qaeda terrorist camps were closed, there would be war. Taliban officials rejected the ultimatum the following day.

Less than three weeks later, NATO launched an invasion of Afghanistan, the stated purpose of which was (1) to remove the Taliban regime; (2) to destroy al-Qaeda's organization and terrorist training camps; and (3) to capture or kill Osama bin Laden. The invasion, named Operation Enduring Freedom, consisted of a twenty-three-nation coalition led by the United States and Great Britain. It began with a major air attack that dropped bombs on al-Qaeda training camps and Taliban military positions, mostly in or near the cities of Kabul, Kandahar, and Jalalabad. Ground forces (joined by troops from the Afghan Northern Alliance) soon followed the air campaign, and by mid-November, most Taliban and al-Qaeda leaders had either been killed or captured, fled the country, or were holed up in caves of the Hindu Kush mountains of northeastern Afghanistan. It was believed that, in the midst of a called truce during the battle of Tora Bora in December 2001, bin Laden and several of his associates escaped and fled to Pakistan. By January 2002, an interim government had been set up under President Hamid Karzai, and coalition ground forces had established their main post at Bagram Air Base, just north of Kabul.

Some historians have marked the launch of the war in Afghanistan as the starting point of the global war on terror, which came to be known in military circles by the acronym GWOT. Others point to President Bush's State of the Union address on January

29, 2002, in which he labeled North Korea, Iran, and Iraq as an "axis of evil," because they threatened the "peace of the world" by sponsoring terrorism and pursuing weapons of mass destruction. It was apparent to senior military leaders that we were going to be in Afghanistan for a long, long time.

Two weeks prior to the President's State of the Union address, the first load of Taliban and al-Qaeda prisoners arrived at the new military prison camp in Guantánamo Bay, Cuba. President Bush even mentioned it in his remarks. "Terrorists who once occupied Afghanistan now occupy cells at Guantánamo Bay [GTMO]," he said.

What the President did not mention was the administration's plan to detain and interrogate the prisoners for an indefinite period of time. Rather, he waited another week, until February 7, 2002, to issue a memorandum stating that he had determined the Geneva Conventions did not apply to members of the Taliban or al-Qaeda because they were "unlawful combatants" and, as such, did not qualify to be labeled prisoners of war. "I accept the legal conclusion of the Department of Justice," wrote the President, "and determine that none of the provisions of Geneva apply to our conflict with al-Qaeda in Afghanistan or elsewhere throughout the world, because, among other reasons, al-Qaeda is not a High Contracting Party to Geneva."

This presidential memorandum constituted a watershed event in U.S. military history. Essentially, it set aside all of the legal constraints, training guidelines, and rules for interrogation that formed the U.S. Army's foundation for the treatment of prisoners on the battlefield since the Geneva Conventions were revised and ratified in 1949. Our current detention and interrogation doctrine had been rendered obsolete and invalid in the war with al-Qaeda. According to the President, it was now okay to go beyond those standards with regard to al-Qaeda terrorists. And that guidance set America on a path toward torture.

In the early days of GTMO, many self-confessed members of al-Qaeda were brought into the detention facility. Still only a few months after the horrendous events of 9/11, tremendous pressure was placed on interrogators to obtain information from these prisoners. Government leaders wanted to gain intelligence for two main reasons: to prevent another possible terrorist attack in the United States, and to identify al-Qaeda cells so they could be wiped out. But that was not an easy thing to do, because these prisoners were hard-core fanatics who were willing to die for their cause. And if forced to operate within the constraints of the Geneva Conventions, interrogators were not likely to gain any substantive information, especially since the enemy was trained to resist interrogation.

President Bush's February 7, 2002, directive did not fix limits on interrogation approaches, nor did it specifically order any new methods to be used. That crucial step was left up to the Department of Defense. More than a month earlier, however, Secretary Rumsfeld had already given orders to the Joint Chiefs of Staff to suspend the Geneva Conventions. Accordingly, in a January 21, 2002, memorandum to GEN Tommy Franks at U.S. Central Command (CENTCOM), the Joint Chiefs passed along Rumsfeld's guidance, stating that al-Qaeda and Taliban individuals under DoD control were not entitled to prisoner of war status as outlined in the Geneva Conventions. Franks immediately codified the lifting of the conventions by sending out his own memorandum on January 24, 2002, entitled "Guidance for PUC (Persons Under U.S. Control) handling," which read:

> The United States has determined that al-Qaeda and Taliban individuals under the control of the DoD are not entitled to prisoner of war status for purposes of the Geneva Conventions of 1949 . . . With the war on terrorism at full speed, the ability to obtain accurate and timely information about the capabilities, intentions, and activities of foreign powers, organizations, or persons may be at odds with the Laws of War and other training our military personnel received.

Prior to the invasion of Afghanistan, the Geneva Conventions and the Laws of War provided limits on authority and prevented abuse of prisoners. President Bush's February 2002 memorandum established new guidance that allowed suspected al-Qaeda prisoners to be tortured. However, the January 21 and 24 memos from the Joint Chiefs and CENTCOM, respectively, had already made it military policy. The next step in the process was for the U.S. Army (Deputy Chief of Staff for operations, G3, in coordination with the Deputy Chief of Staff for intelligence, G2), as the executive agent for military interrogation operations, to establish specific procedures and training related to interrogation operations *before* sending soldiers into battle so they could operate in a structured, controlled environment. Unfortunately, it never happened.

By the spring of 2002, the Pentagon, working with CENTCOM, the lead military element commanded by GEN Franks, had dusted off contingency war plans and began preparing for an invasion into Iraq. And by early summer, LTG Wallace was deeply engaged in the specifics of war planning. Once a strategy and the warplan were agreed upon, Wallace would provide orders to me and his other division commanders so we could begin preparing our division warplans. However, the CENTCOM orders changed frequently—usually on a weekly basis. It turned out that Rumsfeld was micromanaging his generals on issues such as how military forces would flow into Iraq, including their size and composition. Previously, these matters had been the responsibility of the senior warfighting commander. But the Secretary and his staff got so involved, and were so intrusive in the planning stages, that they ended up completely disrupting the process. Constant changes were made to the operation's plans of the warfighting forces. The most devastating impact of Rumsfeld's micromanagement was that warfighting commanders, all the way down to the division level, were never able to plan beyond the basic mission of defeating Saddam Hussein's military. For example, I was initially told that 1st Armored Division would deploy to Iraq as one of the base plan units. Then I was told we would be a follow-on force. Then we weren't going to deploy at all. Finally, we were going to deploy after major combat operations were concluded to handle the Phase IV mission.

Each time our orders changed, we had to stop our planning efforts, rethink and regroup, and then readjust our training programs. The constant changes drained our staff's energy and negatively impacted our mission-specific training regimen. The frustration became so palpable that I finally threw up my hands to my staff and said "Stop! This has got to stop! We cannot continue to change our plans this way. From this moment on, we are going to plan and train for the reserve division mission [the most complex]. That way, whatever they finally decide to have us do, we'll be ready."

The Secretary of Defense was, in effect, involved in the operational decisions of the combatant commander. And I do not believe that was either the spirit or the intent of the Goldwater-Nichols Act of 1986. In my mind, Donald Rumsfeld had changed the doctrine of "civilian control of the military" (when to go to war and for what purpose) to "civilian command of the military" (when to go to war, for what purpose, and *how to wage the war*). That was a very dangerous thing to do, because our national civilian leadership does not have the expertise, judgment, intuition, or staff capacity to make informed decisions or recommendations on the detailed application of military forces. The level of control exercised by the Secretary of Defense went well beyond the establishment of broad policy goals and strategic objectives. He was now beginning a pattern of authoritative and dominating influence over the entire war effort.

The danger itself involves both practical and philosophical issues. Practically, civilian command of the military can cost the lives of soldiers, put at risk vast amounts of resources (such as equipment, supplies, and taxpayer dollars), and endanger our ability to secure national strategic objectives. In addition, political pressures and considerations can lead to ill-advised tactical decisions, which, in turn, may delay, cripple, and result in the eventual failure of a military operation. Civilian command of the military has the potential to breach a larger constitutional issue. The executive branch of the national government has control of the national armed forces, the military leadership maintains command, and the legislative branch has oversight in determining how the armed forces are to be raised and employed. When any one of the legs of this triad gets circumvented, we begin to eat away at constitutional checks and balances, and place the very foundations of our democracy in jeopardy. If the executive branch is allowed to use the armed forces to achieve its own political agenda, our democratic republic begins to approach the status of a fascist dictatorship.

In preparation for the upcoming war in Iraq, the executive branch of the U.S. government (through the actions of Secretary Rumsfeld) was instructing the combatant commander what forces he needed to fight this war, how he ought to deploy those forces, and how he should execute his campaign. In the end, years of deliberate military planning and judgment were discarded to accommodate the desires of the Secretary of Defense. Overall, it was a recipe for disaster.

Meanwhile, an ongoing ping-pong match between Saddam Hussein and the United Nations was taking place regarding weapons inspections in Iraq. Formal talks in Vienna, Austria, in July 2002 failed after Iraq refused to allow the resumption of inspections, which had been suspended since 1998. In August 2002, however, Iraq surprised the international community by inviting the chief UN weapons inspector to Baghdad "to establish a solid basis for . . . monitoring and inspection activities." It was pretty clear that Saddam had seen the writing on the wall.

On September 12, 2002, President Bush addressed the UN General Assembly and outlined Iraq's potential for violence. "The world must move deliberately [and] decisively to hold Iraq to account," he said. "We will work with the United Nations

Security Council for the necessary resolutions. But the purposes of the United States should not be doubted. The Security Council resolutions will be enforced . . . or action will be unavoidable."

Five days after that speech, on September 17, 2002, the Bush administration released a new "National Security Strategy," which called for preemptive action against hostile countries and terrorist groups. "[The United States] will not hesitate to act alone, if necessary, to exercise our right of self-defense by acting preemptively," the document stated. Exactly one week later, on September 24, a crucial British intelligence dossier was made public. It stated that Iraq had reconstituted its nuclear weapons program, and that it would be able to produce nuclear weapons within one or two years if fissile material and other key components could be obtained. This "September Dossier," as it came to be known, also declared that Iraq currently possessed chemical and biological weapons.

Based on this and other key intelligence provided by the Bush administration, on October 10 to 11, 2002, the U.S. Congress overwhelmingly passed a joint resolution supporting the President's diplomatic efforts "through the United Nations Security Council . . . to ensure that Iraq . . . complies with all relevant Security Council resolutions." With this document, Congress also authorized the President to "use the U.S. military to: 1) defend the national security of the United States against the continuing threat posed by Iraq, and 2) enforce all relevant UN Security Council Resolutions regarding Iraq."

By mid-November, Saddam Hussein had accepted a new UN resolution and weapons inspectors were in Baghdad. But when the 12,000-plus-page report detailing Iraq's WMD program was released, Secretary of State Colin Powell flatly rejected the document. As a result, on December 19, 2002, the United States accused Iraq of being in "material breach" of the UN resolution. Through spokesmen, Saddam Hussein quickly responded by accusing the United States of making judgments before UN inspectors could fully complete their inspections. He also stated publicly that Iraq did not possess any weapons of mass destruction, and he invited the CIA to come into the country to see for itself.

DURING THE LAST FEW months of 2002, while the highest levels of the U.S. government were sparring with Saddam Hussein and setting up its case for an invasion of Iraq, there is irrefutable evidence that America was torturing and killing prisoners in Afghanistan. It occurred at the coalition's Bagram collection point located at Bagram Air Base, just north of Kabul. This was the central detention and interrogation center for prisoners captured throughout Afghanistan during the conduct of coalition operations.

Because of the U.S. military orders and presidential guidance in January and February 2002, respectively, there were no longer any constraints regarding techniques used to induce intelligence out of prisoners, nor was there any supervisory oversight. In essence, guidelines stipulated by the Geneva Conventions had been set aside in Afghanistan—and the broader war on terror. The Bush administration did not clearly understand the profound implications of its policy on the U.S. armed forces. In essence, the administration had eliminated the entire doctrinal, training, and procedural foundations that existed for the conduct of interrogations. It was now left to individual interrogators to make the crucial decisions of what techniques could be utilized. Therefore, the articles of the Geneva Conventions were the only laws holding in check the open universe of harsh interrogation techniques. In retrospect, the Bush administration's new policy triggered a sequence of events that led to the use of harsh interrogation tactics against not only al-Qaeda prisoners, but also eventually prisoners in Iraq—in spite of our best efforts to restrain such unlawful conduct.

In concert with this colossal mistake, the administration also created an environment of fear and retribution that made top military leaders hesitant to stand up to the administration's authoritarianism. The result was total confusion within the ranks in the execution of interrogations. The Army, as the executive agent, did nothing to clarify the policy, update doctrine and procedural guidelines, or revise the training programs for interrogators and leaders. Having eliminated the Conventions, it was the responsibility of the Department of Defense and the U.S. Army (as the executive agent) to publish new standards to steer our soldiers away from techniques that could be deemed torture. The fact that this was not done constitutes gross negligence and dereliction of duty.

After 9/11, there was tremendous pressure to extract key information from detainees to locate terrorist cells and uncover their attack plans. That pressure, in turn, led to the implementation of harsh interrogation techniques in Afghanistan. In the most egregiously documented incidents, during the late November to early December time frame, two Afghan detainees were beaten to death with batons. When word reached CENTCOM on December 14, 2002, GEN Franks directed LTG Dan McNeil, the commander on the ground in Afghanistan, to conduct a full investigation into the deaths.

Also in December 2002, CPT Carolyn Wood, the young reservist company commander assigned to the interrogation unit at Bagram, was given a warning order that she and her company might be transferred to Kuwait as part of the force that would invade Iraq. CPT Wood immediately spoke to investigators and fired off a number of communications up the chain of command that requested guidance, clarification of appropriate interrogation procedures, and training for her interrogators. She was taking the appropriate steps to ensure that similar abuses did not take place in Iraq.

At virtually the same time, Secretary Rumsfeld received a November 27, 2002, memorandum from SOUTHCOM requesting his approval for the use of various interrogation methods at the Guantánamo Bay detention facility. Back in October, GTMO had developed a comprehensive list of such techniques. Category I included four methods that were well below the limits of the Geneva Conventions. Category II included twelve techniques that were technically within the limits, but could be considered beyond depending how they were implemented. And Category III listed four methods that were clearly beyond the limits of the Geneva Conventions no matter how they were implemented. On December 2, 2002, Secretary Rumsfeld approved SOUTHCOM's request to use all interrogation methods in Categories I and II, and one method in Category III. In an apparent reference to the request for approval of prolonged standing techniques, Rumsfeld wrote under his signature: "However, I stand for eight to ten hours a day. Why is standing limited to four hours?"

By the third week in January 2003, LTG McNeil, at the request of the Joint Chiefs of Staff, convened a working group to look at all aspects of interrogation operations in Afghanistan. On January 24, 2003, McNeil's recommendations went to CENTCOM and then on to Washington. That report listed current and past interrogation techniques in use (some of which clearly went beyond the Geneva Conventions), and those that had been effective. The report also listed methods that the command desired to implement. In effect, they were asking for authority to use techniques such as "mild physical contact" (hitting prisoners), because they found them effective. Moreover, according to the report, most of the interrogation methods listed in GTMO's Category I, II, and III had been used extensively at the Bagram detention facility.

On January 27, 2003, while his report was percolating in Washington, LTG McNeil received the findings of the investigative panel looking into the Bagram abuses. It was established that a wide array of interrogation approaches had been used at the detention center—those within the limits of the Geneva Conventions, and those that clearly went beyond. In addition, twenty-eight soldiers in the XVIII Airborne Corps were implicated in the torture and death of the Afghan detainees in question.

LTG McNeil immediately fired off memorandums to GEN Franks at CENTCOM, to the Chief of Staff of the Army, and to the Joint Chiefs of Staff at the Pentagon. Basically, he laid out all the problems found in the Bagram investigations, listed the requirements necessary to fix them, and pleaded for help. "We desperately need guidance," he wrote.

By this point, every level in the chain of command (from Afghanistan to Washington) either knew, or should have known, that there was a serious problem regarding interrogation techniques—and that deaths as a result of torture had occurred in Afghanistan. Moreover, the investigations and reports associated with the Bagram offenses meticulously documented every single deficiency in the interrogation and detention operations that would later be associated with the Abu Ghraib debacle in Iraq. Those in authority who were aware of this information included key civilian principals and general officers in Afghanistan, the Department of Defense, the Army, and CENTCOM.

Formal charges were not filed against anyone complicit in the Bagram abuses for fifteen months—not until the Abu Ghraib abuses were revealed in the mass media in April 2004. It could be inferred that the entire chain of command was just hoping it would stay underneath the radar. Because of this, the Afghanistan abuses and, more important, the problems identified with detention and interrogation operations, were never communicated to CJTF-7 in Iraq. After Abu Ghraib broke, however, it became clear that something had to be done. The resulting courts-martial of the Bagram defendants documented everything that had happened, in graphic and horrendous detail.

As the United States struggled with the global war on terror, our forces continued to operate without the constraints of the Geneva Conventions. We had everybody from young officers to three-star generals pleading for guidance and help from the Pentagon. But it never came.

It was the duty of the intelligence community within the Department of the Army to impart specific guidelines and to institute training before sending our soldiers into battle. Such a strategic error can be categorized only as gross negligence and dereliction of duty.

Equally, and perhaps more, culpable were civilian leaders at the highest levels of our government. The Bush guidance and the military memorandums suspending the Geneva Conventions had unleashed the hounds of hell. And no one seemed to have the moral courage to get the animals back in their cages.

CHAPTER 9

The Rush to War in Iraq

On January 28, 2003, President George W. Bush delivered his State of the Union address in front of a joint session of Congress and a national television audience. Near the end of his remarks, he asserted that Saddam Hussein was either in possession of, or in the process of acquiring, weapons of mass destruction (WMDs). Regarding chemical and biological weapons, the President said Saddam had "the material to produce as much as 500 tons of sarin, mustard [gas] and VX nerve agent . . . , more than 38,000 liters of botulinum toxin . . . and upwards of 30,000 munitions capable of delivering chemical agents." On the subject of nuclear weapons, Bush stated "Our intelligence sources tell us that he has attempted to purchase high-strength aluminum tubes suitable for nuclear weapons production," and that "the British government has learned that Saddam Hussein recently sought significant quantities of uranium from Africa." The President went on to label Iraq "a serious and mounting threat to our country, our friends, and our allies." And he stated, that "if war is forced upon us, we will fight with the full force and might of the United States military."

The President's remarks occurred near the end of a months-long national media blitz in which Bush administration officials made their case that Iraq was a clear threat to the United States. Among other things, they stated that Iraq had long-standing ties to al-Qaeda, that Saddam was capable of inflicting death on a massive scale, and even though all information was not in, that they did not want the smoking gun to be a mushroom cloud. The precision of the media campaign had been demonstrated on Sunday, September 8, 2002, leading up to the votes of approval for action in Congress that occurred on October 10, 2002. That September day, *The New York Times* cited "Bush administration officials," in a front-page article stating "Iraq has embarked on a worldwide hunt to make an atomic bomb." To bolster the administration's case for action against Iraq, Vice President Dick Cheney appeared on NBC's *Meet the Press*, Secretary of Defense Donald Rumsfeld on CBS's *Face the Nation*, National Security Advisor Condoleezza Rice on CNN's *Late Edition*, GEN Richard Myers (Chairman of the Joint Chiefs of Staff) on ABC's *This Week*, and Secretary of State Colin Powell on *Fox News Sunday.*

Perhaps the most dramatic performance came eight days after Bush's State of the Union address, when Powell made a televised presentation to the UN Security Council. With CIA Director George Tenet seated directly behind him, Powell showed satellite photos of chemical weapons bunkers and illustrations of truck- and railcar-mounted mobile biological weapons factories. "Leaving Saddam Hussein in possession of weapons of mass destruction for a few more months or years is not an option," said Powell. "Iraqis continue to visit bin Laden in his new home in Afghanistan . . . [The] denials of supporting terrorism take their place alongside the other Iraqi denials of weapons of mass destruction. It is all a web of lies."

Over the course of the next few weeks, however, the international community expressed doubts about the Bush administration's reasoning for action against Iraq, and offered alternative plans to war. Chief UN Weapons Inspector Hans Blix, for example, appeared before the UN Security Council and rebutted many of Secretary Powell's assertions. He not only stated that no weapons of mass destruction had been found in Iraq, but also disputed the existence of mobile weapons labs. France and Germany offered up a peace initiative increasing the number of surveillance flights over Iraq and tripling the number of weapons inspectors. But the Bush administration rejected the proposal as an attempt to disrupt its timetable for invasion. And Turkey's parliament rejected a one billion dollar aid package from the United States in exchange for allowing U.S. troops to be deployed in their country. Such an arrangement would have allowed for simultaneous attacks into Iraq from the north, as well as from the south through Kuwait—and would have provided a second, crucial logistics route.

On February 25, 2003, GEN Eric Shinseki, Chief of Staff of the Army, was testifying before the Senate Armed Services Committee when Senator Carl Levin candidly asked him what kind of manpower would be necessary to keep the peace in a postwar Iraq. "Something on the order of several hundred thousand soldiers would be required," he responded.

But Shinseki suffered an immediate rebuttal from the highest levels of the Department of Defense. In a Pentagon briefing, Secretary Rumsfeld stated "The idea that it would take several hundred thousand U.S. forces, I think, is far off the mark." And Deputy Secretary of Defense Paul Wolfowitz responded by making the following statement to the House Budget Committee: "The notion that it will take several hundred thousand U.S. troops to provide stability in post-Saddam Iraq [is] wildly off the mark." Wolfowitz also told Congress that the Iraqi people "will welcome an American-led liberation force."

Those remarks by Rumsfeld and Wolfowitz sent a chilling message to junior generals like me. They basically said the Army's generals had questionable credibility with senior civilian leadership at the Department of Defense. But GEN Shinseki was right. We knew he was right. And we were very proud that he had the courage to stand by his estimate and did not back down in the face of intense political pressure.

Four months later, on June 11, 2003, a large congressional delegation attended GEN Shinseki's retirement ceremony. But, neither Rumsfeld nor anyone from his office showed up, and that disturbed me. Eric Shinseki served for thirty-eight years in the Army, lost part of his foot in combat in Vietnam, and had honestly answered a U.S. senator's question. The Secretary of Defense should have been there to convey thanks on behalf of the nation. GEN Shinseki had it right when, in his outgoing remarks, he said that arrogance of power is the worst substitute for true leadership.

During the first two and a half months of 2002, as President Bush gave his State of the Union address, and GEN Shinseki testified before Congress, approximately 120,000 U.S. combat troops were deploying to Kuwait and the Persian Gulf in preparation for an invasion of Iraq. An additional 100,000 naval, air force, logistics, and intelligence personnel were sent to the region for support. Part of this deployment included CPT Carolyn Wood and the military intelligence unit from the Bagram detention center in Afghanistan. Along with this unit came all of the interrogation techniques, the procedures, and confusion associated with being transferred from a theater of war where the Geneva Conventions did not apply. No one in CPT Wood's higher chain of command (with the exception of LTG McNeil) had taken any steps to correct the problems that led to the Bagram prisoner deaths. There had been no guidance, no new interrogation procedures, and no training provided by the Army or the Department of Defense. Everybody was too busy focusing on the invasion.

In Wiesbaden, the 1st Armored Division was still struggling with constantly changing force deployment plans. By late February 2003, we still didn't know whether we would be part of the campaign at all. Our standing orders were simply to be prepared to deploy and conduct combat operations in support of a regional combatant commander. That's about as vague as it gets. So, after making a decision to train for a reserve mission, I prepared the division for expeditionary operations in a high-intensity conflict. I was not willing to take any chances.

In February, 1st Armored participated in V Corps's conventional warfighting exercise called Victory Scrimmage. All of the divisions that were expected to deploy to Iraq participated, although certain elements of the corps (including LTG Scott Wallace's tactical command post) were already in Kuwait. At the conclusion of the three-week exercise, we were fully prepared for major conventional combat operations in Iraq. But in terms of the twenty-first-century, 360-degree future battlefield, we were taking a tremendous risk, because we were not prepared to transition from conventional combat to urban warfare, and had not conducted any Phase IV planning.

LTG Wallace recognized this deficiency and called me in. "Ric, I need your help," he said. "I want you to see what you can do about developing some kind of a training package for combat in cities against an unconventional enemy. Obviously, time is of the essence. You've got to take this on and bring it to the entire corps as we prepare to attack."

The Army's doctrine in this area was nonexistent, so we had nothing on which to base a training exercise. But we plowed ahead and developed a simulation exercise that got us into Baghdad, a massive city with a population of more than six million people. The U.S. Army Europe Multi-National Training Center and COL J. D. Johnson, our 2nd Brigade commander, put together a scenario that included attacking, defending, integrating close air support, reconnaissance, and all of the warfighting functions of a battle in urban combat. J.D. and his team took the training package forward to LTG Wallace, but because V Corps was on the verge of attacking, he had no time to do anything with it, except distribute some of the documentation and lessons learned. Fortunately, because 1st Armored had not yet received orders to deploy, we were able to begin training with this new set of guidelines. In fact, I made it mandatory. According to our Cold War fighting doctrine, heavy armor and mechanized forces were supposed to bypass urban areas. But we were now training to defeat residual unconventional forces defending major cities.

In the long run, LTG Wallace's original vision of this task would turn out to be extraordinarily valuable.

On March 4, 2003, I was finally informed that all of 1st Armored was going to deploy. "You're going," they said. "We don't exactly know what you're going to do when you get there. But you're going. It's now solid." Just as our forces were beginning to flow, however, I received a call to stop all movement. Apparently, the entire operation had been complicated by the refusal of Turkey to allow U.S. forces to stage from its territory. So for the time being, Old Ironsides was again on hold.

That interim waiting period gave us not only additional training time, but an opportunity to continue firming up plans to take care of our families once we finally did go to war. Back in 1991, during Operation Desert Storm, we had experienced some major family problems in Fort Benning and, now that I was a division commander, I was determined not to let it happen again. We assigned a quality commander to stay behind in Germany as part of a rear detachment devoted to family care. Next, we set up readiness groups comprised of spouses who had access to resources for any conceivable problem a family might experience (such as finances and emergency leaves). Finally, we established communications centers (called "Yellow Ribbon Rooms") so that spouses and children would be able to reach their loved ones via telephone, e-mail, and video teleconference. After Desert Storm, I became a firm believer in rear detachment operations. Family was one thing I did not want a soldier in battle to be worrying about.

While we were waiting for orders in Germany, America's rush to war continued unabated. In early March, the United States and Great Britain launched an air bombing campaign to soften up Iraqi defenses ahead of the invasion. By mid-March, France's President Jacques Chirac had warned President Bush and British Prime Minister Tony Blair that France would not support any resolution from the United Nations authorizing war with Iraq.

In response, Secretary of Defense Rumsfeld publicly suggested that the United States could go it alone—and Bush and Blair gave the United Nations twenty-four hours to enforce its own resolutions for Iraqi disarmament, or an invasion would be launched within days. Finally, the United States and Great Britain withdrew their proposed UN Security Council resolution, because France, Russia, and China refused to support it. And on March 18, 2003, in a live television address to the nation, President Bush gave Saddam Hussein forty-eight hours to leave Iraq, or else face invasion.

On March 20, 2003, approximately ninety minutes after the deadline expired, coalition forces led by GEN Tommy Franks invaded Iraq. As bombs fell on Baghdad military targets, 170,000 combat troops moved north out of Kuwait. The ground force was comprised of 120,000 American soldiers and Marines, 45,000 Britons, and 5,000 troops from Australia, Poland, and Denmark. This was a much smaller force than we had used during Operation Desert Storm in 1991. However, we made up for it with extensive use of precision high-tech weaponry, including satellite-guided bombs and advanced unmanned aerial reconnaissance vehicles. In his radio address two days later, President Bush stated that the war's objectives were "to disarm Iraq of weapons of mass destruction, to end Saddam Hussein's support for terrorism, and to free the Iraqi people."

Three major offensive lines moved northward out of Kuwait. The Army on the west was led by the 3rd Infantry Division; the 1st Marine Division took the lead in the center; and the British 1st Armoured Division led the way on the east. With the help of coalition air supremacy, Iraq's southern cities were captured first. This move set up protection of the crucial (and only) logistics supply line that would be established out of Kuwait. More than 1,000 paratroopers from the 173rd Airborne Brigade were dropped into northern Iraq to secure key airfields. And by early April, coalition troops had captured Saddam International Airport just west of the capital and renamed it Baghdad International Airport. Shortly afterward, both the Iraq military and the government collapsed, Saddam Hussein went into hiding, and jubilant civilians flowed into the streets of Baghdad to celebrate. The unofficial end of the invasion occurred on April 9, 2003, when a group of U.S. Marines helped Iraqis topple a massive statue of Saddam in a Baghdad city square—an event that was televised around the world.

The U.S. military assumed a significant calculated risk when it launched this invasion. We were using less than half the number of forces we had used in 1991, and the attack was launched while some units were still deploying to Kuwait. Arriving brigades unloaded their equipment, put on their battle gear, and were immediately sent into the fight. We called it a "rolling start." Despite the risk in both size and flow of forces, it turned out to be one of the finest offensive actions that our country has ever conducted—a magnificent performance by all the coalition forces.

Within days of the toppling of Saddam's statue, I was informed that the entire 1st Armored Division would be deployed, and that we should expect to be in Iraq by early May. As I traveled around the division, I heard our soldiers and leaders express disappointment. We had been watching the action on television just like everybody else, and knew that major combat was over. That was what we had trained and hoped for. To only be a part of a nation-building caretaker effort was not "exciting" for the soldiers. Besides, other than our urban operations training, we had not discussed or planned for Phase IV operations or low-intensity conflict. The possibility of an insurgency had been mentioned, but we had not trained or planned to fight one. So a good deal of uncertainty entered the picture. Everybody was asking, "Exactly what are we going to do when we get over there, anyway?"

From that point on, I spent a lot of time explaining, based on my experience in Kosovo, that Phase IV operations usually did involve combat operations. More important, I reminded everybody that if this is what our country asked of us, it was our duty to perform to the best of our abilities. And I knew that everybody would perform well, because they were prepared. The training they had gone through was the best we could offer, and every unit had done exceptionally well. It didn't matter that we were going to be encountering situations we definitely had not seen before or could even have imagined. I still remained confident that our soldiers could prevail. And I was particularly proud of the leadership team we would be taking to Iraq. We had been together for eighteen months, since the summer of 2001, and were a tight-knit group. The brigade commanders of the 1st Armored Division were simply the best in the Army. They were magnificent warriors who loved their soldiers.

As we prepared to deploy to Iraq, I held meetings across the division with all leaders, both officers and noncommissioned officers. Using computer-generated slides, I spoke for about forty-five minutes on combat leadership, and then held lengthy question and answer sessions. Some people might think of it as an inspiring pep talk before the big game. But for me, it was much more than that. The soldiers under my command were going to war, and I was responsible for their safety. So it was my duty to share my knowledge, because I had been where they were going. I also wanted everybody to know that I believed in them, that they were magnificent young men and women, and that I was proud to be part of their team.

I began my presentation with some thoughts on what to expect in Iraq. "We are going to be deployed for an indefinite period of time into an unforgiving environment," I said. "You'll be in a 360-degree, twenty-four-hour-a-day battlefield against a very sophisticated, thinking enemy who adapts daily, is patient, brutal, and has no seeming clarity of purpose. And don't forget that this is a joint operation where some of our coalition partners speak different languages."

I also wanted to set some realistic expectations for soldiers and their families. Our deployment orders were for one year, but I told everybody to plan on being there for at least 365 days, or until the mission was completed. That way, if we were extended, it wouldn't be too big a disappointment. "And don't forget that at the end of the day," I also said, "it is your solemn responsibility to protect the American way of life, to love your fellow soldiers, to accomplish your mission, and to be a warrior." I related my flat-ass rules (FARs) for combat leaders, which was done using the acronym I.R.O.N. SOLDIERS (a play on our division's name, Old Ironsides, and the heavy armor associated with our division).

I was for "Iron Discipline." Prior to Operation Desert Storm, GEN Barry McCaffrey had explained to us what he called the "ten percent rule." No matter what you do, there are 10 percent of your soldiers who are criminals by nature. These troublemakers are kept in line by the discipline standards we impose and the leadership provided by junior officers. "Doing the right thing when no one is watching is the ultimate test of your discipline," I said.

R is for Respect. At all times, we will have respect for our fellow soldiers, for the environment, for the enemy, and for civilians. *Never* forget dignity and respect.

O stands for "On the Battlefield." It's about leadership under fire. In the toughest of times, under the worst of conditions, a leader must be out in front. Soldiers expect it and our warrior ethic demands it. We must be present at critical points on the battlefield. As GEN McCaffrey taught us in Operation Desert Storm, sometimes the worst thing that can happen to you is to not get killed in battle along with your soldiers. We will be ruthless in battle and benevolent in victory. That's what makes us the best army in the world.

N means "Never Run Out of Options." On the battlefield, we must bring to bear all the combat power available to us. It is a leader's responsibility to know what all his options are, and how to mass them on the battlefield. We will never accept defeat.

SOLDIERS means "Taking Care of Soldiers." We must never allow a soldier to go into harm's way untrained. That is a sacred responsibility. The best way leaders can take care of their soldiers is to train them under the toughest of conditions.

After discussing my FARs, I would reemphasize our Army values. I pulled out the chain I have worn around my neck for the last thirty years to display my dogtags and the Army values tag that is issued to every soldier. Holding them high, I would say, "These are the values that you embraced when you swore to be a soldier. Every one of you knows what *L-D-R-S-H-I-P* stands for: Loyalty, Duty, Respect, Selfless Service, Honor, Integrity, and Personal Courage."

Then I spoke about honor and integrity, making it clear to everyone that when it came to these two values, there were no compromises. "Honor is absolute. And the only person that can compromise your integrity is yourself." With regard to personal courage, I tried to ease the lingering doubts that I knew many of our younger soldiers were having. "Don't worry about your courage under fire," I said. "You'll either do your duty, you'll be a coward, or you'll be a hero. Those are the only three options when you're engaged in battle, and yours will be an instinctive reaction." And finally, I told everybody that one of the real tests of a soldier comes when it's time to display moral courage. You must do what is right in spite of personal consequences.

After giving them a little time to digest what I had just said, I offered some personal thoughts. "I had to search for a source of strength in combat," I said. "Friends and thoughts of family provided a certain level of support. But in the end, I had to go back to my roots and dig deep into my beliefs to get the courage and wisdom necessary to do what was right on the battlefield. Each of you has to figure out what your source of strength is going to be. For me, it is my faith and my God.

"After you come back from the combat zone—when all the parades are over and you are back with your family—you will look back on your wartime experience and wonder if you did the right thing. If you have not lost any soldiers due to a lack of training, leadership, or discipline on your part, then you'll be okay."

Finally, I closed out my presentation with the 144th Psalm:

Praise be to the Lord, my rock,
who trains my hands for war,
my fingers for battle.
He is my Loving God and my Fortress.

CHAPTER 10

"Mission Un-Accomplished"

"You Army guys, you have no joint experience and all of you are tied to these byzantine command structures! You don't know how to operate in a joint environment!"

"Mr. Secretary, you're mistaken," I replied. "I have served in two joint assignments as a general—two years at Southern Command and one year on the ground in Kosovo. I also had a joint assignment as a captain. I understand the interagency process and I understand joint operations."

I was speaking with Secretary Rumsfeld in his office at the Pentagon. I knew he was trying to intimidate me, because he had my résumé folder opened on the table right in front of him.

It was April 10, 2003. The statue of Saddam Hussein had just been toppled in Baghdad, and I had been asked to fly to Washington for an interview. I was one of two major generals being considered to take over command of V Corps (a three-star position) from Scott Wallace. Before Secretary Rumsfeld assumed office, such interviews seldom happened. The Army would present the Secretary of Defense with a list of two or three names, one of which was clearly better than the others, and almost always, the Army's recommendation would be approved without an interview. But as part of his micromanagement style, Secretary Rumsfeld had inserted himself into the selection process for all three- and four-star generals. In the early days, senior Army leaders learned that their second or third choice was just as likely to get selected, so the process had changed to where all nominees had to be equally qualified for the billet.

Before my interview, members of the Army senior staff prepped me. "Watch out," they said. "He distrusts Army guys and he'll try to overwhelm you with questions." They were absolutely right, because for thirty-five minutes, Rumsfeld hit me with a series of rapid-fire questions to see if I would focus on the core issues and rebuff his inaccurate statements. Then he abruptly ended the interview, and I had no idea where I stood.

Within forty-eight hours, though, I received a call from the other general being considered for the position. "Congratulations, Ric," he said. "You got the job."

"How did you find out?" I asked.

"I heard it from a guy in the Chief of Staff's office. Apparently, your joint experience was the deciding factor."

On April 16, 2003, GEN Tommy Franks issued orders to withdraw American warfighting units from Iraq within sixty days, and use incoming forces for only up to 120 days. In effect, that order would reduce our presence to fewer than 30,000 troops by the first of August. Additionally, CENTCOM's headquarters element would leave the region in May. As part of this overarching guidance in redeploying forces, GEN Franks explicitly stated that military leaders should take as much risk coming out of Iraq as we did going in—which meant that we were going to try to get by with the smallest number of ground troops possible.

At this point, I did not understand all the conditions in Iraq, nor was I informed of GEN Franks's orders. In looking back, however, there was no doubt that the Army was going to be stretched very thin across the entire spectrum of Phase IV operations. GEN Franks's reasoning, although strategically flawed, was pretty clear. Essentially, the Army was replicating the Desert Storm, Haiti, Grenada, Panama, Bosnia, and Kosovo models. We had gone into Iraq, eliminated the Saddam regime, and liberated the country. The war was now won and the rush to return home had begun.

This order had to have been issued with the approval of the highest levels of the U.S. government. There is simply no way that Tommy Franks would have dared to take such a monumental action on his own authority. Furthermore, all major departments in the U.S. government were aware of the troop movement plans. In association with GEN Franks's orders, for instance, both CENTCOM and Army public affairs issued guidance to be used during redeployment.

Also in April 2003, Secretary Rumsfeld issued a memorandum directing the Department of Defense to proceed with his ongoing military transformation guidance. This was part of the DoD's broader initiative to enable missile defense and transform the military from a Cold War organization to one that was more rapidly deployable, agile, and expeditionary. Rumsfeld's directive was issued *during* the invasion of Iraq and *before* Franks's redeployment orders. It was as if both the war and the postwar operations were secondary. But Rumsfeld was deadly serious about his transformation order, and he put tremendous pressure on the services to comply. For the most part, the services adhered to this strategic shift in priority, because it was natural for them to do so. Before the invasion of Iraq, the Army had been totally committed to creating an irreversible momentum that would allow us to complete the transformation. And now that the "disruption" was behind us, the Secretary's memorandum simply reaffirmed all of our instincts to forge ahead with resetting the force and returning to our main task of preparing for the next contingency operation.

Apparently, our senior leadership really believed that the force levels required for Iraq stabilization could be low. But I had trouble reconciling how the government as a whole, and the Department of Defense, in particular, could shift priority away from the war and just move forward to transform the American military.

Over the course of the next two months, the realities of the Iraq ground situation would create another Vietnam-like Cold War situation in the Middle East. While the executive branch and the Interagency of the U.S. government (the Department of State, National Security Council, Joint Chiefs of Staff, and Department of Defense) were going on with business as usual, ongoing events including de-Baathification of Iraqi society, the disbanding of the Iraqi military, and the reversal of GEN Franks's redeployment orders ensured an extended American presence in the Middle East.

Overall, the concurrence of Franks's drawdown orders and Rumsfeld's transformation directive created havoc throughout the forces. The 1st Armored, the 4th Infantry, and the 101st Airborne (Air Assault) Divisions were still moving to occupy their sectors, and elements of the Marine Expeditionary Force, the 3rd Infantry Division, and a large number of support units were either in Kuwait, moving out of Iraq, or preparing for redeployment. Some units were coming in, some were going out, and others were holding in place. Confusion was the order of the day.

It was difficult to understand why Franks's orders were put in place, and why the senior military and political leaders of our nation so readily accepted them. Had they really allowed themselves to believe that we could go into Iraq, execute regime change, and then come out within sixty days? Had we really learned nothing from our recent experiences in Bosnia and Kosovo—that in order to achieve lasting security and stability, political and economic issues had to be addressed along with the military challenge? America's intervention was supposed to last no more than one year in Bosnia—and we were still there a decade later. I worried about the interactions at the national level between the Department of Defense, the National Security Council, and Congress. Were they all still focused on this war? Did our national political and military leadership know what was really happening on the ground in Iraq? Or was their lack of situational awareness due to indifference and negligence?

I had no answers to these questions, at least not at this point. What I did know, however, was that there was no plan for Phase IV operations in Iraq—certainly not on behalf of the combatant commander. As a matter of fact, before the war, GEN Franks had agreed with Secretary Rumsfeld that CENTCOM would not be involved with Phase IV. He was preparing his battle plan and he did not have the time, expertise, or inclination to work on that phase of the campaign plan. As a result, GEN Franks's original directive was silent on the military's role in Phase IV operations. But subsequent orders directed ground forces to provide only minimal support for postcombat activities. "Offer only transportation, logistics, and some communications capabilities," was the guidance.

It was well known that, before the war, there had been a major philosophical and turf battle between the Departments of State and Defense regarding who was going to be responsible for all aspects of the war. The Department of State, assuming its traditional lead in handling postwar actions and policies, had already completed a massive project called "Future of Iraq." Secretary Rumsfeld, however, went to the White House and obtained a secret presidential directive, National Security Presidential Directive 24 (NSPD-24), in which President Bush assigned DoD responsibility for all Phase IV planning and implementation. That document, signed on January 20, 2003, also established that the Office of Reconstruction and Humanitarian Assistance (ORHA) within the Department of Defense, would handle the postwar effort, including such tasks as dismantling weapons of mass destruction; defeating terrorists networks; reshaping and reforming security institutions (the Iraqi military and police security forces); facilitating reconstruction; protecting infrastructure; assisting in reestablishment of key civilian services (food supply, water, power, and health care); supporting transition to Iraqi authority (establishing new civil administration governance); and providing for its own operational needs.

To lead this massive effort, Rumsfeld appointed retired LTG

Jay Garner, a veteran of the 1991 Gulf War and former Assistant Vice Chief of Staff of the Army. From the very beginning, however, NSPD-24 created a tremendous disconnect between ORHA and everybody else. The Department of State was not involved now. CENTCOM was leaving the region and had issued guidance to provide only minimal assistance. And the Department of Defense had not structured, funded, or resourced the organization properly. ORHA had simply fallen through the cracks of a giant bureaucracy that was led by political appointees who failed to recognize the enormity of the task America was undertaking. The truth is that Jay Garner had been given an impossible task.

LTG Garner, however, performed superbly. He developed a twelve- to eighteen-month engagement plan, took it to President Bush, and advised that ORHA did not have the capabilities to handle some parts of the mission, such as dismantling WMDs and defeating terrorists. The key elements of his strategy included holding elections within three months, selling oil to foreign countries to raise revenue to pay for his plan, using lower-level Baathists to sustain governmental functions, and utilizing between 200,000 and 300,000 former members of the Iraqi Army to help in postwar reconstruction. On April 21, 2003, his first day in Baghdad, LTG Garner also set up an interim Iraqi advisory group (the Senior Leadership Council) made up of key Sunnis, Shiites, and Kurds. His idea was to put a local face on the occupation government—and it was a *good* idea.

Unfortunately, ORHA was doomed from the start. In Washington, there were never-ending arguments over the implementation of policy. The Department of State felt it should be involved, as it had been for the past 200 years in these types of situations. Some of the neoconservatives in the White House and Pentagon wanted to arrest or eliminate members of the former Republican Guard and the highest levels of Saddam Hussein's Baath Party. And others wanted former members of the regime to have nothing to do with a new government in Iraq. It was not surprising then, when after only a few days in Baghdad, LTG Garner received a call from Secretary Rumsfeld thanking him for doing a good job, and informing him that he was being replaced by former Ambassador L. Paul Bremer. I do not know how this change came about. All I knew for sure was that Garner was out, Bremer was in, and the situation in Iraq was not good.

Coalition forces were not yet in control of the entire country. While we had secured most of the southern part of Iraq on our way up from Kuwait, we had not cleared the majority of Anbar province, the Sunni Triangle, or any of the areas in the north—nor did we have any idea what the strength of the enemy was in these places. The ground forces had culminated their attack in Baghdad, but hadn't completely secured the entire city.

There was also significant violence everywhere in Iraq. The fall of Baghdad had ignited an outbreak of tribal skirmishing as everybody began jockeying for power. And with the total collapse of Iraqi police and security forces, there was widespread looting and lawlessness. Government offices, private stores, and palaces were ransacked. Power plants, oil pipelines, and bridges were sabotaged and/or blown up, presumably by Saddam loyalists. And what's more, the coalition combat forces were still focused on their primary warfighting mission of destroying Saddam's military, and were at work clearing the country. It would be some time before they were assigned the mission to confront looting and other criminal activity.

Additionally, there were unsecured arms and ammunition dumps all over the place—not just in Baghdad, but everywhere in the country. We estimated that nearly one million tons of conventional arms and munitions had been distributed around Iraq, and most of it was unguarded and free for the taking. Sure enough, in early May, the Iraqis began holding open-air arms bazaars that sold everything from small handguns to rocket-propelled grenades.

Most of the American public was completely unaware of the situation on the ground in Iraq. To further cloud reality, on May 1, 2003, President Bush declared the war to be over. In the waters off the coast of San Diego, California, clad in a pilot's flight suit and flying in a Navy fighter jet, Bush landed on the aircraft carrier USS *Abraham Lincoln.* When he addressed the nation from the deck of the carrier a short time later, a giant blue and white banner that read "Mission Accomplished" was visible behind him. The President began his speech by saying, "Major combat operations in Iraq have ended. In the battle of Iraq, the United States and our allies have prevailed. And now our coalition is engaged in securing and reconstructing that country." This event, which was televised live and broadcast around the world, immediately caused nearly every service and agency in the U.S. government to declare victory and begin to shift their focus away from Iraq.

At first, I didn't understand the situation on the ground well enough to question or understand the reasons for this declaration. But with time, everything became clear to me. On May 1, 2003, the military coalition was comprised of only five nations, mainly because America couldn't convince anybody else to join. Nobody wanted to commit forces as long as there were "major combat operations" going on. The President's "Mission Accomplished" declaration gave the illusion that things were stable in Iraq. Not surprisingly, then, in the first week in May, a massive U.S. effort was launched to increase participation in the coalition. Back in Washington, governments were offered all kinds of incentives to get them to join up—equipment of all kinds, logistical and transportation support, financial aid, and so on. And it was a successful recruiting effort. The coalition was very rapidly expanded from five nations to thirty-six. But each of those nations would have its own peculiar support agreement, its own national interests to look out for, its own rules of engagement, and its own national chain of command.

Thousands of coalition troops began to arrive in Iraq. But there was nobody working on a plan to bring them into the fold, to help with their predeployment training, or to ensure that America's commitments to them were fulfilled. All of those details were left to the coalition ground commander to sort out and resolve.

On the other hand, by the end of May 2003, the United States and Great Britain had offered the UN Security Council a resolution designating themselves "occupying powers," which would give them control of Iraq's oil revenues. The two nations had also secured a 14–0 vote to lift economic sanctions and confer them with official control of the country.

DURING THE FIRST WEEK of May 2003, I was with the 1st Armored Division in Kuwait and we were 70 percent deployed. With orders to relieve the 3rd Infantry Division and assume responsibility for Baghdad, we launched into Iraq and I arrived in the city on May 8, 2003.

There was chaos everywhere. Looting was more extensive than I had anticipated, and fires were burning all over the place. These images were broadcast in the media, which created tremendous pressure to get things under control. So from day one, 1st Armored Division was consumed with trying to bring stability to Baghdad. As we embraced the task, our soldiers spread out across the city. We set up dozens of small forward operating bases, and then identified 350 to 500 specific sites to protect. These locations included critical government infrastructure, minority neighborhoods, and religious sites. We monitored the areas with drive-by patrols or protected them with soldiers on location serving as guards. After only a few weeks, however, the violence and looting began transitioning to direct attacks against our forces.

At the corps level, LTG Scott Wallace took on the task of trying to get control of the thousands of arms and ammunitions dumps across the country, some of which covered areas that were measured in kilometers. In Baghdad alone, there were hundreds of these stashes all over the city. Every one had to be secured, the weapons demilitarized, and the ammunition destroyed. However, it soon became apparent to V Corps headquarters that the time and resources required to accomplish such a task were beyond the capabilities of our military forces on the ground. As a matter of fact, a quick study estimated that in our current situation, it would take three to five years to complete. Believing that the military's top priority was to quell the violence and stop the looting, higher headquarters made the decision to contract out the entire operation. Civilian companies would guard the sites with Iraqi help, professional explosive ordnance people would demilitarize the weapons and destroy the ammunition, and a minimal number of coalition military units would aid and assist in the effort.

Another immediate task for the coalition forces was to begin work on reestablishing the Iraqi Army, police, and security forces. Believing that the military would have responsibility for this task, the V Corps staff developed a solid plan to create integrated institutions. In addition, V Corps, just like ORHA, believed it was crucial to make the Iraqis part of the solution in reconstructing the security of the nation. Our challenge: how to convince them to come back, and once they returned, whether to bring them in as individuals or units. GEN John Abizaid took the lead on this issue. After Kosovo and a brief stint on the Joint Staff in Washington, he had become the deputy commanding general of CENTCOM under Tommy Franks. He had participated in one of the most successful ground offensives in U.S. history—the invasion of Iraq—and was slated to take over from Franks upon his retirement in July 2003. In late May, Abizaid and I sat down and discussed how to rebuild the Iraqi Army. We knew there were 300,000 to 400,000 regular Iraqi Army soldiers who were jobless and probably willing to return to duty. Realistically, though, we knew they would not fight for us, because after all, we were the infidels. But they might fight for their own leaders. With that thought in mind, Abizaid did some preliminary work in Baghdad and began talks with a few former Iraqi Army generals. "Just what is the possibility of bringing back some of the units of the Iraqi Army?" he asked. The feedback Abizaid received was extraordinarily positive.

In addition to tempering the violence and looting, securing the ammo dumps, and beginning the work to rebuild Iraqi security forces, V Corps started to reopen all the universities and schools across the country so the kids could finish the school year. We also began work on a comprehensive Phase IV plan for all of Iraq that focused not only on security, but also on political self-governance and economic development. As part of that plan, without waiting for guidance from ORHA, we immediately started setting up local provincial governments under the supervision of division commanders. Our strategy was to find the economic centers of gravity in every city, town, and village in Iraq—and then jump-start them into productivity, profitability, and prosperity. In short order, the progress reports from the division commanders gave us cause for great optimism. Things were really looking up, and I was starting to believe that we could really bring stability to this country after all.

All that began to change, however, with the arrival of Ambassador L. Paul (Jerry) Bremer, Garner's replacement. Having received notice that Bremer was flying in, I went down to Kuwait to meet him at the airport. Upon arrival, I linked up with LTG Garner, other members of ORHA, and Ambassador Bremer's advance party. When he got off the plane, he said hello to me, acknowledged LTG Garner, and then asked me to come in and chat with him for a while. Garner waited outside. After exchanging pleasantries while he cleared customs, Bremer said that he had to go meet some people in Kuwait City, and that he was looking forward to working with me. As Bremer left the building and walked to his car, LTG Garner followed along and spoke to him. Bremer then got into the waiting Suburban and left Garner standing on the curb.

Given the culture of his profession, Bremer's attitude toward Garner was fairly normal. Most of his career had been spent in the Department of State, in part as a protégé of Henry Kissinger, as a former assistant to Alexander Haig, and as ambassador to the Netherlands during the Reagan administration. It is normal protocol that when ambassadors transfer responsibilities, they are almost always separated by time and space, and there is no face-to-face transfer. That is counter to the military's ethic and procedure, which is where Garner was coming from. It was his duty, he believed, to provide all the help and information he could to ensure that Bremer had a smooth transition. Ambassador Bremer probably did not understand that.

The next day, Bremer was in Baghdad to begin his tenure as head of the Coalition Provisional Authority (CPA). His first few weeks were a bit rocky, however. When he walked into a meeting with the leadership of V Corps, for example, he announced to LTG Wallace, "I'm the CPA Administrator and I'm in charge. I want you to co-locate your command and control center with CPA in the Green Zone, immediately." That statement completely turned off the military officers, who viewed Bremer's behavior to be arrogant. "Who is this guy and who the hell does he think he is, anyway?" they wondered.

It didn't get any better, either, when in an early staff meeting, Bremer suggested shooting the looters as a way of getting the country's violence under control. That comment got out, made worldwide headlines, and caused him some real problems. Ambassador Bremer didn't really mean it, though. He was reacting to the pressure from the administration and the media, and it just slipped out as a frustrated offhand remark. But somebody in the meeting related his comments to the press and the next thing we all knew, they were blown way out of proportion.

Ambassador Bremer came into the country with some pretty specific priorities from Washington, and they did not coincide with the plan Jay Garner had previously laid out for Secretary Rumsfeld and President Bush. One of the first things Bremer did was to meet with the Senior Leadership Council (SLC)—the Iraqi advisory group that Garner had set up. The SLC was composed of representatives from seven organizations that were supposed to have a "special" place in the new Iraq. More than that, though, these leaders saw themselves as the post-Saddam government. But Ambassador Bremer let them know right off the bat that he didn't see it that way. "You are not the government," he said. "The CPA is in charge." This Iraqi advisory group never met again, and a dejected Jay Garner soon left Iraq, having been in the country about a month. Most of his staff left the country by mid-June and ORHA ceased to exist.

The Coalition Provisional Authority was assigned the mission and given authority to run Phase IV operations in Iraq—with the military in direct support. And while it was clear that Ambassador Bremer was in charge of administering the occupation, details of the command relationship between CPA and the military were never clearly defined by any level of command, all the way up to the Department of Defense. This was a fundamental step that had simply been ignored, and with the lack of a well-defined structure, it was left to the ground commander and Ambassador Bremer to develop the guidelines for how the CPA and the military were going to work together. Initially, Ambassador Bremer believed that the military was going to work for him, and that the CPA could establish priorities and issue orders. No one in Combined Joint Task Force 7 thought that was a good idea. It was civilian *command* of the military, and that was not acceptable.

The presence of Ambassador Bremer and the CPA in Iraq further complicated an already convoluted command structure. Here's how the military chain of command looked on paper during the invasion and in its immediate aftermath:

Department of Defense—Washington, D.C.

Secretary of Defense Donald Rumsfeld

Central Command (CENTCOM)—Qatar

GEN Tommy Franks

Combined Forces Land Component Command (CFLCC)—Baghdad

LTG David D. McKiernan (Wartime ground command)

First Marine Expeditionary Force (MEF)—Iraq

LT GEN James T. Conway

Fifth (V) Corps [Equivalent to MEF]—Iraq

LTG Scott Wallace

(included 3rd Infantry, 101st Airborne, 4th Infantry, and 1st Armored Divisions)

One week after Saddam Hussein's statue was toppled in Baghdad, the military chain of command changed rapidly. On April 16, 2003, as part of GEN Franks's force drawdown order, CENTCOM started planning to leave the theater and move to Tampa, Florida. Its forward command and control center in Qatar ceased wartime operations and was completely gone by May 1. GEN John Abizaid would take over command of CENTCOM in July.

Also on May 1, the Combined Forces Land Component Command (CFLCC), led by LTG David McKiernan, assumed the designation of Combined Joint Task Force 7 (CTJF-7), with responsibility for the activities of all coalition forces in Iraq. First Armored Division completed its deployment into Baghdad within a week and relieved the 3rd Infantry Division by mid-May. But only two weeks later, on May 16, it was formally announced that CFLCC would be departing Iraq and relocating to the United States. This abrupt turnaround was another monumental blunder that created significant strategic risk for America. CFLCC had fought a magnificent ground campaign, and possessed the institutional knowledge, command relationships, and organization to transition the war smoothly from major combat to Phase IV operations. LTG McKiernan had assembled the best staff that the Army had to offer at that point in its history. We called them "The Dream Team." But now, the dream team would be gone. I believe this decision was made by Franks and McKiernan, partly because they thought the war was over, and partly because they did not want to have anything to do with Bremer, the CPA, and Phase IV.

Whatever the reasons for CFLCC's disengagement, the foreseeable consequences were daunting. In country, we would no longer have the staff-level capacities for strategic- or operational-level campaign planning, policy, and intelligence. All such situational awareness and institutional memory would be gone with the departure of the best available Army officers who had been assigned to CFLCC for the ground war. The entire array of established linkages was dismantled and redeployed. Furthermore, V Corps had no coalition operations and ORHA/CPA–related staff capacity, which were departing the theater with CFLCC just at a time when the coalition and civilian administrator support missions were dramatically expanding. Not having these necessary capacities would make it extremely difficult to fight the ongoing war in Iraq, provide much-needed support for the CPA, and bring stability and security to the country. And finally, the loss of our strategic-level national intelligence capacities would cause serious problems that would lead in part, to future problems at Abu Ghraib.

According to the May 16 CFLCC announcement, effective June 15, 2003, V Corps would assume the designation CJTF-7 and all its associated missions and responsibilities. That meant that upon my assumption of command of V Corps, I would become the senior military commander on the ground in Iraq.

Up to that point, I had been working with Scott Wallace to shift the command of V Corps, but now we needed to begin immediate work on transitions for the CFLCC and CJTF-7 missions. Clearly, the current V Corps headquarters staff structure was inadequate for a combined joint mission of such magnitude. Wallace informed me that McKiernan's command had been working on a Joint Manning Document (JMD) that would establish roles, functions, requirements, and personnel expertise for the new headquarters staff. Unfortunately, only two meetings regarding the JMD took place between May 16 and June 15, the last of which was attended by McKiernan, Wallace, and myself. Afterward, I was convinced that a valid and approved transition strategy would not be in place before I assumed command. I knew we would have to do major restructuring when I took over. And we would have to do it while conducting military operations.

As if all that wasn't enough, Scott Wallace gave me more stunning news. "Ric, I've just been told I have to leave Iraq immediately," he said.

"What!" I replied. "But why?"

Scott explained that, during a public briefing, he had made an offhand remark about his concerns regarding a potential insurgency and that we had not planned against such an enemy. Of course, he was absolutely correct. But apparently, somebody in Washington didn't like what he said and orders had filtered down through Franks and McKiernan that he had to leave right away.

"Well, we just can't let this happen," I said. "It is impossible to transition command immediately. I'm still in charge of 1st Armored and there's no replacement designated."

"Ric, we have to find a solution," Wallace responded. "Would you feel comfortable handing it off to one of your assistant division commanders?"

"Well, I guess Doug Robinson can do it. But this is wrong, sir. You don't deserve to be treated this way. We need to go back to CFLCC and have McKiernan delay this for at least two or three weeks."

In the end, LTG McKiernan did, in fact, intervene with GEN Franks to get Wallace extended. However, I was dismayed at the way my corps commander had been treated. We had been friends for years and I knew he was a good man, a magnificent soldier, and a man of tremendous integrity.

Just a few weeks before, LTG Wallace and I had been reminiscing about our time together at Fort Bragg, and I reminded him about the time he called me in for my efficiency report. "You were a captain and I was just moving into the aide-de-camp job," I recalled. "And you mentioned that you had previously experienced problems with Hispanics and didn't want me to join your staff."

"Ric, I don't remember that."

"Sir, it's a comment I've never forgotten. But you also said that I had proved you wrong. And I've respected you ever since for having the guts to admit that to me."

"Did I really say that?"

"Yes, sir, you did."

"Damn, that was stupid!"

"Well, sir, it's nowhere near as stupid as what is going on here in Iraq. Now is it?"

CHAPTER **11**

De-Baathifying, Disbanding, and Dismantling

There was only one real political party in Iraq after Saddam Hussein seized power in 1979. The Baath Party (full name, Arab Socialist Baath Party) was founded in Syria in 1943 and became prevalent in Iraq during coups in 1963 and 1968. When Saddam took over, he became not only the country's president and head of the Revolutionary Command Council, but also secretary general of the Baath Party. A socialist/fascist organization run by a narrowly defined group united through family and tribal ties, the Baath Party became Saddam's vehicle to purge dissidents and rule the country with an iron fist. Nearly the entire civil and governmental infrastructure of Iraq was controlled by party members, and nearly all civil service employees, college applicants, and members of the military were forced to become Baathists if they wanted to keep their jobs and practice their professions. In other words, it became a matter of survival for hundreds of thousands of people. They simply had no choice.

On May 16, 2003, Ambassador Paul Bremer, acting in his capacity as administrator of the CPA, issued Coalition Provisional Authority Order Number 1, an implementation order to disestablish the Baath Party in Iraq. The top four levels of leadership were immediately "removed from their positions and banned from future employment in the public sector." More drastically, however, also "removed from their employment" were full members of the party who held the top three positions "in every national government ministry, affiliated corporations and other government institutions (e.g. universities and hospitals)." Bremer thereby fired approximately half a million people, including 400,000 members of the armed forces, and somewhere near 100,000 civilian workers.

The impact of this de-Baathification order was devastating. Essentially, it eliminated the entire governmental and civic capacity of the nation. Organizations involving justice, defense, interior, communications, schools, universities, and hospitals were all either completely shut down or severely crippled, because anybody with any experience was now out of a job. In one fell swoop, Bremer had created a 60 percent unemployment rate and angered hundreds of thousands of people.

In and of itself, de-Baathifying Iraq wasn't a bad idea. The top four leadership levels of the Baath Party needed to be removed and those individuals with blood on their hands held accountable and prosecuted. It was the unilateral firing of nearly everybody else that created the real problem. Moreover, there was no proper screening and appeals process for the vast majority of average Iraqis to get their jobs back—and this alienated many.

In his initial order, Bremer stated that only he had the authority to grant exceptions "on a case-by-case basis." Nine days later, he established the Iraqi De-Baathification Council, with part of its mission to advise him on which citizens should be allowed to resume their jobs. According to the directive, this new council would report "directly and solely" to Bremer, and he would determine when it would convene and which Iraqi citizens would serve on it.

The obvious problem with this system was that every person who had been fired had to go directly to Bremer (or the De-

Baathification Council), state the fact that they were never in the top four levels of senior leadership of the Baath Party, and justify why they should be allowed to return to their jobs. In order for this to work properly, a massive effort to establish committees in every city and village in Iraq would need to take place. But Ambassador Bremer refused to do it. In effect, he had taken a hands-off approach and made the Iraqis an integral part of the process by giving them responsibility. But he did not provide enough authority, resources, or oversight for them to manage the process. Bremer's actions were later recommended as a primary strategy by Kurdish Prime Minister Barham Salih. "The United States needs to position itself behind Iraqis so it does not get all the blame when things go wrong," said Salih. Essentially, he was telling the CPA to give the Iraqis responsibility, but no authority.

The new De-Baathification Council appointed dissident Ahmad Chalabi's nephew (and lawyer), Sam Chalabi, to manage the process. This might have been the worst possible choice, because from the very beginning Ahmad Chalabi was adamant that no Baathists would ever be allowed back into government service. And that's one reason the whole de-Baathification order became a catastrophic failure.

It wasn't long before Bremer and the military started knocking heads. We went back to him on multiple occasions and told him that his de-Baathification policy was flawed, that it wasn't working, and that no appeals had been processed. But Bremer refused to take any corrective actions. He just listened to us and walked away. By June, we'd had a few major disagreements with CPA. Our soldiers had been working on setting up governing councils, restoring key elements of the infrastructure, and reestablishing some of the schools. Clearly, we had to involve some former members of the Baath Party in the process. Bremer got quite upset by this idea and fired off a number of memorandums addressed to me and my subordinate commanders. "I understand you are not following my orders," he wrote. "You will not allow Baathists to participate in government or civil operations. Any exceptions to the de-Bathification policy will be personally approved by me."

After discussing the issue with him privately, I returned to my commanders and advised them to continue what they were doing. "Figure out as best you can what functions we need to establish," I said. "Get the best people available and let's get these functions back up and running. And for God's sake, keep the schools open and let the kids finish the school year."

I was ordering our soldiers to persist with the work at hand and wasn't particularly worried about repercussions. Somebody had to stand up and do the right thing. Besides, we would likely have already reestablished many of the functions and capacities across the country by the time Bremer would be able to impact our progress.

In thinking through de-Baathification and the lack of an effective process with which to implement it, I had to wonder where the order really came from in the first place. Certainly, it was not what Jay Garner had been advocating before he was removed from his position. And clearly it was not developed in Iraq by any member of the military. It had to have been brought in by Ambassador Bremer, perhaps after it was created by the policy branch in the office of the secretary of defense, with the concurrence of the National Security Council. It appears that they were trying to replicate the de-Nazification of Germany after World War II. Back then, however, the American military handled all the policy, planning, and implementation. To ensure success, specific goals were set and an effective process was put in place to reach those goals. That is basic leadership.

In this case, however, the CPA treated the entire endeavor like they were issuing an academic, theoretical paper. They simply released the order and declared success. But there was no vision, no concept, no experience, and in my opinion, no desire to ensure that the policy was properly implemented. On the other hand, it did look good on paper. When reported on CNN, it gave the illusion that progress was being made, and by the time it became apparent that the policy had failed, nobody knew why it hadn't worked. Besides, the people who put it in place could blame those to whom they had handed it off—the Iraqi Governing Council.

The de-Baathification order constituted a calculated decision. Bremer and the people back in Washington who put it together had to have known that this order would shut everything down. The only other explanation is gross stupidity. And these people were not stupid. What I cannot reconcile, however, is this: When it became obvious that the process wasn't working, why didn't they fix it so that the overall policy had a chance to succeed? The original intended purpose of the de-Baathification policy was to eliminate the top four tiers of the Baath Party leadership, and those with blood on their hands. However, the mechanisms to achieve the objectives of the policy were never clearly defined and were not properly resourced for nationwide implementation; in addition, no system was ever put in place to monitor execution and progress. Actually, the process wasn't even working in the Green Zone, much less in the hinterlands of Iraq. The CPA knew all this, but still refused to take any corrective actions. Why were they willing to let Iraqi society flounder? Didn't they understand that there were hundreds of thousands of disenfranchised people, most of them former members of the military who were going to find a way to express their dissatisfaction? Didn't they realize that this was just going to stoke the already simmering violence across the country? And didn't they care that it was our young soldiers in uniform who were going to bear the brunt of Iraqi retaliation?

A few days after the de-Baathfication order was issued, Ambassador Bremer went out to V Corps headquarters near the airport and visited LTG Scott Wallace. After reiterating that he was in charge, Bremer told Wallace again that he wanted the V Corps headquarters co-located with him at the Republican Palace inside the Green Zone in downtown Baghdad. He also stated that he did not want either LTGs Wallace or McKiernan to attend CPA meetings. Rather, he wanted to deal only with me, because I was going to be assuming command in less than a month.

McKiernan and Wallace both acquiesced to Bremer's demand. Fifth Corps split the staff and started moving part of it into the Green Zone. Wallace then informed me that I now had to attend all the Ambassador's morning meetings. "But, sir, I'm three levels down from Bremer," I said. "Besides, I've got my hands full with the division, and we haven't even settled the occupation of Baghdad yet."

"Well, you don't have a choice, Ric. You've got to go to these meetings. That's what Bremer wants."

"Okay, sir. I'll go. But would you please assign me a liaison officer to help pass on information to you and McKiernan. You're still the commanders on the ground."

For the next few weeks, I wore a "dual hat" as division commander and direct military liaison with CPA. I attended all of Bremer's 7:30-to-8:00 a.m. updates and then passed on all observations, situation reports, issues, and guidance from Bremer. Besides, McKiernan and Wallace still held all the command authority to take action across the country. After the CPA morning sessions, I headed out to the field and continued with my duties to get Baghdad under control. Interestingly enough, during this period, some of Bremer's subordinates began issuing orders to the military regarding location of forces and taskings for small units. Well, I didn't need to speak to Wallace and McKiernan about that issue. I acted on my own authority and immediately put a stop to it.

The process of establishing a V Corps headquarters inside the CPA palace, just as Ambassador Bremer had requested, brought to my attention another significant problem. With both CENTCOM and CFLCC leaving Iraq, V Corps was going to have to operate at the theater strategic level, for which it possessed no expertise, as well as the operational and tactical level across the country. Unfortunately, neither CENTCOM nor CFLCC was planning to provide any help to accomplish that task. Furthermore, even though a Joint Manning Document (JMD) was being prepared, there was very little thought given to mission analysis, or to delineating roles and functions within the new command. In order to get the ball rolling, Wallace and I split the V Corps command structure into two elements—one at the CPA with a theater strategic- and operational-level focus (located in the palace) and another with a warfighting tactical focus (located at military headquarters out at the airport). By doing so, however, we had to temporarily dilute the command and control capacity of the corps at the tactical level. But there was simply no other way to honor the Ambassador's request.

While attending CPA's morning meetings, I noticed a subtle but significant change in the national strategy toward a longer-term political solution. Part of the Bush administration's plan was to place key expatriates in charge of the new Iraqi government. These were people like Ahmad Chalabi, Ayad Allawi, Jalal Talabani, Abdul Aziz al-Hakim, and several others who had been part of the front-end planning for the invasion. Theoretically, by giving them control of the country, coalition forces would be able to leave sooner. Ambassador Bremer, however, determined that this approach was wrong, and he began to push for an interim governing council comprised of a more representative group rather than these already-designated key expatriates.

When Bremer communicated this plan back to Washington, people at the Department of Defense expressed some uneasiness about it, but everybody acquiesced to his wishes. At this point in time, it appeared to me that Washington was distancing itself from all things related to Iraq. No one was focusing, scrutinizing, or analyzing the impact of decisions that were coming out of CPA. Meanwhile, Ambassador Bremer was changing the entire political strategy of the coalition. And it became very clear to me that we were going to be stuck in Iraq for a much longer time than we had all anticipated.

On May 23, 2003, exactly one week after issuing the de-Baathification order, Ambassador Bremer released CPA Order Number 2, which formally dissolved the Iraqi Army, Air Force, Navy, and every other military-related organization, including the Ministry of Defense and its agency counterparts. That part of the order did not become a problem in and of itself, because the military system in Iraq was already disbanded. Bremer just made it official. The problem was that another part of the order stated that CPA was going to create "a New Iraqi Corps." Ambassador Bremer was already aware that we had been working on a plan to rebuild the Iraqi military. He knew that Abizaid had approached former generals, received a good response, and that we had started to bring in small groups to get the process rolling. Our intent was to recall units, train them very rapidly, and put them into the field to help our soldiers.

However, Ambassador Bremer told us to stop what we were doing. CJTF-7 was not going to have anything to do with rebuilding the Iraqi military, he said. I couldn't believe my ears. We were going to build a new military system in Iraq and Bremer wasn't going to allow the coalition forces operating in the country to be involved? Moreover, when we suggested a plan to reestablish the Ministry of Defense with specific goals and deadlines, the Ambassador refused to discuss any timelines. His approach was to work from the bottom up, and he absolutely refused to discuss building Iraqi leadership in the military above the battalion level. Of course, this was exactly the opposite to GEN Abizaid's approach. And it put us in a hell of a bind, because now we had to go back to the Iraqi former officers from whom we had already secured agreement and tell them, "Sorry, this is no longer going to be possible."

By early June, Bremer had brought in Walter B. Slocombe, former undersecretary of defense, as the national security advisor to take charge of rebuilding the military. MG Paul Eaton was selected by the Army to assist Slocombe and oversee the three- to five-year training effort that was to be conducted by civilian contractors. Slocombe and Eaton developed their own plan, went back to Washington to brief it, received approval, and initiated the process. GEN Abizaid and I, however, registered our nonconcurrence at higher levels. It was too slow, we believed. More aggressive, specific timelines needed to be implemented, along with a plan to reconstitute the Ministry of Defense and all senior leadership functions down to brigade headquarters.

Both de-Baathification and this new plan for rebuilding the military substantially slowed down our efforts in Iraq and delayed our exit from the country. At the time, I thought this was going to be the end of it. But we soon got into another major scrape with CPA about reconstitution of the Iraqi police and security forces.

After the fall of Baghdad, nearly all of the Iraqi police abandoned their jobs, which led to soaring crime, looting, and violence. It also hadn't helped the situation that Saddam Hussein had emptied all of his jails just as the invasion was getting under way. Almost immediately, however, LTG Wallace developed an aggressive plan to reestablish police and security forces. He gave orders for his division commanders to call in, train, arm, and integrate former members of the police into our efforts on the ground. Wallace's early attention to police and security forces was critical, because not only did order need to be restored, but power plants and borders needed to be guarded around the country. Wallace's strategy also took the long-term view into account. For instance, he had quickly enlisted the aid of the British 16th Air Mobile Brigade in London to help with strategic planning. Then he assigned the implementation task to our 18th Military Police (MP) Brigade and, by early June, they had already registered, hired, and were training 5,600 Iraqi personnel. Between the British and the 18th MP Brigade, we were going to get this show up and running in six months—at least, that was our plan.

About this time, Ambassador Bremer issued another edict stating that reconstitution of the police and security forces was going to be the responsibility of the Ministry of the Interior, which had just been set up by him. Bernard B. Kerik, the former New York City police commissioner, had been selected by Washington to head the effort. When Kerik arrived in the country, however, I believe he was astonished to find that there were only about a dozen people who comprised the new Ministry of the Interior. He thereby had almost no capacity to achieve a national solution.

On June 3, 2003, the CPA finalized its plan for building the Iraqi police and security forces, and Bremer sent Secretary Rumsfeld a copy of it along with a letter mentioning several key items. The first thing I noticed was that the Ambassador mentioned that the human rights approach of Saddam's police was unsuitable for modern policing, because the previous police force had been a quasi-military institution. Bremer also made it very clear that this was going to be another long-term program, because it would have to involve "professional" training at international standards—and that the decision as to whether it was going to be a national, regional, or bifurcated police force would be left up to the future Iraqi government.

We in the military protested the overall plan, mainly because it was going to take too long, and there was no intent to set up command and control structures or establish senior leadership early on. Bremer, however, refused to back down and made it equally clear to us that he intended to retain control of all policy, resources, training, manning, and equipping of the new police/security forces. Once again, this was not a mission for CJTF-7. Again, we acknowledged his desires, but proceeded with our own plan under the radar. We ordered military commanders to continue the recall of Iraqi former policemen, have them report for duty, and then sort them out. The 18th MP Brigade and the divisions continued working with Iraqi police while we waited for the "professional" training that Bremer had promised. And by the way, every single policeman was a former member of the Baath Party.

We didn't completely ignore CPA, however. We worked closely with Slocombe, Eaton, and Kerik and gave them as much support as we could. All three understood that there was no alternative but to work with the military, because there was no one else available to help them accomplish their missions. The plans that had been put in place for bringing in civilian contractors were not materializing. Ambassador Bremer knew what was going on and he didn't like it, but he had no other choice. Every now and then, though, he let us know he was in charge. In those instances, I took the heat and shielded our commanders. At the same time, however, we bombarded Bremer and the CPA by identifying problem areas and asking them to develop policies for a whole range of issues. That created significant tension, but it was our duty to try to get these issues addressed; and theoretically, CPA was the agency that was supposed to be handling them.

On May 31, 2003, during one of Ambassador Bremer's meetings, he brought up several concerns about U.S. forces in the southern region of Iraq. He took me to task about the military producing councils that, in his opinion, were "uneven, spotty, and in some cases, employing bad leaders that had to be fired." He also demanded that we gather "actionable intelligence" on some of the tribal militias in the south, especially the Badr Corps. Then he told me that he wanted us to take military action and eliminate them as a threat "to show them we mean business." My first reaction was simply to say, "Yes, sir." And then I reported his requests back to LTGs McKiernan and Wallace.

I don't believe McKiernan could have taken Bremer's guidance seriously because there were few American forces left in the southern part of Iraq to tackle such a mission. For the entire month of May, McKiernan had been implementing GEN Franks's force reduction order and, by now, elements of the 3rd Infantry Division and the 1st Marine Division were in Kuwait and the rest of the force that remained in Iraq was preparing to leave. Wallace, who was leaving V Corps in two weeks, was trying to reestablish some level of security and stability across Iraq. And I had my hands full trying to get Baghdad under control. Nobody in the military was interested in mounting an attack on the 10,000-strong, Iranian-trained Badr Corps down in the southern part of Iraq. Ambassador Bremer was dreaming.

Also at the end of May, the Army lifted Stop-Loss and Stop-Move—programs designed to prevent soldiers from leaving their units for virtually any reason (attend training schools, transfer, retire, etc.). These limitations had been in place since before the invasion, but now the Army began to issue move orders en masse, which created a problem at all levels of command—from the division staff all the way down to individual units. In the 1st Armored Division, for example, within forty-five days from the time we actually hit the ground in Kuwait, I was going to see every general officer, every brigade commander, every staff officer (except the chief of staff), and 70 percent of the battalion commanders leave the country. That meant that by June 15, 2003, the day I assumed command of coalition forces in Iraq, all those people would be gone.

As soon as I heard the news, I immediately went to speak with my boss. "General Wallace," I said, "this is absolutely insane!"

"I'm sorry, Ric, but the policy is that we're not going to extend commanders," he replied. "We're going to have to change them out."

"Sir, it is apparent to me that when V Corps is designated CJTF-7 on 15 June, we will not be able to perform what I believe the missions are going to be. V Corps has never operated at a strategic or political level in the theater. We're going to have to interface not only with the field, but with CENTCOM in Tampa and with higher leadership back in Washington. And we will have nobody on our staff who has ever done it. The Joint Manning Document, as it currently reads, is inadequate. Heck, it isn't even approved yet—and there is nobody working on filling our organization. And now we are taking apart our leadership teams across the entire force at the very time we are trying to stabilize the country. As a result, our effectiveness will drop significantly while new commanders adjust to the theater and gain situational awareness. We just can't do it."

"Ric, you're right," Wallace replied. "We need to do something about it."

Over the next several days, the two of us worked together to determine conditions for the transfer of responsibility from

CFLCC to V Corps/CJTF-7. Then Wallace fired off a memo to LTG McKiernan that outlined our concerns. "Conditions for CJTF-7 will not be met by 15 June," he wrote. "Our concerns center on personnel fill, skill sets, and some key-critical must-fill positions which have not yet been resourced, and are unlikely to be addressed in the absence of an approved JMD. Our greatest concerns are at the C-2 (Intelligence), C-3 (Operations), and C-5 (Strategy, Policy, and Plans) levels of the organization."

McKiernan responded by promising to leave a significant amount of personnel behind to provide us the capacity to operate. "However, we're going to leave on 14 June," he said. "We've been ordered out." About ten people arrived prior to June 15, but they were mostly young officers who had less than thirty days left in the theater. When I went back and asked if CFLCC could temporarily lend us its top planner to help with policy and strategy, I was turned down.

LTG Wallace and I also engaged directly with the Department of the Army in Washington in an attempt to stabilize the tenure of commanders and other key personnel. But the Army refused to alter the peacetime rules and directed that changes of command occur as scheduled. Unwilling to accept that as a final answer, I discussed the issue at length with GEN Peter J. Schoomaker (Chief of Staff of the Army) on his first visit to Iraq. "Sir, the Army made a huge mistake by forcing us to change commanders while we were in a critical transition phase attempting to secure the country," I said to Schoomaker. "It has created significant problems for us on the ground. It completely disrupted our operations and definitely delayed bringing stability to the country. We cannot do this in the future."

"Okay, Ric. I see what you mean," Schoomaker replied. "It was a dumb thing for us to do."

As soon as the Chief of Staff returned to Washington, the Army changed its policy for subsequent deployments—commanders would be stabilized so they could train, deploy, and lead their units in combat.

Around this time, I also shifted my attention to the growing detainee problem in Iraq. The numbers were increasing as we continued to weed out pockets of resistance and get the violence and looting under control. First, I met with the person who would be in charge of the military police when I assumed command. In late May, BG Paul Hill, commander of the 800th MP Brigade and his replacement, BG Janis Karpinski, came to my headquarters for an initial office call. All three of us were aware that the 800th would remain in the CFLCC organization and continue to report to LTG McKiernan. He would be in charge of administration, training, discipline, and logistics. As the in-country commander, I would have authority over the MPs for taskings and establishing priorities and movements.

After exchanging pleasantries, we discussed the military situation in Iraq and the capabilities and needs of the 800th MP Brigade. "Well, tell me," I asked, "who is going to be the senior MP in this country that will be responsible for detention operations?"

BG Karpinski raised her hand and said, "I will."

"Good," I responded. "Now, where are you located?"

"Sir, our headquarters are in Kuwait."

"But all your units are up here."

"Yes, sir, but we've been told that we are going to shut down all our operations and go back to the States."

"If you're going to be in charge of detention operations and be my senior military police advisor, I'd like you to move your headquarters to Baghdad. When your orders come through to redeploy, you can leave from here." Three weeks later, BG Karpinski moved the 800th MP Brigade headquarters up to Baghdad.

ON JUNE 10, 2003, in a formal change of command ceremony, I handed off responsibilities for the 1st Armored Division to Doug Robinson. Two days later, on the twelfth, I was given my third star and was promoted to the rank of lieutenant general in the U.S. Army. On June 14, in the rotunda of the Republican Palace in Baghdad, I assumed command of V Corps from Scott Wallace. In attendance were John Abizaid, David McKiernan, and Paul Bremer. The next day, on June 15, 2003, V Corps was designated Combined Joint Task Force 7 (CJTF-7). I was now the senior military commander in Iraq with responsibility for achieving CJTF-7's three-fold mission:

1. Continue offensive operations. Eliminate any enemy forces that are still in the country, including former regime residual enemy groups, terrorists, insurgents, or the like. Defend the nation from all external threats.
2. Provide direct support to the Coalition Provisional Authority (CPA).
3. Provide aid for humanitarian assistance and the reconstruction of Iraq.

We had gotten a few things up and running in Baghdad, but for the most part, everything around the country was still shut down. Some police stations were open but none were effective. Distribution of fuel and electricity was sporadic, at best. A food-rationing system was not yet in place. The political and economic systems of the country were in dire straits. Banks were not open. Commerce was nonexistent. The judicial system had disappeared. There was no national government council yet established, and local councils were few and far between. The mission at hand was daunting.

The Coalition Provisional Authority was seriously understaffed and had no capacity to govern, manage, or sustain Baghdad, let alone the entire country—and certainly nothing was going to be achieved in a timely manner. Moreover, the organization was almost completely focused on Baghdad, to the point where it did not understand the dynamics of the rest of the country. The relationship between CPA and the military was still unclear and undefined, even though we had been ordered to transition much of the work that V Corps had initiated before Ambassador

Bremer had arrived in Iraq. His de-Baathification order had angered hundreds of thousands of people who were already upset that their country was being occupied. We were just beginning to see the rise of an insurgency, presumably by members of Saddam Hussein's former regime. De-Baathification, conceived in the halls of the Pentagon and the White House by neoconservative ideologues, marked the beginning of an incremental dismantling of the original U.S. strategic plan for Iraq. That order assured that the United States would be tied up in Iraq for an indeterminate number of years.

On June 13, 2003, two of the bloodiest enemy attacks since the fall of Baghdad took place. And the next day we were questioning several hundred prisoners who had been captured in the battles. The undeniable fact was that we were still at war. And while Department of Defense leaders dismissed the attacks as "no big deal," or labeled the enemy as "just a bunch of dead-enders," we were identifying targets and conducting offensive operations throughout much of the country. Our forces were moving into the west and north. We really had no idea what was in these uncleared areas and were very leery that there may have been organized elements of the enemy present. After all, Saddam Hussein and his henchmen were still on the loose.

Meanwhile, American forces were rapidly pulling out of the country. GEN Franks's April drawdown order had called for the reduction of our 175,000-man force to 30,000 by September. The combination of increased fighting and rapid force drawdown resulted in rising casualty rates among our armed forces. But few in Washington seemed to care.

When V Corps was designated CJTF-7, every headquarters command above us shut down. Both CENTCOM and CFLCC left, and with them, every single capacity they possessed. The functional gaps were astounding. In the area of intelligence alone, we had no operational or strategic capacity, no interrogation operations capacity, no human intel analytical capacity, and no communications architecture in place. Overall, CJTF-7 had been left unresourced and unprepared to achieve our missions. The lessons of Task Force Smith in Korea had been forgotten. The situation also reminded me of the woeful situation I had experienced with the United Nations in Kosovo. Only this time, it was the U.S. government in charge—and we had done it to ourselves.

On June 15, 2003, I was the youngest three-star general in the U.S. military—only three days in that position, in fact. I had been vaulted up two levels of authority to take command of this situation in Iraq. The burden I felt was unimaginable.

PART III

COMMAND IN IRAQ

★ ★ ★

CHAPTER 12

The Struggle to Stabilize

"Okay, what's the situation?" asked Secretary Rumsfeld.

"Sir, we have a convoy of three vehicles rapidly approaching the Syrian border," I responded. "Our intelligence sources inform us there are high-value targets present. I'm on the satellite phone right now with our on-site tactical commander. If we're going to take them out, we need to issue the order now."

"Who are the high-value targets?"

"Sir, we don't have any names, but our information is that they are former regime members."

"Are there any innocent people in the convoy?" interjected GEN Tommy Franks.

"We are unsure at this time, sir."

"What is our probability of collateral damage?"

"They're out in the middle of the desert, sir. Damage will be limited to the convoy."

"What time is it there, anyway?" asked Rumsfeld.

"Sir, it's about 2:00 a.m."

"Are we sure they're fleeing?" asked GEN Pete Pace (Vice Chairman of the Joint Chiefs of Staff).

"Sir, we have a convoy traveling at high speed headed toward

Syria. Our intel tells us that we have high-value targets that are trying to escape."

"So they're making a beeline for the border in the middle of the night, right?" asked Rumsfeld.

"Yes, sir, that's correct."

"Well, what are you doing right now?"

"Sir, we're tracking them and we're locked and loaded," I responded. "We're ready to take them out."

"Is that your recommendation?" asked Franks.

"Yes, sir, it is."

Upon being awakened about an hour earlier I had gone straight to the command center for a briefing of the situation. My on-scene commander had recommended we take out the convoy and I concurred. However, according to the rules of engagement for Iraq, this decision was beyond my authority. So I had to initiate the conference call to Franks. In turn, he got Rumsfeld on the line, because the Secretary had the final authority to remove high-value targets. While the authority sequence was normal, the detailed questioning about what was specifically happening on the ground was not—and it was an example of Secretary Rumsfeld's micromanagement style.

Sitting next to me in the command center was MG Walt Wojdakowski, deputy commander of Combined Joint Task Force 7 (CJTF-7), who was monitoring the tactical situation on the ground, and BG Bob Williams (director of operations, C-3), who was on a separate phone discussing intel with Deputy CIA Director John McLaughlin. In one hand, I was holding a secure white satellite telephone with Franks, Pace, and Rumsfeld on the line. In the other, I was holding a standard satellite radio handset, patched in with the special operations tactical commander who was flying over the site in a C-130 airplane.

"General Sanchez, this convoy is getting close to the border," said the tactical commander.

"Stand by," I replied.

"Secretary Rumsfeld," I said into the white phone, "the convoy is pretty close to the border and, if we're going to act, we have to do it now."

"Okay, go ahead. You have the authority to engage."

"Very well, sir. I have to get off this line right now, and pass along these orders. Could you please stand by?"

"Yes, but come back as soon as you're done."

I set down the white phone and directed the tactical commander to destroy the convoy. When I was given confirmation that the attack was under way, I got back on the phone with Rumsfeld, Franks, and Pace.

"What happened?" asked Rumsfeld.

"Sir, I don't know what happened," I responded. "I gave the order and we're executing now."

"Well, we need to find out if we killed them. And I want to know who it is we killed."

"Very well, sir."

"Good," replied Rumsfeld. "Get back to me as soon as you can."

In order to comply with the Secretary's wishes, we sent out a special unit to secure the site and police up the area. We loaded debris into dump trucks, brought it back to Victory Base, and attempted to obtain DNA material matches. In the end, we were not able to establish a DNA link with any of the names on the so-called "deck of cards" (the top henchmen in Saddam Hussein's power circle).

This Syrian border incident occurred during my first full week of command in Iraq. Fortunately, I was already familiar with all the division commanders reporting to me. MG David Petraeus, in charge of the 101st Airborne, was located up in the northern part of Iraq, a relatively stable environment compared to the central part of the country. BG Doug Robinson, acting commander of the 1st Armored Division in Baghdad, was soon to be relieved by MG Marty Dempsey. Doug and Marty were in the most complex environment of all, dealing not only with a city of six million, but also interacting closely with the CPA. MG Ray Odierno led the 4th Infantry Division in the north-central region of the country. Because this area included the Diyala province, Ray probably had the toughest neighborhood in the country. MG Buford Blount, commander of the 3rd Infantry Division, was refitting and awaiting redeployment orders. COL David Teeples headed the 3rd Armored Cavalry Regiment, assigned to al-Anbar. LT GEN Jim Conway and MAJ GEN Keith Stadler were in charge of the 1st Marine Expeditionary Force in the southern part of Iraq near Najaf. Also in the south was the 1st UK Division, located in Basra. They would soon begin a major expansion in their sector as coalition forces entered the country.

To honor Ambassador Bremer's request to co-locate headquarters with him at the CPA, we split the CJTF-7/V Corps staff and operated with a dual focus. While continuing to work the strategic level with Rumsfeld, Bremer, Abizaid, and higher military headquarters, I delegated responsibility for the management of day-to-day tactical issues to my deputy commanding general, the brilliant Walt Wojdakowski—and I still remained updated and involved, making all the major tactical decisions.

The major focus of CJTF-7 was to continue our operations to eliminate the potential for any hostilities on behalf of the former Saddam Hussein regime. The week before I formally assumed command, V Corps launched Operation Peninsula Strike, a series of raids using helicopters, small boats, and armored vehicle roadblocks. About 1,000 of our forces from the 4th Infantry and a special force designated Task Force Iron Horse descended on a peninsula alongside the Tigris River in the area near Tikrit. This former haven for regime loyalists and Republican Guard members yielded extensive weapons and munitions caches, along with nearly 400 suspects who were screened for intelligence. Two of those detainees turned out to be former Iraqi generals, but most were released within a few days.

From June 15 to June 19, we launched the broadest offensive effort since the official end of the war (May 1). Operation Desert Scorpion, conducted by elements of the 3rd Infantry, 4th

Infantry, 101st Airborne, 1st Armored, and 2nd and 3rd Cavalry, involved a series of raids across the central portion of Iraq, primarily in the Sunni Muslim strongholds of the Baath Party and in selected areas of Baghdad. We identified many Baath Party loyalists, terrorist organizations, and criminal elements, went after them, and ended up capturing, detaining, and interrogating dozens. Operation Desert Scorpion also tasked the divisions with repairing damaged infrastructure across the country—including bridges, power plants, and oil facilities.

While conducting offensive operations, I was ever mindful that our soldiers were immersed in a 360-degree battlefield environment. We faced multiple enemies, including former regime elements, extremists, criminals, and foreign fighters. And we were constantly taking mortar rounds or rocket fire from just about anywhere in the country north of a line forty kilometers south of Baghdad. This included the Green Zone, Victory Base at Baghdad International Airport, Fallujah, Ramadi, Samara, Sadr City, Mosul, Diyala, al-Anbar, and other areas. You could name any city in the area, and we were probably encountering enemy fire. All soldiers, sailors, airmen, and Marines, no matter where they were located, were subject to getting into a firefight. Almost every day, therefore, soldiers and commanders had to maintain the difficult balance of being ruthless in battle and benevolent in victory. And that required tremendous discipline, focus, and leadership. Never before in the history of warfare was there a situation more dependent on small-unit leadership across an entire battle space. It was a constant, Battle of the Bulge–type scenario.

In this 360-degree environment, casualty rates began to rise. Mortar attacks, rocket-propelled grenades, and sniper fire all started taking their toll. Our troops were being killed or wounded on a regular basis, and I felt we could reduce the casualties by improving our intelligence capabilities, by filling the glaring deficiencies in my staff, and by halting massive troop reductions.

As new missions emerged, my staff developed broad concepts for proposed military operations and provided estimates of the required forces. Over time, some of the missions we addressed included border protection, disposal of enemy ammunition, elimination of the terrorist groups in the north, and securing the oil and electrical infrastructure. Normally, a mission would not be officially assigned to the Task Force until an agreement had been reached between the combatant commander and the Pentagon. Often, however, the Pentagon would come back and ask us to take on those missions without supplying the forces required. I was told to take the required forces "out of hide" (which meant repositioning or reassigning our current forces).

"If I take them out of hide, I incur risk somewhere else," I replied. "Are you willing to accept that risk?"

"Well, no, we are not."

From my earliest days as a general, there had always been an unstated rule that the Joint Staff would never put a combatant commander in the position of having to say, "No, I can't perform the mission with the forces assigned." But by having this discussion and not assigning the mission, the Pentagon avoided a call for additional forces beyond what was on the ground.

In effect, I was told "Do the best you can with the resources available." So that's what we did. To protect the borders, for example, we ran periodic patrols, because we didn't have the manpower to properly secure all borders all the time. Overall, we had to depend upon our intelligence to ascertain where crisis areas could be, and then prioritize and shift our forces around the country accordingly. And it remained that way for my entire tenure in Iraq. I never did have an operational reserve force in the country.

In the meantime, I continued my requests for manning. I asked the Joint Staff for expertise in intel, operations, strategy, policy, planning, detention and legal affairs. "We need lawyers to help us in all of the divisions to address the challenges of detention and interrogation," I told them.

"No, we can't send you any lawyers," was the response. And they never did.

When the Joint Chiefs of Staff finally ordered the various services to fill our requirements, we were questioned endlessly about needs and justifications.

"Why are you asking for all these people? We don't think you need them."

"Wait a minute," I responded. "You have no idea what I need in this country."

"Well, we don't have the forces. You have to give us three months advance notice."

"But I need them now."

"Then put your request through McKiernan's command. CFLCC has to validate your requirements."

"But I don't work for McKiernan."

The bureaucracy within the various services questioned, stalled, and in the end, refused to send the help we requested. To make things worse, there was no mechanism within the Department of Defense to force the individual services to comply with orders issued from the Joint Chiefs of Staff. I simply couldn't believe it. Everybody knew the orders were being ignored, but nobody took the situation seriously enough to fix it. The services were continuing to do their standard bureaucratic dance even though we were still at war. Meanwhile, American soldiers were fighting and dying on the ground in Iraq.

Exactly one month after assuming command, on July 14, 2003, I sent a memo to CENTCOM documenting the status of my requests. "The overall fill rate for CJTF-7 is 37%," I wrote. "[And] only one of thirty critical requirements has been filled."

Manning joint task forces and joint headquarters had been a problem within the services for as long as I had been in the Army. The services balked at it, in part, because they would have to take people out of hide in order to reassign them. And why was that? It goes back to the lack of preparation for a long-duration deployment, like the one we were facing in Iraq. The services were simply stretched too thin. And the only people who had the power to solve the problem were the Secretary of Defense and the four-star generals. But most seemed convinced that the war was over, so they allowed the process to stumble along.

Unfortunately, I could not afford to let the situation on my staff remain the way it was. I simply had too many young people who lacked the experience to get the job done. So I brought up the issue to every senior leader who visited Iraq. And I pushed in every conceivable way for Washington to take some action.

Interestingly enough, Ambassador Bremer also had a lot of young people working for him over at the CPA. And he, too, had a hard time getting more experienced personnel deployed. The State Department had sent him all kinds of young professionals. Fresh out of college, twenty-four to twenty-six years old, they were willing to serve their country and were looking for a little bit of excitement. But they simply didn't have the appropriate knowledge regarding coalition operations, nation building, or planning and execution management, to handle the issues that the CPA had to address.

On the other hand, Bremer had some more experienced people who had been very successful in their fields of expertise back home. They were bankers, educators, lawyers, and various kinds of academicians who understood theories and practicalities, but who had not operated on the scale demanded in Iraq. For instance, when the CPA restored Iraq's Central Criminal Court on June 18, 2003, Bremer placed Judge Donald F. Campbell in charge of rebuilding the entire judicial system of Iraq. But Campbell had almost no staff, and the only people he could turn to for help were members of the military. The CPA was terribly short of people to execute any sort of massive project. But Bremer and the National Security Council had lofty goals for Iraq. They were planning to establish a democracy, draft a constitution, have elections, and set up a duly elected government to run the country—all within a two-year time frame.

Ambassador Bremer's drive to achieve all this led him to question GEN Franks's troop drawdown order. On June 18, 2003, during a video teleconference (VTC) with President Bush, Bremer stated that he thought the troops were being withdrawn too fast. GEN Abizaid came out of that VTC upset that the subject had been raised in front of the President before it was first discussed through the chain of command. "He should have talked to Rumsfeld first," said Abizaid. "Besides, Americans are an antibody in this country. The fewer soldiers we have here, the better off we're going to be." This incident marked the beginning of an ongoing argument between Ambassador Bremer and the military over the amount of forces needed on the ground in Iraq. He never really accepted what Abizaid and I were trying to tell him; that the real answer to fixing Iraq was a consolidated, synchronized strategy, coupled with a rapid stand up of the Iraqi security forces.

Shortly after Ambassador Bremer's VTC with the President, and not long after Judge Campbell suggested he needed help from the military to achieve his mission, I sat down with my staff and we agreed to do everything we could to make Bremer and the CPA successful. It was apparent to all of us that, in the long run, it really didn't matter what we did militarily. If the CPA wasn't successful, we would never get the hell out of this country. Therefore, we went out and canvassed our reservists to find individuals with some expertise that might be put to use. Then we funneled them into the CPA to beef up the staffs of their senior ministry advisors. With time, the number of people we sent over ebbed and flowed. But at the high point, we had approximately 300 men and women assigned to Bremer's organization—most of whom were not initially brought into the country for that purpose. We took them all out of hide in spite of our own dismal manning level.

ON JUNE 20, 2003, GEN Tommy Franks returned to Iraq for his farewell tour. It had already been announced that he was going to retire on July 8 and, in his mind, he had already won the war. When Franks made his courtesy call to the Green Zone, my staff and I gave him a brief situational overview of the country. But he was more interested in talking about the Syrian border incident that had occurred a short time earlier. As Franks began to speak, I got the distinct impression that he felt like he had a brand-new three-star sitting in the room and needed to pontificate a bit.

"You know, that Syrian situation turned out okay, but who did we kill in the end?" he said. "And by the way, I don't need operational discussions in a situation like that. What I need from you is perspective. You should tell me what you think."

"Yes, sir," I said. "I understand."

"Well, it's okay, Ric," he said. "I feel comfortable with everything that went on. You guys are executing at the point of the spear. Sometimes you just can't wait. That's what manifested itself in this case. You just have to make decisions based on the best information you have and take the risk."

Before GEN Franks left, he talked to us a little bit about his perspective on the remaining mission in Iraq. "We are going to be here for another thirty to sixty days, and we've got to help stabilize the situation. We're building a stew with an unknown recipe. Some experimentation will occur and we have to be prepared to abandon the dry holes."

"A stew, sir?" someone asked.

"Yes, that's right. And when the international forces get here, they will outnumber us by about three to one. At least, that's the intent. You should also be aware that Iran has a strategic plan for defeating us." After that statement, GEN Franks adjourned the meeting and went out on his tour.

Later in the day, CNN broadcast images all around the world of Franks walking into the Republican Palace in his fatigues, smoking a big cigar. Many people thought he was celebrating with a "victory cigar." But I don't believe that was the case, because Franks almost always had a cigar in his hand. I was a bit concerned, however, when reports filtered back to me that he had been telling the troops that they'd all be home by September.

A few days after Franks left the country, his future replacement, my old friend GEN John Abizaid came for a visit. We held several long conversations about the current situation and addressed all of our deficiencies. The first item of discussion revolved around our headquarters staff and my frustration in dealing with Washington. "Sir, we just don't have the experience and capabilities within the corps headquarters to accomplish our missions. If they hadn't pulled McKiernan's command out of Iraq and sent them to Georgia, we wouldn't be in this situation right now."

"I agree," said Abizaid. "What do you recommend we do?"

"Well, sir, we need to reestablish a four-star headquarters here that is focused on the strategic and operational levels of war."

"Okay, Ric," Abizaid replied. "This will have to be approved by the SECDEF [Secretary of Defense]. In the meantime, let me see what I can do about getting you some help from the Air Force and the Marine Corps."

Abizaid and I then got into a discussion about how our experience in Kosovo served as a training exercise for Iraq. We were now experiencing just about every single issue we had dealt with there. Tense inner-city operations, strategic communications issues, intelligence challenges, political demands, jockeying priorities, insufficient troop assignments from Washington, and problems with building the security forces—they were all in Kosovo, only on a much smaller scale.

"You know, Bremer was correct when he stated that the Army had already disbanded when he issued his order," I said to Abizaid. "The problem now is that he adamantly refuses to let us rebuild it."

"Yeah, it'll take him forever at the rate he wants to go. He's talking about having only 45,000 Iraqis in uniform three to five years from now. That's bad. We need more people on patrol with us now. Maybe we should do what we did in Kosovo."

"You mean forming the Kosovo Protection Corps?" I asked.

"Yeah, why not?" replied Abizaid. "What have we got to lose? If we don't call it the Iraqi Army or the Iraqi Police, Bremer might help us fund it."

"Well, what should we call it?"

"I don't know. How about something like the Iraqi Civil Defense Corps?"

"ICDC?" I responded. "Sounds good."

"Of course, Bremer will resist the idea. He's already diametrically opposed to our plan to build the Iraq military. But we need to get Iraqi leadership in place by standing up the Ministry of Defense and every level below that. Iraqi-ization is critical."

Finally, GEN Abizaid and I discussed the situation on the ground in terms of forces and, specifically, the troop drawdown order by Franks. After providing a complete review of the ground situation, I pointed out to Abizaid that there was simply no way to achieve our missions given the fact that the troops were pulling out. "If we allow the drawdown to continue," I warned, "we won't be able to fight this war and we'll essentially be giving back the country to the Baathists."

"Well, you're the ground commander, Ric," said Abizaid. "What do you recommend?"

"Sir, I hate to say it, but we just can't allow these troops to go home."

"Boy, that'll cause some severe leadership problems."

"Yes, sir, I know. But what is our alternative? We're still fighting and the country is nowhere near being secure and stable."

"Well, Ric, I can't do anything until I take over."

"I understand, sir. But I'd like to slow this force withdrawal down, so that everybody isn't leaving at the same time."

"Okay," said Abizaid. "Just don't do it so that it rises to the level of the combatant commander. Franks will have a fit. And you better begin to alert the troops that they should be prepared to stay, if we ask them to."

"Yes, sir. I'll also start developing a plan for the Iraqi Civil Defense Corps. I think that has a good chance to work."

"Ric, you've just got to hang on another two or three weeks," said Abizaid, "and then I can change it."

If there was one thing I had learned over the years, it was that I could trust John Abizaid. When he said he was going to do something, he did it. So I immediately began planning for a reversal of Franks's force drawdown order.

While the majority of the military's senior-level commanders were sticking to the illusion that the war was over, a few like John Abizaid figured out what was really going on and stepped forward to help me. There were three other generals, in particular, who also leaped to the rescue. GEN B. B. Bell realized that we had a broken headquarters structure, and volunteered to send his top operations officer to Iraq. When the commandant of the Marine Corps, GEN Michael W. Hagee, visited Iraq, he quickly agreed to send me a Marine MAJ GEN, Jon Gallinetti, to serve as chief of staff for CJTF-7. And perhaps most significantly, Acting Chief of Staff of the Army, GEN Jack Keane, came to Iraq over the Fourth of July weekend to visit our soldiers and to see what he could do to help.

Over several extended one-on-one conversations, GEN Keane probed our needs. "Ric, what is your biggest problem?" he asked.

"Sir, I don't have the right people on my staff," I responded. "This corps can't do the mission. We have young colonels that have no idea how to operate at this level. They're great young officers, but all they've ever done before is fight at a tactical level. There is literally nobody in this headquarters, other than me, who has joint operations experience or who has served at the strategic level."

"What are your most pressing needs?" asked Keane.

"Sir, I am in desperate need of an operations officer, an intelligence officer, and a logistician. GEN Abizaid is already working to get a strategy, policy, and plans officer. I must have at least those three billets filled if I am going to have any chance at succeeding. You've got to help me, sir."

"Well, dammit, Ric. Why did CENTCOM and CFLCC send the dream team home anyway?" he asked. "Did anybody ever give you a reason."

"No, sir. I guess they thought the war was over."

"Well, it appears to me that you have been set up for failure. Now I don't want you to be shy about asking for resources. I'm going to have the Army staff make it our number one priority to reinforce CJTF-7. This phase of operations is more challenging than the war itself. I'll have some names for you by Monday or Tuesday."

And Keane kept his word. A day or so after he got back to Washington, I received the names of four generals to fill some of our staff slots. The one at the top of the list was MG Barbara Fast, who would become our chief intelligence officer.

At that time, there were more then 1,400 CIA and Pentagon intelligence analysts operating as part of David Kay's Iraq Survey Group, but not one of them was assigned to support CJTF-7. They were focused solely on locating Saddam Hussein's weapons of mass destruction. The military cooperated very closely with them and provided extensive support. For instance, any time they had an indication of a suspected WMD location, we escorted them to the sites. We went on hundreds of these missions, but all that was ever found were a few old rusted chemical rounds that had been sitting in a bunker for years.

During this time, GEN Abizaid and I tried to persuade David Kay to help us build the military's intel capacity. "Just give us some added priority, share your resources, and then communicate the intelligence as you gather it," we begged. Eventually Abizaid took the matter all the way to Washington, but was refused. *[On January 23, 2004, David Kay resigned his position in frustration, saying that he believed no WMDs ever existed and that the failure to find them in Iraq raised serious questions about U.S. prewar intelligence gathering.]*

During those early days at CJTF-7, we were having serious problems accumulating sufficient amounts of credible intelligence. It is a fundamental responsibility of any battlefield commander to know who the enemy is and what the enemy is doing. That's how you win wars. So I made our intelligence operation my highest priority. I set up daily meetings with our intel people, limited in number and experience though they were. I asked lots of questions. "Who is carrying out these attacks on us? Is it the Saddam Fedayeen? Is it al-Qaeda? Are there, in fact, any terrorist groups operating in the country? Where are they located?"

I also pressed for results. "We need actionable intelligence at the strategic and operational level," I said. "Right now, we are not getting it. We're only focusing on the tactical level—finding the next target to go after. In addition, we cannot just put information into a report and let it sit. The enemy is probably moving around constantly. We have to shorten the cycle from the time we gather the information to the time we move on it. It will save the lives of our soldiers.

"We're in a tough spot here," I told them. "We have too many prisoners and not enough places to hold them for questioning."

During our early offensive operations, we had swept in large numbers of detainees. People were picked up for a variety of reasons. They might have committed egregious offenses, or been caught planting IEDs (improvised explosive devices), or they might have been innocent bystanders caught up in a cordon and search. Unfortunately, there were no operational prisons in the country. Once again, part of our problem was the lack of Phase IV planning for detainment capacity. So we were forced to start from scratch and improvise. We began by erecting concertina wire (a type of expandable barbed razor fencing) in the middle of the desert. We put detainees inside the wire and provided basic human necessities. We fed them with our own rations, water, and struggled to provide them with shade. Except for the fencing, these conditions were similar to the expeditionary environment that our own soldiers were experiencing.

Our early process for identifying and keeping track of prisoners also had to be done on the fly and was as primitive as you can imagine. We had no computers, so we used written ledgers and makeshift "capture tags," which were nothing more than pieces of paper pinned on the detainees by our front-line units. Usually all that was listed was the person's name, but sometimes the soldiers would also write down place of capture and reason for detainment. One of the first problems we experienced involved the language difference. Most of the Iraqis would give us their names, but because they didn't speak English, they could not tell us the spelling. And sometimes there were four or five different spellings of the same-sounding name, compounded by the fact that our soldiers had to spell things out phonetically.

As our detainee numbers grew, it wasn't long before we started getting inquiries from Iraqis at all levels. "My cousin was captured and we have no idea where he is. Can you tell us if you're holding him?" Part of our problem was that when Iraqis were taken off to prison during Saddam Hussein's regime, many were never heard from again. Before the war, Saddam's people had also told everybody that if the Americans captured them, they would be tortured and killed. When civil leaders began to press Ambassador Bremer for answers, he would call my office and ask why we couldn't answer these questions. "Well, sir," I responded, "we don't know how to spell their names, we don't have a computer data base, and there are no mechanisms in place to perform timely reviews."

"Well, what are you going to do about it?" he asked. "I've got these Iraqis all over my back."

"Sir, we are working to improve our processes internally but it will take time. Our requests for help have been sent to higher headquarters. I'm told help is on the way," I replied.

Our detention capacity was starting to get pretty grim at this point. We had been sending quite a few detainees down to Camp Bucca in the extreme southern part of Iraq. Several hundred prisoners had also been sent to the facility at Abu Ghraib, a neighborhood on the northwestern edge of Baghdad. But that area presented another problem: It was a Sunni stronghold where we were sure to experience continued resistance. In addition, the Abu Ghraib facility was Saddam Hussein's most notorious prison, a place where many political prisoners had been tortured and killed.

We had many debates about whether to use the Abu Ghraib facility at all, due to the very real emotional implications. But it was really the only prison that had an intact physical infrastructure that would allow us to isolate and conduct operations. So in mid-July, Ambassador Bremer made the decision to designate it as a temporary detention facility, until a broader prison capacity was brought online (which was envisioned to be about a year and a half).

A week or so later, CJTF-7 conducted a major cordon and search operation, which we called Victory Bounty. From July 26 to 29, throughout the Sunni Triangle, our forces made a comprehensive effort to round up the remaining elements of the Saddam Fedayeen. We ended up netting nearly seventy former members, including several generals and field-grade officers, nearly all of whom were sent to Abu Ghraib.

After Victory Bounty, however, I took steps to temper the amount of sweeping cordon and search operations. First of all, I was concerned that we were alienating too many Iraqis by taking into custody innocent people who just happened to be in the wrong place at the wrong time. And second, of course, I didn't want to risk overcrowding the system to the point that everything became completely paralyzed. I further asked our division commanders to become more disciplined in their approach to identifying detainees. "Don't just send them back with a name," I ordered. "If you do, we will have no documented reason to detain them, and they will be released immediately."

"But sir," said one of our commanders, "you can't just put these guys back onto the streets."

"General, if any prisoners make it to Abu Ghraib with only a name, we will release them," I responded. "It is your job to provide us the rationale and legal justification necessary to detain them."

The next thing I had to address was how we were going to treat prisoners. My staff and I held extensive discussions about what to do. Are detainees deemed prisoners of war (POWs)? Is their treatment subject to the Geneva Conventions? Or should we heed the Secretary of Defense's guidance and treat them as terrorists? In mid-June 2003, I ended the debate and put out an order to all my units stating that the Geneva Conventions applied to all detainees for all interrogations and handling.

It was obvious to me that avoiding the use of prisoner interrogation approaches deemed harsh or near the limits of acceptability under the Geneva Conventions would require training and supervision. However, the Army had not provided them for either the military intelligence personnel who would be conducting the interrogations or for the military police who would be guarding prisoners.

In Iraq, we had no guidelines whatsoever. So I simply imposed international standards that were consistent with the laws of war. As far as I was concerned, it was the right thing to do.

The first time I visited Abu Ghraib was about a month before Operation Victory Bounty. It had only been two weeks since I assumed command of CJTF-7 when I was approached by Judge Campbell, who, in his job to rebuild the entire judicial system of Iraq, was also senior detention advisor for the CPA. "General Sanchez, I don't have any way of assessing what might be out there in terms of available facilities, and I could use your help," he said.

"Where would you like to go, Judge?"

"Well, I need to go out to Abu Ghraib. I'd also like to see Khan Bani Sadh, which is another prison that Saddam had up on the northeast edge of Baghdad."

"I need to see those two facilities, myself," I said. "Let's take my helicopter and go out there."

We arrived at the Abu Ghraib prison at midmorning, and were greeted by the MPs manning the facility. None of the two hundred or so prisoners were inside the buildings. They were all out in the courtyard in tents that were surrounded by triple-strand concertina wire and towers manned by armed guards. The living conditions were miserable for both the prisoners and our soldiers. There were improvised latrines, no shower facilities, limited hot meals, and relatively little relief from the heat. Nearly everyone was living under very squalid conditions.

When Judge Campbell and I walked into the buildings, we could see why none of the prisoners were inside. The place was completely gutted and trashed out. There was no wiring, no plumbing, no electricity, and every window had been taken out. When we walked into the torture chambers, I saw filth and human waste everywhere. Nooses were still hanging from the ceilings. The cells were as stark and awful as could be imagined. And when we climbed up to the trap door where Saddam Hussein's regime used to hang people, the hair on the back of my neck stood on end. I could practically hear the agony and the screams of the people who had been tortured and killed. It was a terrible feeling, one that I will never forget.

As we left Abu Ghraib, Judge Campbell and I both agreed that we would have to isolate the torture chambers and not allow them to be used in any capacity whatsoever. The future Iraq government would have to make a decision as to whether or not to destroy them. We also agreed that the prison could be used for high-risk, hard-core criminals, although it would take a lot of work before it would be made useable. I also left determined to improve the living conditions of the soldiers and prisoners. We needed to fix the situation fast.

From Abu Ghraib, we flew over to check out Khan Bani Sadh. In this case, there was no need to land, because we could see from the air that there was almost nothing left of the facility. Only five or ten feet of the surrounding wall remained—and there were a bunch of Iraqis chipping away at it with hammers. Some of the buildings had been razed down to the foundation. The two or three left standing were completely stripped—with roofs, doors, windows, and conduits all gone. The facility had been almost totally dismantled by the local residents who used the various parts to construct their own structures. Judge Campbell and I agreed that there was just no way that Khan Bani Sadh was a viable option as a prisoner detention facility.

As we did our final turn over the site, I received a radio call that the bodies of two missing American soldiers had been found. Sergeant First Class (SFC) Gladimir Philippe and Private First Class (PFC) Kevin C. Ott had been kidnapped a few days earlier as they were putting up road barriers to keep people from wandering into an ammunition site. We had undertaken a massive search. Now I was being informed that they had been killed and their bodies dumped on the side of a remote Baghdad road. Turning to the pilots, I said, "Let's fly directly to that location."

After we arrived and landed in a nearby field, I walked up to a group of soldiers guarding the bodies. It was a pretty gruesome sight, and the heat of the early afternoon sun was only making things worse. "Is there a reason we haven't taken them to the morgue?" I asked.

"Sir, we're waiting for CID [Criminal Investigation Division] to come in, secure the scene, and do their investigation," replied the officer in charge.

"Okay," I said. "But why hasn't that happened?"

"Well, sir, they've told us they're busy."

I immediately called our chief of operations back at headquarters and asked him if he knew what the holdup was with CID. "Sir, they're tied up with some reported thefts in one of the units."

At that point, I just went through the roof. "You call the CID commander and tell him he has about thirty minutes to get his ass out here," I said. "I will send my helicopter to get him if necessary. I cannot believe they are screwing around with a reported burglary when we have two of our soldiers lying out here in this hot sun. I want them treated with dignity and respect and I want them attended to now. NOW!"

"Yes, sir. Got it, sir. They're on the way."

CHAPTER **13**

Reversing the Troop Drawdown

While on duty in Iraq, I stuck to a pretty basic daily routine. I rose every morning at Victory Base by 6:00 a.m., and from 6:30 to 6:45, I went into a meeting with my staff where we reviewed the previous night's situation updates. The first thing I looked at was the daily casualty reports of killed and wounded. Afterward, I would either fly or drive to the Green Zone to meet with CPA (Coalition Provisional Authority) officials.

On days that I flew, the helicopter pilots took varied routes for security purposes. So, with time, I got a good aerial view of the "cradle of civilization." Back in 1991, during Operation Desert Storm, I wondered whether the Tigris and Euphrates river valleys would be more beautiful near Baghdad than in the southern part of Iraq. To a degree, they are. As the rivers come together, the land becomes greener and is a marked contrast to the brown, desolate parts of the desert on its edges. There are also miles and miles of palm tree farms running through the interior of the valleys, which can give you the feeling that you're in a paradise. But the water flowing through the banks and into the irrigation systems was filthy in the Baghdad area—a result of the population's large raw sewage dump. At times, I couldn't help but think about the contradictory nature of this place: the abrupt change in color, from green to brown; beauty and filth in the same place; the comfort of history amidst the pain of war; trying to build a democracy in a place where many people apparently didn't really want it.

Upon arrival in the Green Zone, I usually met with Ambassador Bremer at 7:45 a.m. and then we went into his eight o'clock morning meeting with the full CPA staff. Afterward, I spent some time at the Republican Palace or in meetings handling CPA issues, as needed. As soon as my business was finished in the Green Zone, I'd spend two or three hours with a unit somewhere in the country. I could get from one end of Iraq to the other in about an hour if I took a C-130, but most of the time, I'd either fly in the helicopter or drive. Other times, I'd go on patrol with a young tank commander, have lunch with an infantry platoon, or tour a local neighborhood with a group of Marines. My intent was to be "in the box," just as I had always done during training exercises. I wanted to visit the leaders, make contact with our soldiers, and get a feel for what was going on out there. It was very important to assess if my intent and directions were reaching the ground troops so they could improvise their actions accordingly.

By 6:00 p.m., I was back at Victory Base for our regular series of operational updates where we looked at everything that happened in the country that day (battles, intelligence gathered, etc.), and then addressed tactical issues of the task force, issued staff guidance, and decided on future priorities. Normally, I made it back to my quarters about 10:30 or 11:00 p.m., where the first thing I'd do was place a call to Maria Elena. She was the only person I could talk to about certain feelings—and she could always gauge how my day had gone based on my tone of voice. Often, though, she'd be talking to me and there wouldn't be an answer from my end, because I was so exhausted I'd dozed off. "Okay, Ricardo," she'd say. "It's time for you to go to bed."

The next morning, I'd start the cycle all over again. It was a seven-days-a-week routine that just never let up. Time often passed in a blur, but I always knew when it was Sunday, because a church service was on my schedule. Sometimes, I went to mass in the mornings, or we'd squeeze it into the afternoon routine. But the staff knew it had to be fit in one way or another.

In my daily forays to the field, I saw firsthand how our soldiers were living, and the conditions weren't pleasant. Most Americans had the idea that the troops were staying in massive tent cities with all the amenities. It took months, however, before all that could be set up. From May to October 2003, we had as many as 180,000 troops on the ground at one time operating under expeditionary conditions. In the early days, hot meals were scarce, and our soldiers went off and found their own shelter. Some of them lived in amusement parks, in abandoned houses, in gutted palaces, and in zoos. Many used slit trench latrines, stayed in open tents, or lived in combat vehicles. For a time, many lived off what they carried with them—MREs and T-rations—or could scrounge up on their own.

Our problem was low logistical capacity, not lack of funds. We had only one LOC (line of communication) established to transport supplies into Iraq. Only one major road up from Kuwait could be used and every item had to compete for transportation space. The top priority, of course, was to sustain the warfighting strength of the coalition forces. Everything else came next—and I do mean everything. There was virtually nothing in Iraq we could use. Food, supplies, fuel, plywood, you name it, each and every item had to be transported across hundreds of miles of enemy territory via convoys that were vulnerable to attack.

Our logistical problems were exacerbated by the fact that our forces were drawing down. "Most of the troops won't be staying very long, so there is no need to build large stockpiles," was the thinking in May and June. There were ongoing negotiations to open up LOCs through Turkey and Jordan, but neither country was in much of a hurry to help us. The Turks were especially hesitant, because they had already turned down our request to establish a northern staging area and attack route before the invasion.

As CJTF-7 (Combined Joint Task Force 7) and CFLCC (Combined Forces Land Component Command) pushed hard to increase the flow of supplies to our troops, logistics convoys began flowing twenty-four hours a day, through all kinds of terrain. But given the circumstances, it was just a monumental task to improve the quality of life for American soldiers in Iraq. By midsummer, most soldiers had already been deployed for six months. But when we requested a rest and relaxation (R&R) program that would allow them a two-week leave, we were met with considerable resistance from the Army. Eventually, we obtained approval even though most of the leaves came too late for soldiers on the ground in 2003.

One day in July, John Abizaid called to inform me that it was all over Congress that soldiers in Iraq didn't have enough water to drink. "Everybody's up in arms," said Abizaid. "And it's just hit the media."

"Well, I'll check, sir," I replied. "But as far as I know, we don't have a problem."

I was really puzzled by this complaint, because water was the one thing we had plenty of. We had two rivers to tap into, multiple lakes, and we had brought in all our own reverse osmosis purification units. Every organization had been directed to deploy one into the country. For our soldiers to be without water was virtually impossible.

The next congressional delegation that visited Iraq was almost livid. "How can you not provide your soldiers with water?" I was asked.

"Well, sir, I believe they have plenty of water," I replied. "Could you please tell me why you think there's a problem?" After I heard his explanation, I placed a call back to Abizaid.

"Sir, apparently soldiers have complained about running out of plastic two-liter bottles of water," I said. "The congressmen inferred that we weren't providing enough water to our soldiers on the ground. But we have plenty. You're not going to find any soldiers without their canteens full, so long as they're willing to walk down to the water buffalo to fill them up. But we can't provide them all the water bottles they want because of our logistics challenges."

JOHN ABIZAID ASSUMED COMMAND of CENTCOM from Tommy Franks on July 8, 2003. Three days later, on July 11, he reversed Franks's force drawdown order. "The operational environment in Iraq is fluid and continuously being evaluated and addressed," he wrote in his directive. "In light of the current situation [forces previously scheduled to redeploy] will remain in Iraq until replaced by equivalent U.S. or coalition capability." Virtually concurrent with this guidance, GEN Abizaid directed that the troops would deploy for "one year boots on the ground." Abizaid realized that we needed a longer time to gain situational awareness. Besides, there was no rotation plan in place and most soldiers would be on the ground for at least a year before they rotated out.

With these orders, John Abizaid took the first steps in trying to remedy the situation in Iraq. His decisions were courageous and he did the right thing. Abizaid had obtained the reluctant approval of Donald Rumsfeld beforehand, but the Secretary really had no choice by then. It was either that or total chaos in a couple of months.

The immediate reaction in Washington was frenzied. Most of the senior leadership of the armed services had turned away from the war. They were working on getting the troops home, on transformation, and on resetting the force. Abizaid's insistence that nobody could leave without a replacement sent shockwaves through the Pentagon. Everything simply stopped—all the withdrawal plans, all the forces moving out of Iraq—everything just came to a screeching halt. Army leaders were especially upset, because a new force rotation plan would now have to be created from scratch.

When reality set in, shock turned to anger. Some troops (and their families) were terribly upset that they weren't going home as planned. In particular, some of the top generals in the Army were downright mad. Not long after Abizaid's order, I received a visit from GEN James Ellis, the four-star in charge of supplying the bulk of Army forces to Iraq. "Ric, you can't do this," he said. "You and Abizaid have to back off some of your requirements."

"Sir, if we back off, we put the mission at risk," I replied.

"Goddammit, one of these days you're going to be coming back to the Army," he said in response. "You're thinking like a joint warfighter!"

"Sir, I'm just thinking about what I need to do to keep my soldiers alive, and to accomplish the mission I've been given."

Ellis's mind-set was representative of many of our senior generals. They believed that the Army alone should fight and win the nation's wars, that CJTF-7, as a joint operation, was not their responsibility, that the Goldwater-Nichols Act was flawed. And GEN Ellis inferred that if I didn't back off, when I returned to the Army, I'd pay the price for placing joint needs above the Army's needs. In my mind, though, I had never left the Army. I was serving in CJTF-7 and I was doing the Army's mission in a joint environment. That's essentially what Goldwater-Nichols demanded.

Shortly after Abizaid's directives came out, the Army began working to stabilize the active-force component through a detailed plan of rotations. Immediately, however, we realized that the Reserve units were not included in the original one-year boots-on-the-ground orders. But it was obvious that they couldn't leave, either. So DoD had to issue another order referring directly to Reserves. Of course, that sent a second shock wave through the services and resulted in another significant dip in morale.

The emerging rotation plan ran from December 2003 into April 2004. In essence, we were trying to sustain a force of 138,000 (plus 20,000 coalition troops) with minimum impact on soldier morale. The only way we could really achieve that number was to require one-year deployments. If additional troops were needed, we would overlap forces. By bringing in the next scheduled rotation a little early, and retaining the outgoing troops a little longer, we could achieve greater numbers for about sixty days. So that became a pattern we would use for future troop deployments. Unfortunately, based on the realities of the situation in Iraq, we also came to the conclusion that the force, as a whole, was going to have to be deployed indefinitely. There was just no way to tell how long it would take to get the violence under control; establish national security and stability; re-create government systems, infrastructure, and leadership; and rebuild the nation as a whole.

With these actions, GEN Abizaid and I formally implemented the third and final step (along with de-Baathification and disbanding of the Iraqi Army) of the incremental dismantling of the nation's original strategic plan for Iraq. Now, rather than a drastic troop reduction in 90 to 120 days, we would have a significant military presence in the country for an indeterminate period of time. Secretary Rumsfeld, although not pleased, did not reverse our direction. He did, however, state that he would have to personally approve any soldier that was going to stay in Iraq longer than 365 days. On the positive side, once this scenario became policy, a flow of materials that would improve the quality of life for our soldiers began. And as the summer wore on, we shipped in massive amounts of construction material to build living quarters, we established post-exchanges, and implemented a strong R&R program.

By the latter part of July, the Army staff continued to exert tremendous pressure on us to downsize the overall force. "What you want us to do is impossible," they said. "You've got to get your requirements down to a level that we can sustain over time." After some intense conversations about this issue, it became painfully clear to me that the Army did not have the forces available to meet our requirements. We already had 75 to 80 percent of the entire military police and civil affairs units in Iraq. At this point in time, five out of the ten divisions in the Army were fully deployed—and two of the remaining five were partially deployed. That meant that we had more than half of the Army's entire combat force already on the ground in Iraq. How *could* the Army sustain these numbers? It *was* impossible. Accordingly, Abizaid and I agreed to a sustainable force of 138,000.

Even then, the Army struggled to identify replacement units it could send into Iraq. Because we were conducting joint operations at an unprecedented level, and were so desperate to fill the required numbers, the Army simply started pulling soldiers from everywhere, with little regard for long-term implications. We deployed individual soldiers, sent platoons to different companies, and companies to different battalions. We also established "in lieu of" units, which performed roles that they had not originally been organized or trained for. Reserve artillery units, for example, were employed as infantry and military police. Noncombat units were now performing in Iraq as combat units. In other cases, we placed small Army units into Marine brigades, which resulted in a shift of SOPs (standard operating procedures) and warfighting techniques. It also placed additional mental stress on our young officers who had to integrate themselves and their troops into a different culture. This process created a fractured and frayed organizational structure, with inexperienced young leaders operating as independent units. The entire system was becoming a recipe for disaster.

BY MID-JUNE, WITH THE help of CENTCOM planners, CJTF-7 and the CPA came together and created a comprehensive Phase IV plan for Iraq. Ambassador Bremer subsequently published a thirty-page paper on the subject and sent it off to Washington. CJTF-7 immediately developed action plans and began executing the strategy. The CPA, however, never really did develop action plans to accomplish the integrated strategy, largely because it didn't have the staff to do so.

As the command continued our strategic planning for Iraq, we began an in-depth study of the opposing forces. What is this enemy like? Who are they? What are their tactics? We were aware that many of our soldiers were being ambushed while they were out in the cities trying to engage people. In a number of cases, gunmen literally walked up, shot them in the head, and then melted back into crowds. These hit-and-run incidents were becoming more and more prevalent, as was the increased use of IEDs (improvised explosive devices). In addition, our limited intelligence capabilities revealed that there did not seem to be any central command and control organization coordinating the attacks. Rather, it seemed as though former members of Saddam Hussein's regime were acting on their own initiative across the entire country. By mid-July, GEN Abizaid and I concluded that we were facing an insurgency. There was just no other way to describe it.

However, the next week, when Abizaid actually used the term "classic insurgency" at a Pentagon press briefing, the Secretary of Defense immediately rebuffed him. Essentially, Rumsfeld argued that our enemies were simply disparate groups of former regime loyalists, criminals, and foreign terrorists.

The next time I saw Abizaid, he joked, "Well, that didn't go over very well. Maybe I'm going to get fired."

"But you're right, sir," I said.

"Well, there's no appetite in Washington right now to use the word 'insurgency.' And by the way, we're not 'occupiers,' either. We're 'liberators.'"

"That's ridiculous."

"Ric, if you think this is bad, just wait until we start getting close to the presidential election."

According to *any* definition, we *were* up against an insurgency, and we *were*, in fact, occupiers. The Bush administration simply didn't want to use those two words, because they sent the wrong political signals. To those of us on the ground in Iraq, this entire discussion was nonsense. No matter what words were used, we knew what our enemy was doing and we knew our status in the country. Therefore, it was our duty to come up with a strategy to achieve our mission.

On July 28, 2003, GEN Abizaid sent a memorandum to Rumsfeld entitled "Understanding the War in Iraq," in which he outlined who the enemy was, how the enemy was operating, and our recommendations for tackling the problem. Essentially, he listed three solutions: (1) accelerate involvement of Iraqis in security; (2) focus on intelligence, especially HUMINT (human intelligence); and (3) provide reconciliation methods for Iraqis who have no choice but to fight and die. Abizaid also sent a note to Rumsfeld with the Army's formal definition of the word "insurgency."

Rather than dealing with the insurgency on a short-term, one-crisis-at-a-time scenario, we had to prepare a long-term counterinsurgency plan. Because I did not yet have an adequately manned strategy, policy, and plans shop, I was forced to shift the planning priorities around as various crises arose. As we began our detailed analysis of the counterinsurgency mission, it quickly became clear to me that very few of our staff officers had any experience with this kind of warfare. Fortunately, the British had assigned some people with an advanced strategic perspective on dealing with insurgencies. So my chief of staff assigned lead planning responsibility to an outstanding young British officer on our staff, MAJ Pat Saunders. And he did a marvelous job.

We generated a comprehensive counterinsurgency strategy that combined a strong warfighting tactical plan (that included structuring and properly managing our intel/interrogations and detention operations) with an integrated political, economic, and security approach. Through this "hearts and minds" approach, we intended to isolate hard-core insurgents from the general population, provide economic hope, and involve as many Iraqis as possible in the process. GEN Abizaid had often made the point that we Americans were natural "antibodies" in Iraq. If we could achieve neutrality with the average Iraqi, then when it came to the insurgency, perhaps we would be successful.

As we were forging the counterinsurgency plan, Abizaid and I met with Ambassador Bremer several times to discuss our idea for the Iraqi Civil Defense Corps (ICDC). We explained the concept to him and discussed our belief that it should be organized, trained, manned, and equipped separately from the Iraqi Army. We also suggested that details of the ICDC's eventual integration into the national armed forces be left for the Iraqis to decide. But Bremer refused to consider these ideas, and repeated to us that the police and the Army were his responsibilities.

Meanwhile, Ambassador Bremer was aggressively moving forward with his democratization effort. On July 13, 2003, we had an elaborate ceremony to establish the new twenty-five-person Iraqi Governing Council. Live television captured the speeches, the oaths of allegiance, and rest of the pomp and circumstance. Ambassador Bremer had personally chosen each of the members, and he had also given an order that all communications with the new governing council had to be based on his authority and coordinated through the CPA front office.

After the ceremony concluded, Nuri Badran, the new minister of the interior, caught up to me and asked for a meeting. We scheduled an appointment the next morning and, as soon as Badran came in, he said, "General, where's my office?"

"Well, Nuri, I don't know," I replied.

"And where are my people?" he continued. "Where is the money to pay for my people?"

"Have you not spoken to Ambassador Bremer and his advisors about this?"

"I have received no help from them at all, General. Will you help me?"

"Okay, let me see what I can do," I replied.

Later that day, I asked our director of operations to find a building where Badran could set up the new Ministry of the Interior. "And let's see what we can do to find him some furniture and office supplies to get him going," I said. "We can take the money out of CERP [Commanders Emergency Response Program Fund]."

. . . .

IF WE WERE GOING to be successful in our "hearts and minds" approach to solve problems in Iraq and quell the insurgency, we were really going to have to understand the Iraqi people. I had already delved deeply into the history of the region, and my staff and I were learning more every day. Unquestionably, Iraq was still very much a tribal-based society. The tribes, comprised mainly of Sunni and Shia sects of the Islamic religion, were strong and quite active. The Sunnis were predominant in terms of power. They dominated governments in most of the western Middle Eastern nations, such as Jordan, Saudi Arabia, and Kuwait. The Shiites were mainly dominant in Iran.

In Iraq, Saddam Hussein was a Sunni and, of course, he made certain that they controlled virtually every aspect of his regime. The tapestry of the entire nation, however, was quite diverse. Sunnis predominated in the north-central and western regions—particularly in the north from Mosul to the Syrian border, and in the Sunni Triangle, a densely populated area bordered on the east by Baghdad, the west by Ramadi, and the north by Tikrit. The Sunnis were somewhat secular. Although they had their extremist elements, they were not dogmatically subject to ayatollah or imam *fatwas* (calls for action), and often disregarded them. In general, the Sunnis were very upset about de-Baathification and the disbanding of the army, because they had all been eliminated from their jobs and were disenfranchised from all aspects of society.

The center of the Shia religion in Iraq was located in the south, in the cities of Karbala and Najaf. The southern city of Basra was also predominantly Shia, with a heavy Iranian influence. In general, the Shiites were more fundamentalist. The imams and ayatollahs had significant sway over the population, and when a fatwa was issued, large masses of people mobilized across the entire country. With Saddam Hussein and the Sunnis now out of power, there was great hope among the Shiites that they would be able to become the dominant power in Iraq. However, because of the influence of Muqtada al-Sadr and Muhammed Bakr al-Hakim, who both had significant militias, there was a constant danger to coalition forces operating in the region.

Northern Iraq, for the most part, was a very stable environment due to the iron-fisted rule of the Kurdish people, who were predominantly members of the Sunni religion. Kirkuk, the largest city in the region, was a large melting pot of people. Saddam Hussein had put the city through an "Arabization" process, in which he moved many of the Kurds out and thousands of Arabs in. With Saddam's overthrow, however, the Kurds had begun to reassert control and reverse the entire process. Arabs were now being pressured to leave.

There were really no completely homogenous areas in Iraq. The ethnic makeup of the nation changed in virtually every direction, from province to province. Baghdad epitomized Iraq's diversity with separate Shia, Sunni, Christian, and Syrian neighborhoods (to name a few). Sadr City, which lies in the northeast portion of Baghdad, comprised more than one million isolated Shiites. The Sunnis lived around the entire city, but were particularly strong in the southern portions. And just west of Baghdad near Fallujah, the Sunnis were fervently opposed to U.S. and coalition occupation.

The importance of understanding the tapestry of Iraq could not be underestimated from a military, political, or economic perspective. In order to keep our soldiers safe, we had to know where they would most be at risk. And if we were going to effectively stabilize the country, we had to engage the main elements of Iraqi society. Each province had its own story, its own complexity, and its own set of challenges that had to be addressed. So the solutions needed to be specifically tailored to each area.

Within CJTF-7, I knew that if we did not understand the makeup of Iraq, and if we did not take the appropriate actions, we stood a good chance of repeating the same mistakes that the British had made in the early twentieth century. In 1914, during

World War I, British forces invaded Mesopotamia, which included present-day Iraq. By 1917, they occupied Baghdad, and the next year, Britain merged the three former Ottoman Empire provinces of Mosul, Baghdad, and Basra into one political entity. After the war ended, however, the Iraqis rebelled against British occupation and staged an ongoing violent revolt, which had to be put down with indiscriminate force. Despite a solid understanding of the complex mixture of ethnic and religious groups across the country, the British went ahead and imposed a monarchy, influenced the writing of a constitution, and established a constituent assembly, a parliament, and a separate legislature. Even though it *appeared* the Iraqis were in control, the British still ruled the country, and everybody knew it. Eventually, public opinion soured on a military presence in the area. After maintaining a presence in Iraq for fifteen years, Great Britain finally signed an agreement in 1932 that granted Iraq its independence. The terms of that agreement included keeping military bases in the country and having the British train a new Iraqi Army.

In later years, when T. E. Lawrence (Lawrence of Arabia) reflected on lessons learned from his experience in the region, he pointed out the tremendous difficulties the British encountered in dealing with the tribes, the ethnic diversity, and the religious differences of people in the area. Basically, Lawrence warned future generations that, in dealing with Iraqis, success was all about giving them responsibility, putting them out in front, and letting them progress at their own pace. The worst thing to do was to try to occupy the region in a heavy-handed manner. The people were simply too proud to allow such a thing to continue for too long.

IN JULY 2003, THERE were three different wars going on in Iraq. First, Saddam Hussein's former regime was staging a disorganized, decentralized insurgency against coalition forces. They were mainly interested in fighting Americans. Second, Sunni extremists were, at times, rising up and attacking any foreign presence in their neighborhoods. This was especially true in the western part of Baghdad near Fallujah. And third, terrorists were beginning to infiltrate into Iraq through Syria and possibly Jordan. We were never able to satisfactorily quantify the number of terrorists in the country at any one time. However, we estimated—with only a 30 to 40 percent confidence level—that it was probably only a few hundred. They were most likely members of al-Qaeda and were definitely cooperating with former members of Saddam Hussein's regime.

Also looming on the horizon, but not yet quite rising to a level that would cause us much concern, was the Shia-on-Shia conflict in the southern part of Iraq. The two main leaders, Ayatollah Muhammed Bakr al-Hakim and Muqtada al-Sadr were capable of going at each other at the slightest provocation.

Hakim, who had just returned after years in exile, was the leader of SCIRI (Supreme Council of the Islamic Revolution in Iraq), a political party with very strong links to Iran. The Badr Brigade, al-Hakim's militia, was well organized, stretched across the entire southern part of the country, and was completely trained, equipped, and funded by the government of Iran. In early July, there was a big debate within the CPA whether or not to include al-Hakim in the political process. And when Bremer eventually did put him on the governing council, it was with full knowledge that al-Hakim had links to Iran and that he controlled a significant militia.

Muqtada al-Sadr's militia was called the Mahdi Army. But unlike the Badr Corps, it was very loose with almost no discernable command and control structure. Al-Sadr had loyalists all over the South and, with the use of sermons and his propagandist newspaper, he was able to very quickly mobilize tens of thousands of people in Baghdad and most of the provincial capital cities in the region. When he issued a call to arms, busload after busload of Shiites started heading out of Baghdad.

In the Shia-dominated south, there existed tremendous potential for instability. However, the Shiites were motivated to stay peaceful with each other in order to become the main political power in the nation. They were certain that was going to happen, largely because the CPA had eliminated all of the Sunni Baathists from the government. Under the surface, it was also important to know that al-Sadr and al-Hakim were fighting for control of the holy cities and mosques. Why? Because with control of the cities and mosques came the big money. And as it does everywhere, money means power.

If the U.S.-led coalition was going to win the three wars in Iraq, temper the simmering hostility brewing in the Shia community, and stabilize the nation as a whole, military success had to be accompanied by political and economic progress. John Abizaid and I had learned that lesson from our experience in Kosovo. Any other approach would result in prolonged conflict and increased resistance. Abizaid and I repeatedly met with Paul Bremer and urged him to work with us in the process of tribal engagement and reconciliation. But he adamantly refused to do so. The reason, I believe, was more philosophical than practical. The Bush administration's—and therefore CPA's—grand vision for Iraq was to create a democratic state where tribes had minimal-to-no influence in running the government. Unfortunately, Iraq was rooted in its tribal customs, and that wasn't going to change any time soon.

With Bremer's reluctant acquiescence, GEN Abizaid and I began the process of tribal engagement, which included the Sunni, the Shiites, the Kurds, and whoever else was relevant in the various sectors. We identified key leaders across the country and invited them to talk with us. Throughout the summer, we conducted separate meetings in each province, with Sunni leaders and with Shia leaders. We invited the CPA along and, in most cases, they sent representatives to attend, monitor, and report back.

Some meetings began with a huge meal prepared by the hosting tribe or coalition unit. For the main course, we always had our choice of fish, goat, or chicken. Some of our officers wouldn't go near the goat, but I relished it. And the Iraqis noticed. As a matter of fact, it usually provided a starting point for us to begin conversations. Another common bond revolved around their greeting *"In cha'Allah"* (if God wills). "Where I come from, in South Texas, we say, '*Si Dios quiere*,'" I would tell them. "It has the same meaning—if God wills." It seems trivial, but eating goat and using similar language helped me relate to the Iraqi leaders and that, in turn, led to more productive talks.

Once we got down to serious discussions, the tribal leaders, whether Sunni or Shia, always began with similar complaints that coalition forces were not respecting their traditions, their culture, and their families. "Why can't you improve our lives?" they also asked. "Saddam could give us electrical power. We had fuel when Saddam was in charge. Now we can't get fuel for our vehicles or even for cooking."

After all the complaints were aired, it turned out that nearly everybody was willing to cooperate with us. All they wanted were fundamental rights: government representation, participation in the new security structures (police and Army), and a rise in economic opportunity. Those were the common demands among everybody across Iraq with whom we met, and I thought they were quite reasonable.

During this period of time, I interfaced quite often with President Bush via video teleconferences. Ambassador Bremer was usually in the room with me, and GEN Abizaid and Secretary Rumsfeld were patched in from wherever they happened to be at the time. The President was always sincere, focused, and aware of the key issues we were discussing. Frequently he would inquire about the status of our engagement and reconciliation process. "How's our Sunni engagement strategy coming along?" he would ask. "Are we making progress?"

Ambassador Bremer would usually say, "Oh, Mr. President, our engagement strategy is coming along. We're focused on it."

"Well, this is kind of important," Bush would say. "We've got to get these folks to start cooperating."

"Yes, Mr. President. We're working on it," Bremer would reply.

. . . .

THREE NOTEWORTHY EVENTS OCCURRED in July 2003. Two were international in scope, but remained largely out of the view of the American public. The third was more local in nature and received massive news coverage and worldwide attention.

On July 9, the Iranian army seized a disputed southern border post inside Iraq and, from a diplomatic standpoint, all hell broke loose. We had to be prepared to reestablish the recognized international boundaries of Iraq. As I issued orders to begin planning a military option, my mind flashed back to Germany and the studies we undertook of GEN Eisenhower's coalition challenges during World War II. I immediately called our British commander in the southern region and urged him to take the lead in planning for a military solution to the crisis. He quickly elevated the issue all the way to London. It was as if I had ordered an immediate attack to occupy Iranian territory, which wasn't the case. I always considered diplomacy our first option and the diplomatic options were being executed by London because the United Kingdom still had formal diplomatic relations with Iran. After securing an agreement to maintain the status quo on the disputed southern boundaries between Iraq and Iran, the British government persuaded Iran to destroy the border post and pull its troops back into Iranian territory.

In late July, another international incident took place when we captured a group of Turkish special forces operating in northern Iraq. They had entered the country to search out and destroy members of a Kurdish terrorist group, KADEK, which had been conducting random bombings inside Turkey. The whole situation quickly escalated to Washington and Ankara, where the Departments of State and Defense had to engage with their Turkish counterparts. In the end, the administration decided to release the prisoners and turn a blind eye to the fact that a U.S.-recognized terrorist group was operating inside northern Iraq.

Turkey, however, never stopped pestering the United States to get KADEK under control.

The event that garnered worldwide attention happened on July 22, 2003, when Saddam Hussein's sons, Uday and Qusay, were killed in a gun battle at their hideout in the northern Iraqi city of Mosul. In a very efficient operation, U.S. Special Forces and the 101st Airborne, upon receiving solid intelligence, surrounded the house and employed armored firepower to quickly end the fight. By early afternoon, I received word that four people had been killed and the probability was high that Uday, Qusay, and Saddam's fourteen-year-old grandson were among the dead. The bodies were transferred to our morgue at Victory Base, and we began the process of identifying the bodies. X-rays were taken, DNA samples were secured and sent off to the lab, and the Iraqi Governing Council was asked to put together a delegation to view and, if possible, identify the bodies. Some in the group had gone to school with Saddam's sons, knew them well, and were able to make positive identifications. The boy was confirmed to be the grandson, and the fourth person turned out to be a known bodyguard.

Next, we needed to decide on the best way to make the news public and convince the Iraqi people that these bodies were, in fact, who we said they were. Because the Geneva Conventions did not allow us to broadcast pictures of the corpses, we had to take the issue all the way to Washington. After some discussion, the administration made the decision to override the Conventions. So the day after the killings, we went public with the news and released photos of Uday and Qusay laid out in plastic body bags. Forty-eight hours later, DNA lab results confirmed their identities.

Shortly after the media hubbub died down, our staff began planning for the eventual capture of Saddam Hussein. We created a notification plan, a detention plan, a plan to manage consequences, prepared a media package, and determined methods for positive identification. We also debated whether a confrontation with Saddam should be a capture or a kill mission. A capture would present long-term problems. We'd have to try him and, if found guilty, sentence and execute him. On the other hand, if we killed Saddam, the pain would be over very quickly, but he would become a martyr and we might never be able to prove to the Iraqis that he was really dead. In the end, we determined our most likely scenario was that we'd wind up in another firefight, which would probably result in his death. If we were lucky, we might capture him.

Iraqi public reaction to the deaths of Uday and Qusay was mixed. While elated, people were also very worried that Saddam might exact revenge. And Iraqi leaders were not shy about expressing their concern that until we got the former dictator, there was always a possibility he might come back into power. These fears were heightened when Saddam released an audiotape declaring his two sons to be martyrs and urging his people to press the fight. And over the course of the next week, three U.S. soldiers were killed during an ambush in Mosul, one died in a grenade attack south of Baghdad, and 10,000 young Iraqis came forward to join an "Islamic Army" in the city of Najaf.

CHAPTER 14

The Insurgency Ignites

As July turned to August, the CJTF-7 staff received a temporary boost when the Joint Warfighting Center (JWFC) out of Norfolk, Virginia, came in to help us. Scott Wallace, who had previously served at the center, had requested their assistance back in May when the corps first learned it would transition to a joint task force. I was very grateful for the help, because I knew what they could do. Five years earlier, a team had showed up at SOUTHCOM in Miami for a forty-five-day immersion in joint operations. I had learned that a tactical warfighting victory can turn into a strategic defeat if transition and postcombat operations are not properly addressed. During those exercises, it became obvious that some U.S. government agencies were so small and stovepiped that they were virtually incapable of conducting the interagency operations necessary to succeed. I believed I was now experiencing that scenario with the CPA.

Back in 1998, the SOUTHCOM experience had been largely a training exercise. But in this case, we were training while fighting. We worked on real-time issues and made the most of the JWFC's "best practices," which had been gleaned from operations observed around the world. The JWFC visit made a big difference across all staff functions—knowledge, ability, and operational performance. And it was just in time, too, because the coalition national forces were starting to arrive en masse.

By mid-July, only ten nations had deployed to Iraq. In addition to the United States, Great Britain, Australia, and Poland, which had participated in the invasion, there were also Iceland, Latvia, Lithuania, Bulgaria, Denmark, and South Korea. As the Departments of State and Defense forged separate bilateral agreements with new nations, the Pentagon directed CENTCOM to accelerate the flow of coalition forces into the country (specifically mentioning Tonga, Croatia, Nepal, Bosnia, and Uzbekistan). Within a couple of weeks, the floodgates had opened and August became an extremely busy month for integrating thousands of coalition troops from other countries, including Italy, the Netherlands, Spain, Hungary, Ukraine, Slovakia, Macedonia, Thailand, Mongolia, Nicaragua, Honduras, the Dominican Republic, El Salvador, New Zealand, the Philippines, and Fiji. Others that would straggle in during the coming months included Portugal, Norway, Kazakhstan, Moldova, Singapore, and Japan.

Trying to integrate this mass of troops into the Iraq theater of operations presented an extraordinary challenge. We had to assign each nation specific areas of the country, get them situated on the ground, and make sure they received the proper flow of logistical support. Nearly all the nations were expecting to receive equipment as promised by the U.S. government. But we had nothing to give them. So we wound up taking material and equipment from American soldiers and redistributing it to the multinational forces. Had we not done so, the forces would not have been able to perform their duties or assume responsibility for their area of operations.

Most of the nations were deployed to the Marine and British sectors in southern Iraq, from a line thirty miles south of Baghdad and below. I went down there regularly to make sure operations were proceeding smoothly and on schedule. When I encountered the Spanish brigade, they were absolutely delighted not to have to use a translator during their briefings. I also gave the Dominican Republic, which struggled with English, permission to present their briefings in Spanish and provide translated paper copies to our staff. Language, overall, turned out to be a problem for the coalition. There was no requirement inside any of the national units to speak English. In some cases, we had to use three-way translations to communicate effectively. For example, in the Mongolian contingent, we had to go from English to Polish, Polish to Russian, and Russian to Mongolian. Overall, our language problems significantly delayed our planning and execution times.

Nearly all of the nations also had caveats attached to their service in Iraq. There were some things they simply would not do, and it varied from country to country. The restrictions were so complex that I had to carry a five-page spreadsheet listing all the countries, their rules of engagement, and who was authorized to do what. After a while, I learned that an order I issued would be carried out either in due time or not at all. In some cases, they were not willing to obey the order. In others, they were not able. And almost always, they had to go back to their national political leadership to ask permission.

The entire thirty-six-nation coalition, in and of itself, was a massive effort to demonstrate international support and cooperation in the stabilization and rebuilding of Iraq. That is certainly what the media portrayed and, as a result, it's what the American public believed. Below the surface, however, the coalition was fraught with complexity and lack of commitment. Most of the nations had signed up because they were offered all kinds of economic incentives by the U.S. government. However, the majority had no intention of deploying their forces into a combat environment. In the end, the entire southern half of Iraq operated under very restrictive defensive rules of engagement. Other than the British and Italians, none of the international troops were authorized to impose order. So every time the use of offensive force was necessary to maintain security, we had to send American troops in from the central and northern portions of the country.

There was always a possibility of violent flare-ups in the south, and in August 2003, Muqtada al-Sadr started to become a problem. Although not an imam or ayatollah, he was a young, fiery, influential cleric who had behind him the credibility of his late father, the Grand Ayatollah for whom Sadr City was named. Muqtada had been speaking out against the coalition and was the chief suspect in the brutal mob murder of Ayatollah Abd al-Majid al-Khoi, a pro-U.S. member of the Shia ruling council.

After a newly established Iraqi judge reviewed all evidence in the murder, a warrant was issued for al-Sadr and twelve of his top lieutenants. Initially, it seemed as though the most feasible option was to make the arrests, so we asked the Marines in the area to develop a plan. They studied al-Sadr's daily moves, the number of forces that would be required to make the arrests, and the potential blowback we might receive. Our assessment was that there would be demonstrations and potential violence from his followers in Najaf, Karbala, Al Kut, Baghdad, and possibly all the way down to Nasiriyah and Basra.

When the plan was briefed to Ambassador Bremer, he made a big push to take action, even speaking with Condoleezza Rice and gaining approval from the National Security Council. "Ric, we have to execute and we have to do it now," Bremer said to me.

"Okay, sir, let me take a look at it and I'll get back to you on the right timing," I replied.

When my staff and I evaluated the situation, we agreed with the Marines, who didn't like the idea of trying to arrest al-Sadr. They were only three weeks away from going home and they didn't want to create any instability. In addition, all the multinational forces, led by the Poles and Spaniards, were actively flowing into the region. They were about ready to assume responsibility for the south-central area that included Najaf and Karbala, which would be the center of unrest. The fact that most of the coalition nations would not engage in offensive operations also created a problem. I concluded that we were simply too vulnerable while in transition, and that this was the wrong time to undertake the mission.

So I prepared a memorandum addressed to GEN Abizaid and handed a copy of it to Bremer. "I understand the criticality of this mission," I wrote. "I understand that it is necessary for the stability of the country. But my best military judgment is that the timing is absolutely wrong. It would be a strategic blunder for us in terms of the coalition effort. We would be setting them up for failure. Therefore, my recommendation is to defer the operation. Once we have the right timing and situational awareness, we'll be able to execute."

Ambassador Bremer was very upset with my recommendation. GEN Abizaid, however, agreed with me. "You're absolutely right, Ric," he said. "We can't fracture the coalition even before we put it together." Abizaid forwarded his concurrence to the Pentagon, and on August 18, 2003, Washington ordered Bremer not to arrest Muqtada al-Sadr.

Over the next few months, our plan was refined and briefed to the Ambassador. We determined that an operation to arrest al-Sadr would most likely result in a firefight, because he traveled with a large contingent of armed guards. Any mission to capture the volatile cleric would most likely result in his death. So we recommended to Washington that the mission be designated "kill or capture." Then we told Bremer to give us two weeks' notice and we would be able to execute. In the meantime, al-Sadr's movements were continuously monitored so that his location could be pinpointed for an eventual attack.

At this time, one of the things I was particularly concerned about was the inadequacy of our intelligence operations. Obviously, we needed detailed and reliable intel to execute such a crucial mission, as we did for just about everything we were doing in Iraq. Fortunately, we began to see some improvement concurrent with MG Barbara Fast's arrival in the country. I was elated with her deployment, because she had visited with me several times back when I was in Kosovo and she had a reputation as one of the few intelligence officers with the appropriate strategic-to-tactical intelligence background for this situation. Her deployment was a direct result of GEN Jack Keane's visit back in early July when I asked him for help in our G-2 (Intel) shop.

As soon as word got out in Washington that Barb Fast was being transferred to Iraq, Stephen A. Cambone, the Deputy Undersecretary of Defense for Intelligence, called her in and gave her a very specific charter. She was to look at the entire intelligence architecture at CJTF-7, assess and define the existing challenges, and then make recommended changes so that we could accomplish our mission. Cambone instructed her to submit a findings report directly to him. At the end of August, we completed the assessment and submitted it through the chain of command to Washington. As a result, by September 10, 2003, the highest levels of the Department of Defense were well aware of our intelligence deficiencies in Iraq.

When MG Fast arrived in Baghdad, my guidance to her was pretty straightforward. "We don't know what we don't know on intelligence," I told her. "I want you to ramp up that which has been shut down [when McKiernan's command left]. You have free reign to establish the intel structures and capacities necessary to make CJTF-7 successful." I also asked her to focus specifically on building our capability in human intelligence gathering, and to set priorities for our interrogation operations.

Because increased violence and our counterinsurgency operations were now resulting in additional detainees, I issued orders requiring the movement of long-term prisoners to Abu Ghraib prison. This move definitely increased the detainee burden on the MPs and interrogators at the facility. However, it was necessary to take advantage of high-intel-value prisoners, and Abu Ghraib was really the only place we could put them. To make sure that detainees were properly treated and registered as required by the Geneva Conventions, I walked through the processing of prisoners myself. In addition to securing all key information, we searched each detainee upon arrival, gave them medical exams, and made a written inventory of all personal items.

In the meantime, Barb Fast and I both were attempting to

persuade higher headquarters and all the elements in Washington to give us detailed guidance and assistance on the entire spectrum of interrogation operations. While my June memo issued instructions to stay within the limits of the Geneva Conventions, our policies did not provide any specific guidance on techniques, training, or oversight. We were in desperate need and it was the Department of Defense's, CENTCOM's, and the Army's duty to provide it.

Because Abu Ghraib had been designated a temporary facility, we weren't putting huge amounts of money into upgrading the infrastructure, but we were making it livable and safe. That decision was tied to rebuilding the country's judicial system, which included the planned construction of permanent new prisons for the entire country. Unfortunately, the money necessary for that program was supposed to be allocated from the major supplemental request going to Congress. But when it eventually came under consideration by the House and Senate, lawmakers balked at funding the initiative, because they needed prisons for their districts back home. So the supplemental was delayed, and, when eventually passed, it contained insufficient funding for the new Iraqi detention facilities that CPA planned to build over the course of the next year.

While we were having trouble securing money to build new prisons, the CPA was moving forward to build Iraq's new security and police forces. Unfortunately, they had no capacity for strategic-level planning, development, or execution. As a matter of fact, the months of July and August revealed some areas of the CPA operation to be nothing less than a series of reckless efforts.

For starters, Bernard Kerik had gone out and found a former Iraqi police commander that had operated in Baghdad during Saddam Hussein's regime. After bringing back some of the former policemen, and running them through a quick training program, they began conducting police raids across the city without coordinating efforts with the military. When a related incident resulted in a formal investigation by the 1st Armored Division, I went to see Kerik and asked him to knock it off. "You're going to wind up in a firefight with our soldiers," I said. "We've got troops patrolling the neighborhoods, and if they see a group of unknown armed Iraqis show up, they're going to engage." It also became apparent that Kerik's security forces were almost completely Baghdad-centric. But we didn't need them conducting raids and liberating prostitutes in the city—we needed a strategy to build a police force across the entire country.

There was a lot of talk around this time about training and securing equipment for those Iraqis signing up. In Ambassador Bremer's meetings, Kerik would tell us that he was going to send the Iraqis to Hungary or Jordan for training, and that he had contracts out for all kinds of equipment, including 1,000 inbound cars. However, after listening to many excuses, I gave our commanders permission to utilize CERP (Commanders Emergency Response Program) funds to purchase uniforms, cars, and radios so the new Iraqi police could at least operate with some credibility.

At the end of August, just before Kerik was slated to leave the country (without a replacement being named), I directed my staff to coordinate with CPA and review records to find out exactly what had been contracted and when it was due to arrive. Now that there was sure to be a void in leadership, I wanted to plan ahead so we didn't lose any momentum. The results of our assessment, however, were shocking: "Sir, the only thing we can document for sure is a contract for 50,000 Glock pistols," I was told.

"What?" I replied. "Is that all?"

"Sir, we just cannot establish orders or delivery dates for anything else. Nothing."

"Do you have the paperwork on these pistols?"

"Yes, sir."

When I was informed of the exorbitant prices that were being paid for these pistols, my first reaction was that there had to be some impropriety, but I had no evidence to substantiate it.

However, I did go directly to Ambassador Bremer and reported our findings.

"Mr. Ambassador, there is only one contract in place," I said. "And there are no programs for anything else, at least not that we can find."

"How can that be?" he asked.

"Sir, I don't know. We delved into the status of all contracting for the police and it is just not there. It's all talk."

Bremer was visibly shaken. "I don't believe it," he said. "It can't be true."

From his reaction, I believe Paul Bremer, indeed, did not know what was going on in Kerik's shop. However, Bremer had adamantly refused to give up or share responsibility for building the new security and police forces, and with that came control over contracting. Unfortunately, the entire process failed miserably. Shortly after we discovered the extent of Kerik's actions, Abizaid and I sent a memo to Secretary Rumsfeld requesting that responsibility for building Iraq security forces be turned over to the military. But Rumsfeld would not become embroiled in the controversy. "Why don't you all talk about it down there," was his only response. *[Bernard Kerik left the country on September 3, 2003. Fourteen months later, he was nominated by President Bush to succeed Tom Ridge as Secretary of Homeland Security, but the nomination was withdrawn the following week. In November 2007 Kerik was indicted on sixteen counts of federal criminal charges, including corruption, fraud, and income tax evasion.]*

ONE DAY DURING THE summer of 2003, I walked into Ambassador Bremer's office and saw stacks and stacks of U.S. currency sitting on his desk. "Holy cow," I said. "What's this?"

"We wanted to see what a million dollars in cash looked like," he replied. "It's probably the only time any of us will ever see that much money in one place."

Back then, the CPA was flush with cash. As a matter of fact, there was a room downstairs in the Republican Palace where millions of dollars were kept. It was commonplace for CPA regional coordinators to come in, get what they needed, and walk out with bags of money to take back to their respective parts of the country, for use in ongoing reconstruction projects. I was aware of this operation, because coalition troops often acted as guards to protect the regional coordinators from being robbed.

The CPA's cash was not American taxpayer money. Rather, it was Iraqi money that had been obtained from the United Nations. Billions of dollars of Saddam Hussein's money had been frozen before the war. And now, through UN initiatives, such as the oil-for-food program and the Iraq Developmental Fund, the cash flowed back into Baghdad. And the Coalition Provisional Authority was responsible for all of it—the intake, the handling, and the disbursement. We also seized quite a bit of American cash during operations like Victory Bounty. One Republican Guard stronghold, for instance, was hoarding $8.5 million.

When it came to Iraqi and U.S. military funds, the corps comptroller kept meticulous records of every dollar that came into and went out of our coffers. I insisted on it, not only for moral reasons, but because I knew somebody was eventually going to question what the money was used for. Kosovo had taught me this lesson. I never got the feeling, however, that any consistent accounting procedure was firmly in place within the CPA to effectively track expenditures.

Money, people, and supplies were in short supply for the building of the Iraqi Army. Early on, MG Paul Eaton approached me about providing him with planning assistance, because he had been assigned only a handful of people. "This is all I've got," he said. "And I'm supposed to stand up the Army? You've got to help me, General Sanchez."

I had no authority over Eaton's efforts with the Iraqi Army. He did not report to John Abizaid or me. But he was an Army general and, as far as I was concerned, part of the overall team.

Besides, I knew his efforts were critical to the success of our mission in Iraq. "Okay, Paul," I responded. "We'll loan you some people to get you going, including a few of our planners. Let's put together a manning concept to justify your needs and then send it back to Washington to see if we can get some help."

We proposed an interim structure and sent off a plan to allow him some temporary help. But it had been in Washington for only a few days when we received a call from the Joint Chiefs turning us down flat. "We're not going to task the services to support your request," was the response. "It's too short notice. Man it out of hide."

"Okay, the decision has been made," I said to Eaton. "We go out of hide."

Our staff canvassed all the people in CJTF-7 who had the necessary expertise and found that everybody was tied up. The only choice I had was to task the 3rd Infantry Division, which was in Kuwait, getting ready to go back home. After calling GEN Abizaid for approval, I issued the order to extend about twenty people for another forty-five to sixty days. So we had to drag these poor guys back up to Baghdad and they became the starting point for Paul Eaton's efforts to build the Iraqi Army. With time, Eaton would redefine his efforts and get things going, but his organization never really did get manned at the appropriate level.

After Ambassador Bremer disbanded the Iraqi Army, he also refused to pay its former members any salary, including retirement benefits. But that soon became a major economic and security issue. When people started rioting across Iraq, CJTF-7 had yet another source of violence to contend with. At one point, I received a report from the Marines in Najaf that read: "Temperature in our AO [Area of Operations] is past boiling point on the issue of payment to former Iraqi soldiers." In an attempt to solve the problem, my staff and I (with the help of Walt Slocombe) persuaded Ambassador Bremer to set up a temporary plan to pay the Iraqis with money from the Iraq Development Fund. Bremer didn't like doing it, and he later told two U.S. senators that "it simply does not sit well with me."

It turned out to be the right thing, however, as the riots immediately stopped. An added benefit came when all the former members of the military who wanted to get paid came in and wrote down their names and addresses. That provided MG Eaton with a fairly comprehensive list from which to work. The list also became useful in establishing the Iraqi Civil Defense Corps.

Because things were moving so slowly with the police and army, GEN Abizaid and I decided to make another run at establishing the ICDC. This time, Abizaid started by briefing the concept to the Secretary of Defense, who immediately embraced it. With Washington now on our side, we reengaged Ambassador Bremer, and he reluctantly gave his formal approval on July 11, 2003. When we asked for a long-term commitment, he agreed to let us try it for a year. When we asked for permission to recruit and build eighteen battalions, he gave us approval for six. When we asked for the CPA to fund our effort completely, he said he'd pay for the cost of training and basic supplies. We'd have to foot the bill for recruiting costs and major equipment. But that was progress, and it was certainly enough to help us get going.

We decided to form the ICDC by region, place the division commanders in charge, and deploy the Iraqis in their own cities and provinces. The effort took hold very quickly. Actually, we had more people than we knew what to do with, as all across the country, Iraqis waited in long lines to get back in uniform and start receiving regular paychecks. When I saw the response, I immediately issued orders to apply large amounts of CERP funds to purchase weapons, vehicles, and uniforms. After a short three-week training program, we put the new recruits right to work patrolling and performing basic security in the cities. The feedback we received was immediate and positive. We had Iraqis working with Iraqis, and Iraqis working with coalition forces. Before long, all of our local military units wanted to have an ICDC contingent. So we went back to Bremer and asked for permission to build more battalions. Seeing our success, he agreed to increase the number from six to fifteen. By the end of the year we had twenty-eight, and in another four months' time, we would be looking at upwards of forty-five battalions. That was approximately 36,000 Iraqis in uniform.

So now we had a parallel effort going on in the country to equip, train, and get Iraqis into uniform: CJTF-7 had the ICDC, and CPA had the new Iraqi Army. Predictably, Walter Slocombe and Paul Eaton became concerned. They viewed the ICDC as competition to the Iraqi Army, rather than as a complementary capability.

At this point, Abizaid and I began to press Bremer to set faster timelines for the Iraqi Army and build its national command and control structure in parallel. These organizations (Ministry of Defense, Joint Chiefs of Staff, Army and division headquarters staffs, etc.) were truly necessary to build the Army's operational competence. Eventually, Bremer asked Slocombe and Eaton to revise their plans and expedite the fielding of the Iraqi Army. The success of the ICDC had simply been too impressive to ignore.

As Eaton stepped up training of the Army in August, their contractors came in, but wound up doing a poor job. So the 4th Infantry Division provided trainers, security, housing, and other assets to get the program moving. MG Eaton eventually graduated his first battalion of soldiers in late September. And by the end of the year, two more battalions were trained, in uniform, and ready to fight.

ON AUGUST 19, 2003, a suicide bomber parked a truck loaded with explosives next to the wall surrounding the UN compound. The resulting explosion blew up the wall and collapsed the UN headquarters building. As soon as I received word of the attack, I flew to the scene and walked amidst the chaos and devastation. Scores of soldiers and other rescue workers were combing the rubble for survivors. The bomb had exploded just below the office of Sergio Vieira de Mello, head of the UN mission in Iraq, who was in his office at the time. Soldiers crawling through the rubble were talking to him, but after about twenty minutes, they reported that he was no longer responding. Twenty people were killed and 100 were wounded in the attack. Sergio Vieira de Mello was among the dead.

One week earlier, a car bomb had exploded outside the Jordanian embassy, killing eleven. Threats were also made against both the Russian and Italian embassies. It was now clear that the enemy was targeting the international community for its support of coalition efforts. More than that, however, during the month of August, the enemy had ignited its low-level insurgency, extending all across Iraq. On August 14, a British soldier was killed and two others injured as they drove an ambulance along the outskirts of Basra. On the fifteenth, an oil pipeline in northern Iraq was blown up, resulting in a huge fire and a halt in all oil exports to Turkey. Two days later, a water main exploded in Baghdad. On the twenty-third, three British soldiers were killed and one wounded in a Basra attack. The next day, three Iraqi security guards were killed in Najaf. And these were just *some* of the acts of violence occurring across the country.

Notable in the August attacks was a new sophistication and synchronization of tactics. In the past, explosives were planted and left to go off on their own. Now we were seeing wires with command detonation devices. Prior to August, IEDs and small-arms fire were separate events. Now both were happening at the same time. A device exploded and people shot at troops in the vicinity. We also began to see the increased use of VBIEDs (vehicle-borne improvised explosive devices), which were trucks, vans, or cars filled with explosives and set off by either suicide bombers or remote devices. The UN bombing that killed Sergio Vieira de Mello was a VBIED.

It was no coincidence that an escalation in violence occurred during the same month that the majority of coalition forces entered the country. The insurgents were very clearly going at the will of the coalition. And it was beginning to work, because a number of nations started to back away from their commitments. During this time, I kept thinking about GEN Tommy Franks's statement that the ratio of internationals to U.S. troops was going to be three to one. Even if we had not reversed the American troop drawdown, we would never have been able to sustain a ratio like that. Franks had been widely off the mark.

Along with everything else that was going on in Iraq during the month of August, we also began to see attacks on the Shiites, who were now assuming power. Extremists and elements of the former regime staged these incidents, because, in their minds, the Shiites were just as guilty of cooperating with Americans as the coalition nations. Perhaps the worst single attack occurred in Najaf on August 29, 2003, when a bomb exploded near the Tomb of Ali, one of the holiest of Shia Muslim shrines. Weekly prayers had ended and a mass of people were just leaving when the bomb exploded, collapsing the front entrance to the shrine. At least 100 people were killed and many more wounded. Among the dead—and the likely target of the attack—was Ayatollah Muhammed Bakr al-Hakim, the leader of SCIRI.

In the immediate aftermath of al-Hakim's death, tens of thousands of Shiites took to the streets to stage demonstrations and riots. They believed this was the work of coalition forces, but after a while, it became apparent to them that the perpetrators were Sunni extremists. So we were not only seeing an escalation of the insurgency, but for the first time, we were witnessing signs of civil war in Iraq.

At the end of August 2003, CJTF-7 conducted a formal ninety-day assessment of the Iraqi theater of operations. With the igniting of the insurgency, our battlefield had clearly expanded all across the country—to Mosul in northern Iraq, to Baghdad and Najaf in central Iraq, and to Basra in southern Iraq. The enemy force was so unstructured that we were unable to specifically identify objectives against which we could maneuver our forces. There was also an increased intensity in the fighting, and casualty rates spiked as our soldiers fought and died every day.

Militarily, we took aggressive steps to beat back the enemy. Politically, we stepped up our engagement and reconciliation meetings. However, the escalation of the insurgency confirmed that we were inadequate not only in terms of headquarters staff capabilities and our linkages to the CPA, but our ground forces were wholly lacking in sufficient armored protection, and we simply didn't have enough armored Humvees for our troops.

We didn't have much hope of fixing the economic problems anytime soon. The U.S. supplemental funding bill had not been passed in Congress, and no action was being taken to free up any substantial dollars. We did not know it at the time, but such funds would not be approved until the end of November 2003, and they wouldn't be available until February 2004. In the interim, we had no reconstruction funding from the United States or from any other member of the coalition.

As if all that wasn't enough to deal with, I received a call right around the first of September that the Spaniards, who were shortly due to replace the Marines in southern Iraq, were refusing to assume command of their sector. We immediately flew the Spanish brigade commander and the leaders of the Latin American battalions to Victory Base for a meeting. As soon as they walked into the office, I began conversing with them in Spanish.

"Gentlemen, I've been told that you are not willing to take over your sector," I said. "Is that true?"

"That is correct, General," came the reply. "America hasn't given us our equipment."

"Your equipment?" I asked.

"Trucks, weapons, body armor, communications—all of the equipment that was promised. We haven't received any of it, and neither have the Central American countries [Honduras, the Dominican Republic, El Salvador, and other smaller units] that report to us."

"You are putting us is a major bind, General," I said.

"Sir, I am sorry, but this is the political guidance from our minister of defense. The United States has to fulfill its commitments."

"Well, that's pretty definitive," I replied. "Let me see if I can get your minister of defense on the phone."

After speaking to Spain's minister of defense in Madrid, I understood why he was taking such a firm stance. In the rush to get other nations to join the coalition, the U.S. government had promised all kinds of equipment and other incentives, but had not delivered. Spain's key concern was that none of the Latin American countries, which would be operating under Spanish command, had received any of the promised equipment. They didn't have it now that they were in Iraq and, as far as I knew, there was nothing on the way. In my mind, the Spanish had every right to be upset. It was an outrage. And I promised the minister of defense that his troops would have their equipment.

We decided to delay the transfer for a few weeks while we launched a mad scramble to acquire everything necessary to fulfill our obligations. After checking with GEN Abizaid and Army headquarters in Washington, I realized that our only option was to take the matériel from U.S. forces on the ground. This was equipment that our American troops couldn't afford to give up. But we had no choice. We had to stand the Spaniards up in southern Iraq, because, after all, they were replacing the Marines.

When the Spanish minister of defense realized that we were taking matériel away from our own soldiers to give to his troops, he sent me a note of thanks. I think he was stunned, in part, because he was in possession of signed agreements from the U.S. government that had not been honored. But I had made a commitment to him and given my word. And as my father and uncle had drilled into me years before: "Once you make a commitment, by God, that's it. Your word is your bond."

CHAPTER 15

The Lead-Up to Abu Ghraib

In early August 2003, I went out to the Abu Ghraib detention facility and gathered together all the interrogators on duty. There were about a half dozen of them—a captain, a warrant officer, and several specialists.

"What kind of training do you all have?" I asked.

"Well, sir, basic course," they responded. "Just the fundamental training that the Army gives."

"What experience do you have?"

"Panama." "Desert Storm." "Haiti." "Afghanistan," came the responses.

"How are we monitoring the interrogations?"

"Well, that's our responsibility as interrogators, sir."

"Who approves your interrogation plans?"

"We do, sir."

"How do you know whether you're getting close to the limits of our authorities as defined in the Geneva Conventions?"

"Just experience, sir."

"Well, that is a pretty gray area," I said. "How do you know if you're getting close to crossing the line? Have you ever been trained in the specifics of the Geneva Conventions as they apply to interrogations?"

"No, sir. We have not."

"Okay, look. You all are performing a very critical function. We have a desperate need for human intelligence. It is the only source we have to find out *who* these people are, *how* they're organized, *what* their plans of attack are, and *where* and *when* the next acts of violence might be. So we must push our interrogations to the limits of our authorities as defined by the Geneva Conventions. Each of you must know when our techniques are approaching the limits allowed, but you cannot cross those boundaries. And knowing when you're getting close to the limits is crucial. From what you've just told me, however, I don't believe we properly understand what those limits are."

MG Barb Fast and COL Marc Warren (CJTF-7's chief attorney) were with me on this visit and, as soon as I finished speaking with the interrogators, I pulled them aside. "We've got to put out some guidance for these young men and women," I said. "I'm just not comfortable at this point. We have no standards, our training is abysmal, and our interrogators don't have clear limits. We also need to do something about oversight mechanisms to improve our interrogation plans and approaches."

"I agree, sir," replied Fast. "Let me get with Washington again. I'll press them this time until they give us some sort of answer."

"Okay, good. But let me make sure I'm straight here, Marc. As long as we're at or below the authorities of the Geneva Conventions, we'll be fine, right?"

"Yes, sir, absolutely correct," said Warren. "That's why they were established in the first place."

Two key issues had spurred this particular visit to Abu Ghraib. First, we needed additional and better intelligence to effectively combat the growing insurgency. I was looking for long-term answers to such questions as: What's the insurgency doing? What is their organization? Who are their leaders? How do we attack them? At Abu Ghraib, we were trying to establish a strategic intelligence capability that was monitored on a regular basis.

Second, I had received some worrisome reports from the field that prisoners were being treated too harshly (in terms of physical contact and threats) upon capture and during tactical questioning. Greater than 90 percent of these abuses happened on the battlefield while units were trying to obtain tactical intelligence. They wanted answers to such questions as: Where is this guy's buddy? Where is the IED stash? Where is the enemy hideout?

With the benefit of hindsight, we now know that some improper interrogation techniques had migrated into Iraq due to the deployment of soldiers from Afghanistan, where a totally open universe of techniques caused by the lifting of the Geneva Conventions in the war with al-Qaeda had caused confusion in the ranks. From April 2003 to February 2004, our division commanders initiated a large number of investigations that dealt specifically with this problem. Early on, there were no trained interrogators at the tactical level, and I wanted to make sure that potential prisoner abuse was eliminated. We had to have the tactical-questioning problem solved by the time formal interrogation operations were initiated at Abu Ghraib. That's why I decided to speak directly with the interrogators.

Because of the August spike in violence by former regimeists targeting U.S. forces, coalition forces, and Iraqis who supported the coalition, our detainee numbers had swelled to somewhere between 6,000 and 7,000. We were experiencing overcrowding, plus identification and documentation problems. If we didn't take quick action to fix things, we were going to be in real trouble. For starters, we consolidated a number of our temporary facilities around Fallujah, Baghdad, and several other high-impact areas—and then we began conducting regular inspections of all detention operations at the division, brigade, and battalion levels. I also personally monitored the daily division detainee reports and the weekly Abu Ghraib population snapshot lists.

Walt Wojdakowski and Barb Fast then came up with the idea to hold meetings of all leaders and staff involved in detention and interrogation operations. Over the course of a month, the group met three times at Victory Base for approximately one to two days at a time. Chief among the dozen or so participants was BG Janis Karpinski, the senior MP responsible for detention operations, whom I had first met back in May. MGs Wojdakowski and Fast conducted the meetings and, while I was not present for any of the working sessions, I was provided a summary of each day's discussions. I also personally approved the group's proposed actions and requests for assistance.

In excruciating detail, we determined requirements and deficiencies across the entire spectrum of detention operations in Iraq. It was not good news, either, as we confirmed that we were lacking in just about every conceivable resource and procedure—medical capacity, intelligence infrastructure, quality of life, identification, communications, information technology and database support, strategic debriefing, interrogation procedures, oversight mechanisms, checks and balances, and so on. After each of these "detention summits," we issued orders to the entire command, which then began to establish specific solutions, procedures, and timelines to fix our detention problems. Initially, we created formal review boards to examine and speed up the release of detainees. The boards met once a week, but then we asked them to meet every day of the week so we could expedite the process.

Our detention summits also focused on CJTF-7's intelligence capacities, and we determined that we needed help in organization, personnel, analysis, collection, training, experience, reachback to national agencies, and integration of interagency assets. In an effort to improve these intel capacities, we consistently requested help from Washington. Once in a while, people in both Army intelligence and the Joint Chiefs of Staff would express concern about our situation, but rarely did we receive any specific guidance from them. However, after the murder of Sergio Vieira de Mello in the UN bombing, Ambassador Paul Bremer aggressively engaged with the principals in Washington about our intelligence deficiencies. He had been a good friend of de Mello, and became concerned about our inability to anticipate such heinous acts of terrorism. Ambassador Bremer told me that he was going to put pressure on Washington and push for interagency cooperation among the CIA and all other national intelligence agencies to address our current intelligence shortcomings. And he was true to his word. Bremer told Washington that he considered this issue "a crisis of enormous proportion . . . ," one that both "stands to undermine the progress we have made to date" and "endangers the successful achievement of U.S. objectives for Iraq . . . It must receive the highest priority."

Additionally, on September 10, 2003, CJTF-7 submitted the detailed assessment of the intelligence architecture that Deputy Undersecretary of Defense Stephen A. Cambone had requested of MG Barb Fast before she entered the country. Copies of this report went up through the chain of command, including CENTCOM, the Joint Chiefs of Staff, and the office of the secretary of defense. I was optimistic that some major changes would be made, especially now that Ambassador Bremer was engaged.

Eventually, the CJTF-7 requests for assistance made it through the Joint Chiefs and wound up squarely in the Army's lap, which is where it should have been all along. Within the Department of Defense, the Army is responsible for interrogation operations. It writes the manuals that are used by the other services. However, by the time the Bush administration suspended the Geneva Conventions, the Army had not trained for or conducted interrogations on this scale in decades. They had only performed tactical questioning immediately upon capture in places like Panama, Haiti, and Bosnia, and during Operation Desert Storm. I was worried the Army had validated the interrogators only in the most elemental tactical questioning techniques, which were inadequate for what was going on in Iraq. They had provided no training for the kind of interrogation procedures we needed to conduct at Abu Ghraib. In the meantime, CJTF-7 was struggling with a complete lack of guidance and doctrine. We didn't even know what to train, and neither did the Army. The Army *Field Manual* was not sophisticated enough in 2003 to ensure that our interrogators were employing all the necessary authorities accorded under the Geneva Conventions while ensuring the proper treatment of detainees in a disciplined, supervised interrogation environment.

A few days after we had met with the Abu Ghraib interrogators, COL Marc Warren came back to me with an update. "Sir, I have talked to CENTCOM and Washington and pressed them," he reported. "However, this is a real hot issue and nobody wants to put out any guidance."

"Well, that means they don't have the experience and they're afraid to take on the administration."

"Yes, sir. Either way, they're choosing to do nothing."

"Well, we're just going to have to do it ourselves!" I said. "We're in desperate need of standards and oversight mechanisms. We must impose some kind of controls to make sure our soldiers are properly trained. Marc, I want you to assemble a team and come up with a plan. We have to find out who the national-level experts are in interrogations. And I don't care who they are, we've got to find them and get their help. The people at Guantánamo Bay [GTMO] are probably the most experienced at these kinds of operations."

"Yes, sir, you're right," replied Warren. "GTMO is probably the most experienced. Maybe we can model their system and adapt it to comply with the Geneva Conventions. I'll get on it."

Over the next several weeks, Marc Warren put together a team that canvassed a variety of experts on the issue, and came up with a proposed set of interrogation rules of engagement for Iraq. The lawyers on his staff had taken every suggested interrogation approach and pared them down by determining which were allowed by the Geneva Conventions. Marc then brought them in to me, and walked me through the entire document. "Sir," he said, "internal to our command, we have unanimous agreement that every one of these is authorized by the Geneva Conventions."

"Good," I replied. "But what about higher headquarters?"

"Sir, they won't answer, won't tell us anything, won't give us any guidance on this issue."

"All right, I'm fed up. I can't wait for them while our soldiers are executing detention and interrogation operations every day without guidance. Prepare a memorandum that says, unless otherwise directed, these interrogation rules of engagement are now implemented for this theater. Our objective is to formalize a plan, lay out specific guidelines, institute checks and balances, and provide oversight mechanisms with approval chains. Please keep in mind when you draft the memo that there has been no guidance issued prior to this whatsoever. So it's not like we're superceding anything. And let's make sure to specifically state that the Geneva Conventions apply here. We just can't have it any other way."

On September 14, 2003, via memorandum, I issued the CJTF-7 Interrogation and Counter-Resistance Policy. As I wrote in the cover letter, the program was "modeled on the one implemented for interrogations conducted at Guantanamo Bay, but modified for applicability to a theater of war in which the Geneva Conventions apply." Before providing a list of techniques, I reemphasized that "CJTF-7 is operating in a theater of war in which the Geneva Conventions are applicable. Coalition forces will continue to treat all persons under their control humanely." The final page of the memorandum was a list of "General Safeguards," to be used when conducting interrogations.

I was pleased that we had finally made an attempt to bring standards and a sense of discipline into CJTF-7. One of the problems, of course, was that this memo didn't apply to everybody in Iraq, but only to those units under my command. The Special Forces units, which are under a separate operational command, received the memo, but they didn't have to comply. And the CIA, as usual, was operating under its own secretive rules.

When this memo reached CENTCOM, the senior lawyer there read it over and thought it looked fine. However, one of the young majors on his staff said she was uncomfortable with some of the interrogation procedures. Marc Warren then came back to me with her reactions.

"Marc, are you convinced that these methods are legal?" I asked.

"Sir, absolutely, they are," he replied. "There's no question. These are authorized by the Conventions."

"Well, I don't want any disagreements with CENTCOM, the Joint Chiefs, or DoD on this memorandum. Can you get with Abizaid's headquarters, sort it out, and come up with a solution?"

"Yes, sir. Will do."

After some significant debates with CENTCOM, Warren came back to me and reported that the approaches in question were too close to the line and that consensus would be reached if these particular techniques were eliminated.

"So CENTCOM will concur if we drop them?" I asked.

"Yes, sir. Everybody will be in agreement."

"Okay, let's republish the memorandum without those approaches. Also, I am not comfortable with the approval authority at the colonel level. Bring it up to my level."

So on October 12, 2003, we sent out the new document specifically stating that it superceded the September 14, 2003, memorandum, and that "use of techniques [listed] must be approved by me personally prior to use," that any "request for approaches not listed must be submitted to me through CJTF-7 C-2 (MG Barb Fast)," and that "a legal review from the CJTF-7 Staff Judge Advocate (COL Marc Warren) must accompany each request." By implementing this approval process, I was ensuring the review of all interrogation plans by the most senior experts in the command. I did not want interrogators to be unclear about what techniques they could and could not use.

The Interrogation and Counter-Resistance Policy was an important first step for the Iraq theater of operations, but alone, it would not solve all our problems. With the increase in violence related to the insurgency, we were still receiving complaints at the tip of the spear. So it was necessary to reinforce our commitment to humane treatment of all detainees under our control whether in the field or at a detention facility. I periodically sent out memorandums that reinforced the obligation of coalition personnel to comply with the laws of war and the Geneva Conventions. With the expanded battlefield and the escalated intensity of the fighting, the message had to be repeated constantly.

BECAUSE ABU GHRAIB WAS the center of detention operations in Iraq, I made a number of trips there from July through October 2003. I focused on issues of base defense, quality of life, intelligence architecture, and the status of interrogation and intelligence infrastructure. I did not get into the details of how a cellblock or interrogation booth was run. On one of those visits, BG Karpinski was with me, and it was apparent that while the quality of life had improved somewhat, it was not at a satisfactory level given the resources now available. As we walked through the facility, Karpinski told me that the reason for lack of progress was due to the CJTF-7 staff's failure to fill the many requests she had made of them. After questioning why she had not pressed harder or brought the matter to my attention earlier, I got on the phone and called BG Scotty West, my logistics officer, who said that he had no pending requests from the 800th MPs for anything related to Abu Ghraib. At this point, I was seriously beginning to question BG Karpinski's ability to command her brigade. I directed Scotty to do whatever was necessary to upgrade the living conditions to acceptable levels. "Don't worry about the fact that Abu Ghraib has been designated a temporary facility," I said. "This has gone on long enough. We need to improve the situation out here for both our soldiers and the prisoners. It looks like we're going to be here for a year or two, so do whatever is necessary to get this place fixed." Within forty-eight hours, eight of the eleven platoons from the 94th Engineer Battalion were on site making improvements in shelter, showers, latrines, and medical and sanitation facilities.

I made another trip to Abu Ghraib right after a mortar strike killed a number of prisoners and a couple of soldiers. I was concerned about protecting the facility, because the surrounding area was a rat's nest of insurgents with constant fighting that included drive-by shootings, regular mortar, and rocket-propelled grenade attacks.

While on the site, I walked the perimeter and went up in a couple of the guard towers, only to have the MP sergeant major tell me that he had been asking for .50-caliber machine guns and that nobody could provide them. "What do you mean nobody can provide them?" I asked.

"Oh, yes sir, we've been turned down by higher headquarters."

So I again called BG West and reported what I had just been told. "General Sanchez, I've never heard of this before," he replied.

"Well, you have now," I said. "Let's fix the problem."

Again, it was apparent that the 800th MPs were not prepared to defend Abu Ghraib. For a period of time, the detention facility was getting mortared and rocketed every day, as were other sites around the country. However, it is a fundamental responsibility of every commander to defend his position and be prepared to fight, if necessary. That just wasn't happening. There was no aggressive patrolling, no plans for fire support or counterfire, and mortars were lying around in the compound unprepared to fire. In addition, there was no coordination with the 82nd Airborne Division, which had operational responsibility outside the gates of the prison.

Part of the problem stemmed from the fact that the 800th MPs were reservists who did not have training or experience in combat operations of this nature. The U.S. Army had not prepared them properly. To make matters worse, the reservists in Iraq had an ineffective pay system that created morale problems, and a cumbersome troop-replacement system forced some of their units to operate at less than 60 percent strength. The Army had been ignoring its Reserve component problem for years, and now it had become a real nightmare. These conditions compounded

BG Karpinski's leadership challenges, and I knew I had to cut her some slack because of it.

The highest-ranking officer at Abu Ghraib was Tom Pappas, an active-duty full colonel, whom I had previously assigned to run the intel and interrogation process. The MPs were responsible for detention operations, defense, and force protection. COL Pappas was a pretty good young commander and I was confident in his ability to lead. He was my V Corps military intelligence brigade commander. So based on the total lack of progress toward defending Abu Ghraib, I made an on-the-spot decision. I pulled him aside and said, "Tom, as the senior commander on the ground, you now have responsibility for the security and coordinated defense of Abu Ghraib. You are now in charge of all warfighting requirements and you will retain the authority over intel and interrogation operations."

This move was later second-guessed, due to the fact that I had placed reservist MPs under the command of an active-duty officer who happened to be a military intelligence colonel. There is nothing in military regulations, though, that prevented me from taking such an action. In fact, placing a senior officer in charge of the defense of a military installation is a fundamental military principle. And given the fact that something had to be done right away to protect the lives of our soldiers and the detainees, I believed it was the right thing to do.

Immediately upon assuming command in Iraq, I requested assistance for our detention and interrogation operations from any available experts within the Department of Defense. After being ignored for months, I received notice out of the blue that "MG Geoffrey D. Miller with his team from GTMO is inbound to Iraq to assess and assist CJTF-7." My initial reaction was mixed. I knew there was risk in the GTMO visit, but I was finally going to get some help. "Okay," I thought, "GTMO has the latest on detention and interrogation operations, and is viewed, internal to the Army and Department of Defense, as the best in the world." Besides, Geoff Miller and I had known each other since 1978 when we were both young captains at the Armor Officer Advanced Course in Fort Knox, Kentucky. He had been in command at Guantánamo Bay since November 2002.

During Miller's initial office call, the CJTF-7 staff and I explained the Iraq theater of operations. At GTMO, he was not operating in a combat environment, he had a stable prisoner population that included some hard-core terrorists, and he had significant resources. Here, we were in a serious combat environment, we had a dynamic detainee situation with rapid turnover, our prisoners were a mix of insurgents and criminals, and we had very few resources. "We're in desperate need of help," I told him. "There is nobody here that has expertise in how to conduct proper detention and interrogation operations on this scale."

I asked MG Miller to help us become efficient and effective in our overall operations. "Identify problems, train our interrogators, establish priorities, provide sample operating procedures, and link the intelligence gleaned to the overall task force," I said. "But whatever you do, just make sure you're in compliance with the Geneva Conventions."

Miller's team hit the ground running and, within a couple of days, Geoff came back to me with an initial assessment. "There are some huge problems out at Abu Ghraib, especially with the MPs," he said. "If it's all right with you, I'd like to implement some immediate fixes."

"Absolutely," I responded. "If there are things that are wrong and can be fixed immediately, then by all means, go ahead and take action. And don't bother to ask me again. Just go ahead and fix the problems as you find them. That's exactly what I need you to be doing."

During the month that Miller's team was in the country, changes were made to detainee segregation methods, interrogation booth setups, and interactions between MPs and the military intelligence (MI) personnel, to name a few. At the end of the assessment period, Miller also provided a recommended set of new standard operating procedures (SOPs) and training methods that had to be modified for our theater of operations. Just about every one of those recommendations was approved and distributed to military police and military intelligence commanders on the ground for immediate implementation, including BG Karpinski and COL Pappas for Abu Ghraib. And finally, MG Miller agreed to provide continued training in interrogation operations if approved by DoD. In late September and October, GTMO "Tiger Teams" comprised of psychologists and veteran interrogators came into Iraq and worked directly with our own MI and MP personnel in on-the-job training.

In the end, Miller gave a fairly damning assessment of BG Karpinski's entire operation. And in the SOPs left behind, he specifically addressed the role of military police in the interrogation process. He cited an inextricable link between MPs and MI interrogators that had to be managed properly. Basically, the MPs were to provide support to military intelligence. They would transport the prisoners to and from their cells and monitor them while in the cells. They would watch for behavior patterns, clues, and any other indicators that might be useful to the MI personnel, and then pass on that information on a regular basis. In so doing, they would help set the conditions for productive interrogations. MG Miller was also very specific that under no circumstances should MPs ever perform interrogations.

Coincidentally, in October 2003, the top MP in the Army, Provost Marshal MG Donald J. Ryder, arrived from Washington with a large team of detention experts. While Geoff Miller focused primarily on interrogations and related procedures, Ryder concentrated on detention operations. During our initial meeting, I asked him to assess how we were running our facilities, identify our requirements for personnel, equipment, and training, and then provide a list of recommended improvements. I further gave him the authority to immediately implement fixes to problems and train MPs, just as MG Miller was doing with the interrogators. I also asked him to personally evaluate BG Karpinski and provide a recommendation as to whether or not she should stay in command of the 800th MPs.

After several weeks of intense work, on November 6, 2003, MG Ryder gave me an outbriefing as to his team's findings and recommendations. He offered new SOPs regarding in-processing of detainees, capture cards, databases, and release mechanisms. We accepted nearly all his recommendations and believed they would lead to significant improvements in our detention operations. Ryder also pointed out that he did not completely agree with Miller's position that MPs and interrogators should have a close relationship, preferring rather to have a strict separation between the two. This was an age-old doctrinal and training issue that had never been adequately resolved.

When I asked him for his recommendation about BG Karpinski, he said that she was a weak leader and should probably be relieved. Her unit's performance, he noted, was really hurting operations at Abu Ghraib, she was not effectively prioritizing and allocating her MP assets, and despite some areas of good performance, there were problems across her organization. However, for a number of reasons, Ryder did not recommend that she be relieved. He cited unusually difficult challenges for the overall mission and, now that new SOPs were being put in place and training provided, some of the problems would be fixed. In addition, Ryder pointed out that the 800th Brigade was only about thirty to forty-five days away from going home. "It would probably be best to leave Karpinski in command and allow her to take her brigade home," said Ryder. "Besides, the Army will not be able to provide you with a replacement commander in the timelines we're facing."

I thought long and hard about this decision. Don Ryder had some valid points in his assessment. I realized that the "difficult challenges" Ryder referred to went beyond what met the eye. The fractured command relationships increased the complexity in Iraq. Karpinski had responsibility for MP detention operations in

Iraq, and CFLCC retained responsibility for all her support. She was also a reservist who had not been given the training for a job of the magnitude we faced. And finally, the Army had refused to deploy an active-duty senior officer to help oversee detention operations. In my opinion, her failure was, in large part, the Army's failure. Therefore, with only a short time left on her stay, I decided to follow MG Ryder's advice and not relieve Karpinski.

On October 15, 2003, GEN Abizaid at CENTCOM received a memorandum from the Chairman of the Joint Chiefs of Staff, GEN Richard Myers, entitled "Intelligence Architecture in Iraq." Essentially, this document acknowledged that the Pentagon had full understanding of our intelligence deficiencies and that we needed immediate assistance, and pledged to get us help. The message was a direct response to the intelligence assessment we sent to Stephen A. Cambone on September 10, 2003, through the military chain of command, and to Ambassador Paul Bremer's reaction to Sergio Vieira de Mello's murder.

Specifically, the Myers memorandum offered a variety of steps to remedy our intelligence deficiencies. For example, a multiservice interagency team was being built to address our needs, the Iraq Survey Group (under David Kay) was going to be tasked to support CJTF-7, the integration of CPA and CJTF-7 intelligence operations would be enhanced, the Army was going to organize a strategic counterintelligence detachment specifically for our needs, and additional interrogators were going to be sent to Iraq. In addition, the Army would be taking steps to improve the training of intelligence personnel and enhance the capabilities of interrogators; our request for 5,200 native linguists to help with translations was being expedited; and a series of technological solutions was forthcoming.

At CJTF-7, we all had high hopes after receiving the Myers memorandum. However, most of these promised items never materialized. Rather, they just spun around Washington in the normal bureaucratic process. They were discussed, understood, validated, and, eventually, forgotten.

Despite the lack of help from Washington, we did begin to see some improvements in our detention and interrogation operations. Barb Fast had done a thorough review of the CJTF-7 intel architecture and had defined the structure that we needed to pursue. MGs Miller and Ryder had come in, both validated and complemented Barb's assessment, and then provided new procedures and training. We had also received a brief visit from the number two intel leader in the Department of Defense and the top military intel officer at the Pentagon. LTG William G. Boykin (Deputy Undersecretary of Defense for Intelligence) and LTG Ronald L. Burgess (the J-2 Intelligence Director for the Joint Chiefs of Staff) came to Iraq and spent the day with our staff in an effort to clearly understand our needs. We provided them with a detailed briefing of our intel program, including challenges, deficiencies, priorities, and the actions needed to improve our intelligence systems overall. By the time they left, my feelings were mixed. I knew they now understood what our challenges were on the ground, but I still wasn't sure about their level of commitment.

Unfortunately, some scandalous abuses occurred at Abu Ghraib that would forever obscure our hard-fought efforts to fix the intel and detention systems in Iraq. These events happened below the radar while experts in the Army were assessing Abu Ghraib. Unfortunately, we would not know about the extent of the abuses for another couple of months—and only then, because one brave MP had the courage to bring them forward.

Over the course of two nights, October 18 and 19, 2003, four enlisted military policemen of the 372nd MP Company and three intel specialists of the 325th MI Battalion pulled about a dozen Iraqi prisoners out of their cells at Abu Ghraib and abused them. The event was spurred by a small prison riot in which a few prisoners got into a scuffle with their guards. One of the prisoners, who had a pistol smuggled in by an Iraqi guard, tried to shoot SSG Ivan Frederick, an MP. Frederick and Specialist (SPC) Charles Graner (also an MP) then decided to retaliate. They, along with the other MPs and MI specialists, ordered the detainees to strip naked, tied them up in cells, piled them into pyramids, humiliated them sexually, and threatened them with unmuzzled guard dogs. As several of the offenders later admitted, they were "playing a game," "screwing around," and otherwise, just doing it "for the fun of it." There were no interrogations involved during this "party." In fact, none of those involved were ever interrogated. Unquestionably, the abuses constituted criminal behavior by soldiers in my chain of command, and clearly violated the Geneva Conventions and CJTF-7 policies prohibiting such behavior.

A few weeks later, on the night of November 4, 2003, a death occurred at Abu Ghraib—apparently caused by torture at the hands of a CIA interrogator. The morning after the incident occurred, November 5, 2003, Barb Fast walked into my office with her hair on fire. "Sir, the CIA came in last night and left a dead body at Abu Ghraib!" she said.

"A dead body?" I replied.

"Yes, sir. I just received the call. They told me it was brought in last night and iced down to preserve it. In fact, they're calling it 'Ice Man.'"

"Why in the world would the CIA drop off a dead body there?" I asked. "What's going on?"

"They want us to take care of returning the body," said Barb. "What do we do, sir?"

"First, let's try to identify the body, bring it in for autopsy, and then see that it is properly returned to the family. Then let's send a memorandum to the CIA. We've got to put this back in their channels and make sure it gets investigated properly."

About a month later, the CIA station chief (who reported to Ambassador Bremer) and his deputy left the country unannounced. When the new station chief walked into one of our early-morning meetings with Bremer, I asked what had happened to the other guys. "Oh, they had some issues and had to leave," was all he said in reply. I later heard some scuttlebutt that the CIA was considering charging them with abusing and killing people in Iraq. But I never heard anything more about it.

With time, however, more information came out about "Ice Man." He was identified as Manadel al-Jamadi, a suspect in a VBIED bombing who had been abducted from his Baghdad apartment by a group of Navy SEALS. According to eyewitness testimony, he had been dropped off at Abu Ghraib as a "ghost" (unregistered) prisoner while still alive. The CIA interrogated him at the prison, and apparently killed him with blunt-force injuries to the head. A Department of Defense autopsy ruled the death a homicide.

It is important to note that the nation's experts in detention and interrogation operations—the GTMO Tiger Teams and MG Ryder's team—were conducting their assessments in Abu Ghraib when these abuses occurred. However, there is no evidence to suggest that they were involved in any way. As a matter of fact, the Tiger Teams were not working with the MPs at all, but only with the MI interrogators. In addition, the fact that the abuses were not interrogation-related (except for the "Ice Man" incident) precludes them from having had any direct participation.

However, it is entirely possible—and even likely—that members of the Tiger Teams discussed with MPs and interrogators some of the harsher GTMO interrogation techniques that might have indirectly influenced the abuses. Furthermore, Special Forces did, at times, conduct interrogations across the country with the support of CJTF-7 interrogators, whom we provided at their request. Looking back, I have little doubt that the techniques employed by Special Forces influenced the CJTF-7 interrogators. At one point, I had received at least one credible allegation of abuse at the hands of a Special Forces unit. But because we did not have the authority to investigate the incident, we sent it to CENTCOM for referral to Special Operations Command, in accordance with the chain of command. I never received word of an outcome in that particular case.

Although no one in the highest levels of military leadership in Iraq (Karpinski, Fast, or I) knew of the abuses at Abu Ghraib when they occurred, we all became aware a few months later when they were reported to the Criminal Investigation Division (CID) in mid-January 2004. Within twenty-four hours, the entire echelon of American senior leadership was made aware of the situation, all the way up to the Secretary of Defense. And three months later, in April 2004, the Abu Ghraib abuses would explode across the world, driven by an international media that fed on the fact that the young people who had perpetrated the abuses were brazen enough to have taken pictures—lots of pictures.

CHAPTER 16

The Decision to Transfer Sovereignty Early

By late October 2003, the Iraqi Governing Council was putting pressure on Ambassador Bremer to increase traffic flow through and around the Green Zone. One major thoroughfare they wanted reopened ran over a bridge across the river and right past the al-Rasheed Hotel. We had originally closed down the road for security purposes. But with the approach of the Muslim holy month of Ramadan, and now that the hotel had been upgraded and converted to quarters for CPA and coalition personnel, the council wanted to send a message that life was getting back to normal in Baghdad. When Ambassador Bremer and I discussed the matter, I pointed out that it would split the Green Zone, create some serious security challenges for us, and that MG Marty Dempsey was adamantly opposed to the move. "But we've got to do this," he responded. "It's time to show some progress and a return to normalcy."

"Okay, sir, we'll give it a try," I said. And over the next few days, we worked very hard to compartmentalize the road in order to minimize the risk to the Green Zone. We opened up the flow of traffic at midday on October 24. Less than forty-eight hours later, at 6:10 a.m. on Sunday morning, October 26, 2003 (the day before Ramadan began), the al-Rasheed Hotel was attacked by insurgents. They launched eight or ten rockets from a makeshift platform hidden in a trailer that had been towed to a cloverleaf intersection nearby. The rockets blew out windows, went through rooms, and destroyed part of a side of the eighteen-story hotel. One American colonel was killed, seventeen people were wounded, and Deputy Secretary of Defense Paul Wolfowitz, who was staying on an upper floor, narrowly escaped injury.

The very next day, on the official beginning of Ramadan, a suicide bomber rammed an ambulance filled with explosives into the Red Cross compound near the Green Zone. And in coordinated attacks, four recently opened Baghdad police stations were attacked in a similar manner. With thirty-five killed and 244 wounded, October 27, 2003, would go down as the bloodiest day since the fall of Saddam Hussein. Unfortunately, it was just the beginning of a new wave of violence, possibly spurred on by release of a letter from Saddam that urged a holy war against "the hated invaders" and those who cooperated with them.

From late October to early November, there was a huge spike in violence all across Iraq, and our military forces became involved in the toughest fighting since May 1. We were seeing attacks on a greater magnitude than anything we had previously experienced. Over a five-day period, three U.S. Black Hawk helicopters and one Chinook helicopter were downed, resulting in thirty Americans killed and thirty-one wounded. The next week, a suicide bomber struck the Italian military headquarters in Nasiriyah, killing fourteen Italian soldiers and eight Iraqis. And less than a month after that, forty-one U.S. soldiers and six Iraqis were wounded in a suicide car bombing outside a barracks in Mosul.

Also during this time, Sunni insurgents were making significant strikes against the Shiites in southern Iraq. The worst event occurred during a mass bombing that killed 100 people in Karbala. The Shiites responded with uncontrolled and indiscriminate violence. In Sadr City, for instance, between 300 and 600 al-Sadr supporters attacked a convoy of U.S. trucks, which led to more frequent and widespread confrontations with the Mahdi Army.

Now we were beginning to see the very real threat of civil war in Iraq. Sunni-on-Shiite violence was already happening, Shiite-on-Shiite was looming, and the U.S. and other coalition forces were right in the middle of it all. For a short period of time, our combat engagements across the country skyrocketed to between 75 and 100 a day. The instability was widespread, but there was no panic because ongoing troop rotations had actually resulted in an increase of forces on the ground.

All of these factors caused huge concern within the Bush administration and led to a series of high-level meetings in Washington about how to bring the crisis under control and, just as important, to provide a thorough reconsideration of Iraq policy. We were only one year away from the 2004 presidential election and a disaster in Iraq could not be allowed, because in all likelihood, that would cost President Bush a second term.

The first action that came out of the White House meetings was an order to obtain a thorough assessment of what was really happening in Iraq. As a result, three high-level players, one each from the Departments of Defense and State and the NSC, came into the country during a three-week period from October 23 to November 11. Deputy Secretary of Defense Paul Wolfowitz was the first to show up. In fact, he arrived just when the al-Rasheed Hotel was attacked. Deputy Secretary of State Richard L. Armitage arrived next. And finally, Ambassador Robert D. Blackwill, the National Security Council Deputy for Iraq, came into the country along with an assessment team.

A major part of their assessments focused on Ambassador Bremer's "7-step plan," which he first announced in an op-ed in *The Washington Post* on September 8, 2003. Some members of the Bush administration felt it was both unfeasible and would take too long, because it implied a multiyear commitment. Essentially, Bremer's seven steps were:

1. Creation of a twenty-five-member Iraqi Governing Council.
2. Appointment of a committee to prepare to write a constitution.
3. Turning the day-to-day operation of the Iraqi government over to Iraqis.
4. Creation of a new constitution for Iraq.
5. Ratifying the constitution by popular vote.
6. Election of a permanent government.
7. Dissolution of the Coalition Provisional Authority (CPA).

The Iraqis were also dissatisfied with this plan and constantly demanded more autonomy from Bremer. "You need to let us handle our own problems," they said. "You need to let our militias impose security. We need to have Iraqi leadership, not American leadership." They wanted everything to happen *right now*—and the CPA just wasn't moving fast enough for them.

However, putting Lawrence of Arabia's advice into practice did not prove to be such an easy task for either the CPA or the military. During his late October meetings, Paul Wolfowitz was hammering Bremer about turning over greater responsibilities to the Iraqis. But for the most part, the Iraqis were unable to do the job simply because they had no experience or training. The coalition not only had to help them build their governing capacity, but it was also our responsibility to guide them through the processes of running their own country.

Ambassador Bremer, however, never embraced the critical task of building Iraqi capacity for self-governance. On November 4, 2003, for instance, he published a memorandum transferring all responsibility for de-Baathification to the Governing Council. Also during this time, Bremer directed his regional coordinators across the country to execute what the CPA called "a refreshing of Governing Councils." In essence, his orders were that any local governing council that had been established by the military without explicit approval from the CPA was going to be reviewed and refreshed with people they approved—and it didn't matter whether it was operating effectively or not. But CPA coordinators in the field were tightly linked with our military commanders out of necessity. There was simply no one else who could help them accomplish their mission, so we integrated them into every aspect of our operations. Within a very short period of time, Bremer's own people were reporting to him that the military was doing a good job with the local governing councils, and that the CPA needed to pursue tribal and Sunni engagement, reconciliation, and economic opportunity more aggressively. Unfortunately, it never happened.

Part of the problem was Ambassador Bremer's firm belief that progress in Iraq had to be done sequentially; first the security mission, then the political, and then the economic rehabilitation of the country. But that was flawed thinking. Our recent experiences in Bosnia, Kosovo, and Haiti had taught us that the political, security, and economic tasks all have to be coordinated and executed simultaneously. Furthermore, it was becoming apparent that Ambassador Bremer himself was part of the problem. The CPA's own Finance Ministry senior advisor had warned me that the Ambassador was not "a trusting soul" and that he did not tend to share information or decision-making authority outside a small circle. This "inability to manage properly," he said, would become a particularly urgent issue once the CPA was responsible for managing the $18 billion supplement that would be available in February 2004.

When Richard Armitage's assessment confirmed what GEN Abizaid had been saying for months—that there was, indeed, an insurgency in Iraq—I believed it would help set the stage for some significant changes, especially if he told the NSC what he told me. "Real insurgencies take years and years to defeat," he had said.

I'm not sure exactly when it happened, but by early November the CPA was placed under the authority of Condoleezza Rice and the National Security Council. I believe it was done, in part, because Ambassador Bremer didn't like answering to Donald Rumsfeld and the Department of Defense. Certainly, Bremer had lobbied hard for such a change. However, the overriding reason probably had to do with the fact that Armitage and Blackwill were finding that the CPA was failing in Iraq, that the situation was growing worse by the day, and that the NSC was the only organization that had the authority to bring all of the agencies of the government together to develop and execute a new grand strategy for Iraq.

On November 8, 2003, Ambassadors Bremer and Blackwill were back in Washington for meetings with the highest-level government principals (Bush, Cheney, Rice, Rumsfeld, Powell) and their deputies. I was not part of those discussions, but I did receive a copy of a memo that was written by Secretary of State Colin L. Powell on November 10, 2003, the day that Bremer and Blackwill were flying back to Iraq. Powell addressed his memorandum to Vice President Dick Cheney, the President's Chief of Staff (Andrew Card), Dr. Rice, and CIA Director George Tenet. He addressed the memo in handwritten form, which, for security purposes, ensured the smallest possible circle of distribution. Basically, the Secretary of State argued that Bremer's 7-step plan would not work fast enough, and that the current Iraqi Governing Council's organizational setup was not sustainable. Accordingly, Powell offered an alternative plan:

1. Form a Constituent Assembly (CA), including the twenty-four members of the Governing Council, twenty-five ministers, and 150 to 200 additional delegations from the various Iraq provinces.
2. The CA chooses an interim government to be in power for two years.
3. The CA chooses a commission with a two-year deadline to draft a constitution.
4. The interim government of Iraq will be sovereign.
5. The Coalition Provisional Authority will become a large U.S. embassy.
6. The coalition forces will stay under a United States unified command.

On November 11, 2003, the day after Powell's memorandum was written, I was involved in a video teleconference with President Bush, Secretary Rumsfeld, GEN Abizaid (all out of country), and Ambassadors Bremer and Blackwill (who were back in Baghdad). Of significant interest to me was the fact that President Bush again raised the issue of Sunni engagement. "It might make sense to pour massive amounts of money into the Sunni areas of Iraq," said the President.

"Yes, sir," replied Ambassador Bremer. "We have been pursuing a Sunni strategy for the last two months." Bremer had, in fact, appointed two people to work on the issue but, as usual, they did not have sufficient resources to get the job done. The fact is that a Sunni engagement strategy was never a top priority for the CPA.

Not long after the teleconference concluded, Bremer and Blackwill were summoned back to Washington for additional high-level discussions. After only a day there, they again flew back to Baghdad, and Bremer spent most of November 14 in closed-door meetings with members of the Iraqi Governing Council, including Jalal Talabani, the council's president.

The next day, November 15, 2003, America announced that by July 1, 2004, Iraq would become a sovereign nation, the CPA would officially dissolve, and in its place, the United States would establish a new and fully functioning embassy. Other steps on this timeline included: drafting and approving a fundamental law and a new constitution; appointing and approving a new organizing committee; convening a governance selection caucus across Iraq; and seating a new and permanent Iraqi government. All this was to be completed by the end of 2005, just a little over two years away. The entire plan seemed very similar to the proposal made by Secretary of State Powell in his November 10, 2003, memorandum. Of course, all this information was preliminary and subject to change. However, the plan was in place: Sovereignty would be handed to Iraq within seven and a half months, rather than the original two and a half to three years that had been planned.

I realized that the decision to transfer sovereignty to Iraq by July 1, 2004, amounted to a calculated political decision. The Bush administration knew that things were going poorly in Iraq and were sure to get worse. And with the Wolfowitz, Armitage, and Blackwill assessments now confirming that fact, the administration knew something had to be done immediately so that the November 2004 presidential election would not be impacted. Giving Iraq sovereignty by the first of July would create the illusion that significant progress was being made. It would also provide a full four months for voters to be convinced through the media that America's mission in Iraq had been a success. If things got worse in the interim, the administration could simply blame it on the Iraqis and a bit of bumpy transitioning. The politics of a presidential election year were beginning to unfold. It was all about winning the presidential election and maintaining power. And as John Abizaid said that summer, "We just don't know how ugly it's going to get."

ON NOVEMBER 16, 2003, the day after the announcement, there was a noticeable change of direction for the CPA in Baghdad. All I kept hearing were two key buzzword phrases: "Transition everything to the Iraqis" and "Do it as fast as possible." Nothing else seemed to matter, and I could certainly understand why. Hardly any work had been done on the new initiatives related to the transfer of sovereignty—fundamental law, a new constitution, elections, a new government, a new embassy. Bremer had a lot of work to do and he knew it. So he focused almost entirely on these initiatives, paid little attention to other issues, and was careful not to allow *anything* to delay his timelines or jeopardize the first of July deadline.

Meanwhile, the CJTF-7 staff was fit to be tied. I believe that for an action-oriented group, their reactions were entirely appropriate. Some of the comments made by various staff officers included: "There's no reason for us to go this fast, sir. We're not ready." "Iraq is in no position to be able to assume sovereignty. It doesn't have any judicial, economic, security, intelligence, or communications structures in place!" "We've got a totally incompetent governing council, and we've made the decision to hand everything off? It doesn't make any sense." "Doesn't this eliminate any shred of hope that we might be able to synchronize the military, political, and economic elements?" "How are we ever going to be able to counter the insurgency now?"

After letting the staff vent, I said, "Okay, we're going to have to go into a crisis-planning mode. Along the lines of operation, we'd better help the CPA figure out what critical tasks need to be accomplished. What resources are going to be required? What are the most critical actions that need to be completed to make the transfer of sovereignty happen? That's what our focus has to be in order for us to have any chance of achieving security and stability. The CPA is leaving on July 1, and everything that isn't completed will wind up in our laps. We're going to be left holding the bag, so we better get ready."

I placed my chief of staff, Marine MAJ GEN Jon Gallinetti, in charge of the crisis-planning effort and he did a magnificent job. In a very short period of time, he had outlined all the critical steps to get us to the July 1 deadline. As soon as our plan was completed, we sat down with Ambassador Bremer and briefed him on it. He seemed to appreciate our effort, as he certainly needed help.

That initial briefing led to another major gathering between our staff and Bremer's senior executives for the purpose of assessing our current state of affairs. Up on the wall was a big "stoplight chart," with each functional area and objective evaluated with a red (long way to go), amber (partway there), or green (on track) circle to indicate the status of progress. The CPA leaders made their reports first. The senior advisor for the Ministry of Justice, for example, assessed the status of his efforts to establish the rule of law, install the Iraqi court system, build detention capacity, build police forces, and train lawyers, judges, and everything else on his to-do list. "We have some challenges in general," he said. "But everything is going great. It's all on track." And it went that way for every one of the CPA advisors. Nearly everything they presented had a green circle next to it. Of course, I realized it was complete BS. Bremer obviously didn't know the real status of things and he wanted to believe what they were telling him.

My staff presented a much more realistic assessment of the security line of operation and its impact on the CPA areas of responsibility. All the political and economic items were the same as those the CPA had presented, only we were relating the progress of our own specific responsibilities. And just about everything on our chart was red or amber. When the presentation was finished, Bremer, who was sitting right next to me, turned and said, "Well, Ric, what's the problem with the military?"

"Sir, I just don't believe your staff assessments," I replied. "I think I have a pretty good understanding of where we are across this entire country on all of these issues—and we are not where your guys say we are. We're just not there."

With that, Ambassador Bremer ended the meeting and I walked outside, totally dumbfounded at what I had just witnessed. Once we got far enough down the hall that we were out of earshot, I started laughing, turned to the members of my staff, and said, "What's wrong with us, guys? Why are we so incompetent? We're the only ones who have ambers and reds in the whole country. We've got more than 160,000 service members out there. Why can't we work harder?" My quip broke the tension among the staff. They had worked so hard and done such a magnificent job that I felt the only way to react was to make light of the situation.

Now that we had a plan in place to help the CPA achieve its goals for the transition of sovereignty, we focused on our own needs. The CJTF-7 staff was still operating far below our required manning level (1,000) at only about a 60 percent fill rate. However, with CPA leaving on July 1, 2004, we all knew it would be necessary for the military to take over many of its functions.

So we recalculated our projected staffing needs to account for the stand up of the four-star headquarters, and on November 30, 2004, I sent a memorandum to GEN Abizaid pointing out that our Joint Manning Document currently called for a 1,000-person staff. "As of July 1, 2004," I wrote, "we will need 1,700."

Another major issue for CJTF-7 at the time was troop rotation. November was a critical time to finalize plans for all forces in Iraq. From December to April, we would be rotating more than 330,000 service members, including all U.S. and all international troops. This would be the largest U.S. troop movement since D-Day. But the resistance we received from Washington was never-ending. First, they had tried to eliminate our requirements on a case-by-case basis. If they couldn't eliminate, then they tried to reduce the numbers. And if they couldn't get us to reduce, they pressured us to reduce the training requirements. It was abundantly clear that the Army did not have the force capacity to provide the one-to-one replacement that we needed to accomplish the mission. With more than half of the total U.S. Army divisions in Iraq during the summer of 2003 (five of ten, plus elements of two more), it was a physical impossibility.

Another reason the Army needed to delay was to allow more time to train the troops coming into Iraq. However, because it took so long to receive individual mobilization approvals from the Secretary of Defense, they were forced to shorten the training timelines for some reserve units. When the Department of the Army informed me that they would be sending us units that had not been trained adequately for combat operations, I objected. "We will not accept under-trained units in the theater of operations," I informed the Pentagon. "Incoming Army units must be properly trained for combat operations at the battalion level. Nothing less is acceptable."

In the end, we had more than 100 units that were extended, or what we referred to as "gapped." These were troops that had reached their 365-day, boots-on-the-ground time limit, and the Army and Department of Defense were not ready to deploy their replacements. So either they had to stay for as much as sixty days longer or we would have a void until the replacement unit arrived. That may not seem like much to civilians, but in a war-torn environment, it could mean the difference between life and death for soldiers on the ground.

In Iraq, there was always some sort of combat going on somewhere in the country, and often on more than one front. Coincident with the October–November increase in violence by Sunni insurgents, for example, Muqtada al-Sadr's supporters and militia began conducting significant attacks on their Shia rivals, which, in turn, created tremendous instability in southern Iraq. And when the Mahdi Army seized the al-Mukhayyam mosque in Karbala, CJTF-7 headquarters received notice that the Spanish command refused to take any action. "We don't consider al-Sadr a threat," they said.

"Not a threat?" I responded. "They just took over a holy shrine by force. Of course, they're a threat."

"No, no," responded the Spanish commander. "According to our rules of engagement, we don't have the authority to act against them." It then became apparent that not only were we going to have to move expeditiously to deal with Muqtada al-Sadr and the Mahdi Army, but U.S. forces were going to have to take the lead, because the international forces would not.

Before the transfer-of-sovereignty announcement, the CPA pressed us to move. "If we do not get al-Sadr now, the situation will get worse," they said. "Not only does he have to be taken, if we do not act, the Iraqi people will castigate the coalition for inaction."

In response, we pulled out our detailed plan to take out Muqtada al-Sadr, gathered key intelligence, and prepared to move a major force to the south. A day or two after the announcement to transfer sovereignty, we briefed the plan to Ambassador Bremer and sought authority to execute the mission. But the decision came back in short order. "No, absolutely not. This would be the worst thing we could do, given that we just said we are transferring sovereignty to the Iraqis. Don't even contemplate it." So we backed off, put the plan back on the shelf, and simply continued to monitor the al-Sadr situation.

Right on the heels of this decision, Ambassador Robert D. Blackwill and his team came back into the country. His charge was to assess the current conditions, actions, working relationships, plans, and capacities of the U.S. mission in Iraq. Over time, I gained a great deal of respect for Blackwill and his people, because he was very clear on his intentions and kept providing me with feedback that confirmed what I had been telling Washington all along.

Blackwill's final report was sent back to Washington at the end of November 2003. From my discussions with the Ambassador and from the feedback I had received from various other sources (such as staff officer interactions and draft reports), I believe the final assessment stated:

> *Regarding conditions on the ground.* The insurgency is clearly worsening, especially in the area of the Sunni Triangle and it will get worse before it gets better. The current de-Baathification policy is a disaster, because former Baathists now have little choice but to join the insurgents. The conditions are right for a civil war between the Sunnis and Shia, and between the Shia elements themselves.
>
> *Regarding the Coalition Provisional Authority.* The organization is highly stove-piped and displays very recognizable bureaucratic behaviors. CPA administrators are incapable of coordinating across organizational lines, unwilling to share resources, and refuse to shift priorities in a frequently changing environment. Ambassador Bremer demands complete control and makes all major decisions.
>
> *Regarding CJTF-7 and the military, in general.* The military has been highly effective and is clearly taking the lead in political elements of the counterinsurgency effort, such as establishing local and provincial councils, which

> are working well. It has performed considerable outreach to tribal leaders, and made good use of CERP funds to reinvigorate community life and generate employment. The Iraqi Civil Defense Corps has been particularly efficient in balancing effective security operations without alienating the Iraqi populace. The military has put in place a strong plan and has done everything possible to fight the growing insurgency.
>
> *Conclusions and recommendations.* The insurgency can be mitigated only through a synchronized approach of military, political, and economic elements, which must be coordinated aggressively and immediately. However, nothing can be done by the United States to completely eliminate the insurgency. The U.S. presence in Iraq is as much a part of the problem as it is a part of the solution.

As far as I know, Ambassador Blackwill's report was officially shelved. Certainly, there was no new action taken as every political move in Iraq stayed focused on transferring sovereignty to Iraq by July 1, 2004.

DURING LUNCH ON THANKSGIVING Day, November 27, 2003, I visited as many of the dining facilities in Baghdad as I could. I spoke with the cooks, the soldiers, and spent some time roaming around Baghdad. At one of our field artillery units, I noticed four Hispanic soldiers sitting together and enjoying their meal, so I pulled up a chair and joined them. We talked about where they were from, how they were doing, and, right before I left, I asked, "What do you guys need? What can I do for you?"

I expected they might mention that they needed some equipment or perhaps some better quality of life support, or something along those lines. But their response surprised me. "Sir, the only thing we need from you is to help us become American citizens," they said. "We've been trying for years, but the bureaucracy is just impossible. And now they want us to go back to the United States for immigration hearings. But we're here fighting in this war, and we can't do it. You need to help us, sir."

I was flabbergasted. I knew it was common for people who are not citizens to join the American military as long as they have a green card and are legal immigrants. But I did not realize it was so difficult for soldiers in a combat zone to obtain their citizenship. So here I had several of my soldiers who had applied to be U.S. citizens. They were caught up in the bureaucratic process, and yet they were unflinching in their commitment to defend our democracy. There must have been others, because the Army opened up a way for them to get an education, advance themselves, and serve the country. What's more, they were as committed, if not more so, as anybody else in uniform.

"Well, men, this is absolutely wrong," I told them. "I'm going to do whatever I can to change the system and help you. Happy Thanksgiving and God bless you."

Right after that, I asked our staff to look into the matter, and we pressed every single congressional delegation that came into the country to take up the matter. In turn, they started an initiative to expedite the citizenship process for our soldiers who had been fighting in Iraq, but were caught up in the government's web of bureaucracy. It would take a number of months, but eventually we would hold several very large swearing-in ceremonies.

Later that same evening, President Bush paid a surprise visit to Baghdad to have Thanksgiving Day dinner with the troops. I had received word three days earlier that he was coming and immediately began preparations, which included the construction of a huge tent at Baghdad International Airport, with a podium, backdrops, tables, and arranging a turkey dinner with all the trimmings. An hour and a half before the President's scheduled arrival, we brought in several hundred U.S. troops representative of the divisions serving in Iraq. For security purposes, they were all searched for weapons; none of them knew who was going to be there. The only information we had released was that a VIP was going to be there in a show of appreciation. Most people thought it would be Secretary Rumsfeld.

When Air Force One landed, I was in an armored Suburban waiting to meet the President. We drove up to the aircraft, parked directly at the base of the stairs, and waited. During the intervening minute or two, I heard several explosions in the distance that appeared to be coming from the neighborhood of Abu Ghraib, some five or six kilometers away. "Oh, great," I thought to myself. "This is all we need right now, a mortar or rocket attack at the airport." Because of the secrecy surrounding the entire operation, and the security we had provided, I realized such an attack was highly unlikely, so I calmed down a bit. Sure enough, the background explosions stopped, the doors to Air Force One opened, and Bush walked down the stairs, took two steps, and got into the Suburban next to me.

"Hi, Ric, good to see you," he said.

"Happy Thanksgiving, Mr. President," I replied. "It's great that you're here. I know the troops will be very glad to see you."

As we drove over to the tent, I briefed Bush about the details of the event and how we were going to get him in front of the troops. Ambassador Bremer was waiting at the tent to greet the President, and then he and I walked out on the stage while Bush waited behind the backdrop. We welcomed everybody to the event and then I casually said, "Gee, I wonder who outranks me back there and wants to come out and talk to the troops."

At that cue, the President walked out from behind the backdrop and the effect was electrifying. Our troops went absolutely wild. They cheered, screamed, applauded, and jumped up and down. It was a fantastic demonstration that lasted for four or five minutes. After everybody calmed down, Bush came up to the podium and spoke, but not for very long. He wished everybody a happy Thanksgiving, thanked them for their service to our nation, and made several other impromptu and heartfelt remarks. Then he walked out among the crowd and started shaking hands with everybody. Eventually he made his way over to the food line and began serving the troops as they passed through. Then he filled his own plate and went out and sat down at the tables with everybody else.

President George W. Bush was at his best that day. His concern for the troops and his sincerity in thanking them sent a powerful message to our men and women in uniform. For the moment, I completely forgot about all the politics of Washington. It really didn't matter what the President's policies were, or whether he was a Republican or Democrat. What mattered most was that he was there showing his appreciation to these fine young people who had been enduring such hardships in service to their country.

Over the next several weeks, word spread across the country about this very inspiring visit from the President of the United States, and morale among the troops soared. Just about the time the talk was beginning to fade, another president made headlines, which served to further boost morale. Only this time, it was the former president of Iraq.

On December 13, 2003, at about seven thirty in the evening, I received a call from Ray Odierno, commander of the 4th Infantry Division. "General Sanchez, I think we've got Saddam," he said. "He's scraggly, dirty, and has a huge beard, but we think it's him because of the tattoos on his hand. They match up with descriptions provided by Special Forces."

Earlier in the day, a key facilitator for Saddam Hussein who was picked up and interrogated by Special Operations Forces had revealed that the former president might be hiding at a small farmhouse near the town of Adwar, about ten miles south of Tikrit, Saddam's hometown. This man was immediately given a lie detector test, which he failed. But a savvy warrant officer thought there might be something to this tip, and that it should be pursued.

Just after dark, about 600 troops descended on the location, but initially found nothing. After cordoning off the area and conducting an intensive search, however, a couple of soldiers noticed the entrance to an outside "spider hole" covered by Styrofoam and a carpet. They were just about to toss a hand grenade inside when Saddam Hussein, hiding in the hole, identified himself. Our soldiers quickly grabbed him, dragged him out, and threw him to the ground. Inside the hole, which was about six feet deep and eight feet wide, they found an AK-47 rifle and about $700,000 in U.S. cash.

As soon as I heard the news, I directed the CJTF staff to execute the Saddam postcapture contingency plan we had developed in August after the deaths of Uday and Qusay. In order to give us time to make a positive identification, we shut off all communications to the outside world. I asked MG Odierno to keep the news under wraps, and we quickly put the clamps down on everything inside Iraq, including his troops on the ground, his headquarters, and my headquarters. We were also fortunate that it was past dark, because reporters usually were not around at night. I was confident we had about six to eight hours before the press suspected that something major was taking place.

Shortly after hanging up with Odierno, I placed a call to John Abizaid and gave him the news. In turn, he called Secretary Rumsfeld. Then I notified Ambassador Bremer and told him that we were going to get Saddam to Victory Base as fast as possible. I believe Bremer subsequently called back to Washington to inform the President. Over the next four or five hours, there was some confusion regarding Saddam Hussein's status: Was he a prisoner of war and subject to the Geneva Conventions? After the issue was resolved between CENTCOM and Washington, GEN Abizaid called and directed me to treat Saddam in accordance with the Geneva Conventions.

Initially, we kept our new high-value detainee at Victory Base in a facility where we were holding several other former Iraqi officials, such as "Chemical Ali" (Ali Hassan al-Majid), Tarik Aziz, and Saddam's half brother. First, we gave Saddam a short physical exam. I was present in the room along with a security guard, a physician's assistant, an interpreter, and a cameraman (to document the prisoner's physical condition).

Saddam arrived at Victory Base around 2:30 a.m. with his hands zip-tied behind his back and a sandbag over his head so that he remained unaware of his location. We walked him into a small room, stood him up against the wall, and pulled the sandbag off his head.

"Who are you?" asked the interpreter.

Slightly disoriented, he looked around. "I'm Saddam Hussein, the president of Iraq," he said. "Why are you doing this to me?"

"We are going to give you a physical," said the interpreter. "Is there anything wrong with you?"

"I think I have an injury to the side of my head and a cut in my mouth," replied Saddam, referring to the fact that he had been thrown to the ground upon his capture.

"What other problems do you have?"

"I have high blood pressure and I need my medicines."

"We'll get your medicines for you," I said.

Over the next ten or fifteen minutes, the physician's assistant administered the physical exam. He particularly checked inside Saddam's mouth and beard looking for the bruises and cuts he'd complained about. There weren't any. A couple of bruises were documented on the side of his body, but otherwise the prisoner was in pretty good physical condition.

Later, when we released a short video clip of the exam, the press immediately began speculating that the physician's assistant was checking Saddam's beard for lice and looking inside his mouth as if he were checking out a slave or an animal. That was not the case, however. He was simply checking for the cuts and bruises that Saddam had mentioned. We were also informed by key members of the Iraqi Governing Council that the clip had conveyed the message that Saddam was off his lofty pedestal and under the coalition's control. Actually, there was no intent to send such a message. Our sole intention was to convince the Iraqi people that we had, indeed, captured Saddam Hussein. Of course, we had to get permission to release that video clip because, technically, it violated the Geneva Conventions. However, the decision was made in Washington that it was necessary to do so in order to convince Iraqi citizens. We also had to get a similar exception to shave off Saddam's beard in order to compare him with pictures taken prior to the war.

Our next step was to move Saddam into a room with a one-way mirror and then bring several of the other high-value detainees in to get a visual identification. We brought his half brother in and asked, "Can you tell us who that is?"

"Why, yes," he replied. "That is Saddam."

When we brought in Chemical Ali and Tarik Aziz and asked them the same question, each responded that they didn't recognize him. "No, I don't know who that is," they said.

"Take a closer look," said the chief warrant officer who had brought them in.

"Oh, yes, that looks like Saddam," said Aziz.

"Yes, yes, that's Saddam Hussein," replied Chemical Ali.

The next day, Ambassador Bremer brought in four members of the Governing Council who also made positive identifications. To confirm, though, we sent out samples of his DNA to a lab; they later came back positive.

While we were preparing for the press conference to announce the capture of Saddam, I received a call from Secretary Rumsfeld who stated matter-of-factly that he did not want Paul Bremer to make the announcement on Saddam's capture. He wanted me to do it.

So I went to see Ambassador Bremer, informed him of our timelines for the press conference, and told him what Rumsfeld had said. Initially, he simply heard me out. But he called me back a short time later and said, "I'm going to speak first."

"Well, okay, sir," I replied. "I can't tell you not to, but I did communicate the Secretary's wishes."

"I know," he said. "But I'm in charge here and I'm going to speak first."

"Okay," I said.

Then I got on the phone to GEN Abizaid. "I can't comply

with Rumsfeld's direction, because I cannot control Bremer," I said. "You'd better call the Secretary and tell him what's going to happen."

"Let me call Bremer first," he replied. "I'll get back to you."

After speaking with the Ambassador, Abizaid called me back. "Just let Bremer do it, Ric," he said in a defeated tone. "Just let him do it."

"Well, okay, sir. We need to notify the SECDEF."

On December 14, 2003, at 3:00 p.m. (7:00 a.m. Washington time), Ambassador Bremer kicked off our formal announcement of Saddam Hussein's capture.

"Ladies and gentlemen," he said, "we got him."

NINE DAYS LATER, ON December 23, the 1st Armored Division opened up a small rest and relaxation area inside the Green Zone just in time for Christmas. I was invited to attend the opening reception with my former unit and really enjoyed interacting with the soldiers. One of the men I spoke with at length was Command Sergeant Major (CSM) Eric Cooke, whom I had known for about fifteen years. He had a reputation for being a frontline leader, always out with his troops, day or night. We had a good talk about what was going on in the field, the morale of the division, and how he was doing, in particular. As we parted, I shook his hand and said, "Take care of your men and be careful, Sergeant Major."

"Aw, don't worry, sir. We're out there every day. We'll be fine."

Spirits were high among the troops as we all did the best we could to celebrate the Christmas holidays, even though we were far from our families. On Christmas Eve, I called home to Maria Elena and the kids, and read "The Night Before Christmas" over the phone—a Sanchez family tradition. I told them I loved them and was sorry I couldn't be with them. The children told me they loved me and that they understood.

On Christmas Day, after morning mass, I went around and visited as many of the units in the Baghdad area as I could squeeze in. People were watching football games, singing Christmas carols, and sharing gifts and Christmas cards from home. Even in the most austere and remote outposts, the troops had put up improvised decorations, such as little lights and stars fashioned out of metal and cloth. It reminded me of when I was a kid and my parents did the same thing at our house back in Rio Grande City.

When I returned to my quarters that night, I received a phone call from COL Mike Tucker, one of the brigade commanders in 1st Armored. "Sir, I'm sorry to have to tell you that CSM Cooke went out on a patrol with his men this afternoon and his vehicle was hit by an IED. He was killed instantly."

My eyes welled up with tears, and I couldn't speak for a few moments. Finally, I thanked COL Tucker for the call, hung up, and then sat down and cried.

I was bereaved that my friend had been killed, but I was also angry. And I began to think about what I had always told our troops, that they should be ruthless in battle, but benevolent in victory. At that moment, it was brought home to me just how hard that was to do, especially when your people are actually fighting, bleeding, and dying on the battlefield. And so I prayed. I prayed for the strength to keep from displaying my anger in front of our troops. I prayed that this war would be over soon. And I prayed for the soul of my fallen comrade.

A few days later, I attended the memorial service for Eric F. Cooke, Command Sergeant Major of the 1st Brigade, 1st Armored Division (Old Ironsides); Bronze Star; Purple Heart. He was from Phoenix, Arizona, had joined the Army at the age of eighteen, and had left behind a loving wife and children. He was forty-three years old.

Meeting with President George W. Bush in the Oval Office on May 20, 2004. Secretary of Defense Donald Rumsfeld is standing behind the President. *(Official White House Photo)*

Collage of early years (late 1940s to mid-1950s) in Rio Grande City, Texas. *Left to right:* Grandmother, Mom, and her sisters; family, friends, and I, aged four *(at lower right)*; the first house I lived in is in the background on cinder blocks. *(Courtesy, Ricardo S. Sanchez)*

College graduation from Texas A&I University, 1973, with my mother, father, and Maria Elena. *(Courtesy, Ricardo S. Sanchez)*

In Saudi Arabia in the fall of 1990, planning movement routes during Desert Shield training. *Left to right:* CPT Geoff Ward, MAJ Doug Robinson, MAJ John Tytla, and I. *(Courtesy, Office of the Commanding General, CJTF-7/V Corps)*

Washing clothes in Saudi Arabia just before the Desert Storm attack into Iraq, in early 1991. *(Courtesy, Ricardo S. Sanchez)*

In Germany, speaking to the 35th Armor Battalion in March 2003, setting expectations for deployment to Iraq. *(Courtesy, Ricardo S. Sanchez)*

Greeting GEN Tommy Franks at Baghdad International Airport (BIAP) upon arrival for his farewell tour, June 2003. *(Courtesy, Office of the Commanding General, CJTF-7/V Corps)*

President Bush on the deck of the USS *Abraham Lincoln*, announcing the end of the war on May 1, 2003. *(Associated Press/J. Scott Applewhite)*

In a Blackhawk helicopter preparing to fly over Baghdad with Secretary of Defense Donald Rumsfeld in the summer of 2003. *(Courtesy, Office of the Commanding General, CJTF-7/V Corps)*

With GEN John Abizaid, my boss, at BIAP during one of his many visits to Iraq. *(Courtesy, Office of the Commanding General, CJTF-7/V Corps)*

Surveying the devastation after a suicide bombing on January 18, 2004, of Assassin's Gate, a key entrance to the Green Zone in Baghdad. *(Courtesy, Office of the Commanding General, CJTF-7/V Corps)*

Late June 2003, on the white satellite phone with Rumsfeld, Franks, and GEN Pete Pace during the Syrian border incident. *(Courtesy, Office of the Commanding General, CJTF-7/V Corps)*

Assessing Iranian infiltration routes into Iraq with the Polish Division, at the Iranian border in December 2003. *(Courtesy, Office of the Commanding General, CJTF-7/V Corps)*

On patrol in Sadr City with elements of the 1st Cavalry and 82nd Airborne in March 2004. *(Courtesy, Office of the Commanding General, CJTF-7/V Corps)*

One of dozens of my meetings with regional leaders—part of our tribal engagement strategy. Here I'm interacting with the Shia leadership in Basra. *(Courtesy, Office of the Commanding General, CJTF-7/V Corps)*

Discussing intelligence and interrogation operations with military intelligence personnel inside Abu Ghraib prison, summer 2003. *(Courtesy, Office of the Commanding General, CJTF-7/V Corps)*

With the MP battalion at the south wall of Abu Ghraib, developing plans for the prison's defense, summer 2003. *(Courtesy, Office of the Commanding General, CJTF-7/V Corps)*

With Ambassador Paul Bremer as President Bush addresses the troops during his surprise visit to Baghdad on Thanksgiving Day 2003. *(Courtesy, Office of the Commanding General, CJTF-7/V Corps)*

At a press conference with Ambassador Paul Bremer on December 14, 2003, providing details of the operation that captured Saddam Hussein. *(Courtesy, Office of the Commanding General, CJTF-7/V Corps)*

Eluding sniper fire on the roof of CPA headquarters in Najaf, on April 4, 2004. *(Courtesy, Office of the Commanding General, CJTF-7/V Corps)*

April 20, 2004, just north of Najaf with the 3rd Brigade, 1st Infantry Division, only hours before they are committed to battle against Muqtada al-Sadr's Mahdi Army. I'm emphasizing the Army values of L-D-R-S-H-I-P (on my dogtag chain) and being "ruthless in battle, but benevolent in victory." *(Courtesy, Office of the Commanding General, CJTF-7/V Corps)*

Testifying before the Senate Armed Services Committee, May 19, 2004. *(Scott J. Ferrell/Getty Images)*

November 2004, greeting children at the dedication of General Ricardo Sanchez Elementary School in my hometown of Rio Grande City, Texas. *(Courtesy, Ricardo S. Sanchez)*

The Sanchez family at my retirement from the U.S. Army on November 1, 2006. *Left to right:* Maria Elena, Michael, Rebekah, Lara, Daniel, and I, Fort Sam Houston, San Antonio, Texas. *(Courtesy, Ricardo S. Sanchez)*

CHAPTER 17

A Window of Opportunity Lost

On January 13, 2004, Sergeant (SGT) Joseph M. Darby, a twenty-three-year-old U.S. Army MP who worked in the office at Abu Ghraib, left a note with a CD containing photographs of prisoner abuses on the desk of a Criminal Investigation Division (CID) agent. Darby received the material from Corporal (CPL) Charles Graner and agonized about whether or not to bring it to the attention of the authorities. In the end, as he later said to a congressional committee, his decision was made because "[the abuses] violated everything I personally believed in and all I'd been taught about the rules of war." SGT Darby did the right thing. At great risk to himself, this young man had the moral courage to stand up and turn in members of his own company. For that, he was both ostracized and praised.

The next day, January 14, 2004, I received a call from the CID commander informing me that he needed to come in right away and brief me about a new major case. A couple of hours later, the commander was in my office relating details of the case to COL Marc Warren and me.

"Sir, yesterday a CD was dropped off in our office that contained several hundred pictures of prisoner abuse at Abu Ghraib," he began. "Clearly some of the things that were done are illegal and definitely violations of the Geneva Conventions. We have pictures of naked prisoners, some of which are pornographic in nature. Others show the use of unmuzzled dogs, and there are even pictures of MPs posing with a dead body."

"How many people were involved?" I asked.

"Sir, at this point, it looks like there were six MPs, several unidentified soldiers, and about a dozen prisoners. We have located and seized computers and other CDs that may have copies of the pictures, and are holding them as evidence. We have identified some of the MPs in the pictures, have conducted initial questioning, and already have a few confessions."

"What in the world was going on? Were they interrogating prisoners?" I asked.

"No sir, these guys were just screwing around," he responded. "They were angry about several inmate disturbances and were getting back at them. The prisoners were herded into a cellblock late at night."

"Why the pictures?"

"Sir, these kids took pictures with their digital cameras like they would at a party. One of the young women told us that they did it for the fun of it."

After the CID officer talked us through his timelines for the rest of the investigation, he left the office, and I met privately with COL Warren. "Marc, we need to request an immediate investigation from LTG McKiernan's command [CFLCC], because the 800th MPs report to him, not me."

"Yes, sir, I'll prepare the request immediately," he replied.

"I also want these guys completely out of contact with prisoners. Get the names of everybody CID has identified and prepare the paperwork to suspend the chain of command. Let's move any suspected individuals out of Abu Ghraib and into temporary office jobs someplace else."

"Sir, I'm not sure you have the authority to do that," said COL Warren. "Technically, you only have tactical control over the MPs."

"I don't care. I want them out of there now. This is outrageous."

"Yes, sir."

"I also want you to prepare a list of immediate administrative actions for the leadership of the 800th MPs. Let's make sure they've been informed of what I intend to do, and hear what they have to say. Let's look at the entire chain of command. Suspensions, letters of reprimand—prepare the documents based on what the evidence supports. I think we have enough to initiate administrative actions, but make sure I'm within my authority. Okay?"

"Yes, sir. I'll get on it."

"And one more thing. There might be other incidents like this that have not been reported. I want to initiate a broader internal investigation to find out if there have been other abuses of which we are unaware."

After Marc Warren left, I called GEN Abizaid to inform him of the situation.

"Have you seen the photographs?" he asked.

"No, sir. The CD is being handled in accordance with evidentiary rules. But the pictures have been described to me in detail by the CID commander."

"Pretty bad, huh?"

"Yes, sir. This is going to be really ugly when it breaks and we better be prepared for it. We definitely need to make sure that Washington, the Secretary, and the Army leadership are aware of this."

While Abizaid called Washington to inform Secretary Rumsfeld, I briefed Ambassador Bremer. Within a few days, I sent a letter of admonishment to BG Karpinski with specific instructions on whom I wanted suspended from duty at Abu Ghraib. As a result, most of the chain of command was moved out. In addition, we formally requested an investigation from CFLCC, and McKiernan assigned MG Anthony M. (Tony) Taguba to lead the effort. At the time, MG Taguba was the Deputy Commanding

General for Support in CFLCC with responsibility for oversight of the 800th MP Brigade. We also received an e-mail on January 18, 2004, from the CENTCOM deputy commander advising us that Secretary Rumsfeld was very concerned that the scandal would break before the President's State of the Union address, scheduled for January 20, 2004.

On January 16, at the request of GEN Abizaid, I flew to Bahrain to have a dinner meeting with him and Rumsfeld. As soon as I walked in the door, the Secretary lit into me. "Why in the hell did you let Bremer make the announcement on Saddam?" he thundered.

"Mr. Secretary, I don't control the Ambassador," I answered. "I gave him your message, but he said he was going to speak first and that was it."

"I talked to both Ric and Bremer about this," Abizaid said, jumping to my rescue. "Bremer just would not relent."

"Well, dammit, you guys ought to listen to my guidance," replied Rumsfeld. "Now, sit down."

It wasn't a great start to our meeting, but the Secretary eventually calmed down and we discussed a variety of subjects. In addition to the ongoing status of operations in Iraq and the timeline for transition of sovereignty, Rumsfeld seemed particularly interested in Saddam Hussein. So I gave him a detailed update.

After the decision came down to keep Saddam in Iraq, we built a special detention site at Victory Base to house him. For security purposes, the public did not know where he was being kept. And to keep Saddam from knowing his own whereabouts, we blindfolded him in the middle of the night, flew him around in a helicopter for twenty minutes, and then landed him right back at Victory Base. In reality, we had moved him only about 200 yards. In his new holding area, he was monitored twenty-four hours a day and allowed visits from the Red Cross. He complained about his shower stall, the lights being turned off too early, and not having enough fruit to eat—all of which we changed to his satisfaction.

Washington provided very specific instructions that only the CIA would be allowed to interrogate him, but I was fairly certain that Saddam was not subjected to any severe treatment. From information contained in captured documents, it turned out that he was not directly controlling the insurgency. In addition to recording a couple of audiotapes for public release, about all he was doing was passing information back and forth to key operatives. Apparently, the insurgency was being run by separate cells across the country, and Saddam would be periodically updated.

After we finished discussing Saddam, Secretary Rumsfeld made a cursory reference to Abu Ghraib. Then he surprised me by asking if I would be willing to extend my stay in Iraq. Fifth Corps was due to rotate out in mid-February, and Abizaid and I had previously discussed the possibility of my extension.

"Well, how long?" I asked Rumsfeld. "Are we talking another year?"

"It could be that long. But at least through the fall, which would get us through the entire transition to sovereignty. Tom Metz [a three-star] will come in to be your deputy. After that, we're looking at promoting you to four stars and putting you into command of SOUTHCOM in Miami. So what do you think?"

"Sir, I'd like to speak to my wife about it," I replied.

"Okay, that's fair," said Rumsfeld. "But I want you to stay."

"Mr. Secretary," interrupted Abizaid, "we really need to give Ric a couple days to talk to his wife and think about it. This is a big decision."

The meeting had started with Rumsfeld chewing me out and ended with him assuring me of a promotion and offering an opportunity to extend my command. Later that night, I spoke with Maria Elena and gave her the news. "Do you want to continue?" she asked.

"Yes, I think I owe it to my soldiers and to the country."

"Well, if that's what you want to do, okay," she said. "I don't like it, but I will support you."

Within a couple of days, I told GEN Abizaid that I would

agree to Rumsfeld's request and stay on. The only concession I asked for was to be able to attend my son Daniel's high school graduation in May 2004.

It couldn't have been more than forty-eight hours later that I received an e-mail from *Washington Post* reporter Tom Ricks stating that he was the bearer of bad news for me. "I'm hearing that you are about to be relieved and that you will be replaced by LTG Metz," he wrote. "I'm sorry, I don't make up the news, I just report it." He then asked me for my comments before he wrote an article.

"Tom, don't believe everything you hear from your sources," was all I wrote back.

Ricks was not happy with my response, but I simply wasn't going to give him much information, because of a previous interaction I'd had with him in May 2003, when I was still a division commander.

Back then, he was embedded with us in 1st Armored for about four or five days and was attempting to get me to buy into a story that the 3rd Infantry Division was only putting a finger in the dike with regard to maintaining stability and keeping the Iraqis from looting the city. The day after one of our briefings by the commander in the Green Zone, an article appeared in the *Post* in which he seriously misquoted me. So I called Ricks and asked him to come over to my office right away.

"I never said this, Tom," I protested, laying the paper down in front of him. "You're fabricating comments. This was your agenda from the very beginning."

"General, you did say that," he responded. "I don't like the fact that you're questioning my integrity."

"Show me where I said this, Tom."

"It's in my notes. You said it in the briefing."

"No, sir, I did not. You need to go review your notes. If you don't print a retraction, I will never allow you into one of my units again."

The next afternoon, Ricks came in and apologized. "General, you were right," he said. "I got confused." The paper then printed a two-sentence retraction that attributed the quote to someone else.

Overall, this incident with *The Washington Post* exemplified to me how the media manipulated information to fit their predetermined agendas. As I would later find out, it was just a prelude to what would happen when the Abu Ghraib abuse story was presented by the mainstream media.

AT 8:00 A.M. ON January 18, 2004, Ambassador Bremer's morning meeting was just beginning when a massive bomb exploded at the north gate of the Green Zone about a kilometer away and shook the entire building. My instinct to ride to the sound of the guns kicked in, and I immediately told my aides, "Let's go! Let's see what happened." We stopped our vehicle about fifty yards from the main gate and, as soon as I started walking, I could see bits and pieces of human remains scattered about. Up at the scene, we encountered mass confusion, with people screaming and running around. Our soldiers had responded quickly and were scurrying to secure the scene. One young captain had pulled out his 9-millimeter pistol and, with a glazed look in his eyes, was waving it around and issuing orders. I grabbed him by his web gear and said, "Look at me! You're doing the right thing, but put that pistol away before you shoot somebody or yourself. Help your soldiers get on with doing their jobs."

As I walked outside the Green Zone, I observed that the explosion had occurred about thirty or forty yards beyond the gate. Dozens of wounded people were laying about begging for help. Fifteen or twenty cars were completely destroyed. Some were charred black, still on fire, or smoldering—and there were skeletons in the front seats of many vehicles. At the site of the detonation, there was a huge crater in the middle of the road. We later determined that a suicide truck bomber had tried to bypass a line of cars at the security checkpoint, but the thousand pounds of explosives he was carrying had gone off prematurely. The blast killed twenty and wounded sixty-three, most of whom were Iraqis. Two Department of Defense civilians waiting in line were among the dead. Because our gate fortifications performed as designed, none of our soldiers were killed.

With this shocking attack, the insurgents sent us the message that Saddam Hussein's capture didn't mean that the violence and brutality were going to stop. It was an attempt to penetrate the Green Zone through the Assassin's Gate, which was right at the doorstep of the CPA. *[The gate was named for the military unit that manned the gate, the Alpha Company Assassins of the 2nd Brigade of the 1st Armored Division.]* To my mind, this terrible incident underscored the extent of the horror the enemy was willing to inflict.

Although there were other spectacular acts of violence similar to the Assassin's Gate explosion after Saddam Hussein's capture, there was, for the most part, a tremendous lull in activity across the country. Our number of combat engagements dropped drastically, the overall amount of insurgent attacks went way down, and there were really no major areas of instability anywhere in the country. Shortly after Saddam's capture, John Abizaid and I talked about having a huge window of opportunity. "Ric, this is kind of scary," said Abizaid. "But it looks like there may actually be an opening to stabilize things. We need to figure out how to take advantage of this window. What actions do you think we need to take to advance reconciliation, security, and stability?"

The CJTF-7 staff went into an almost round-the-clock effort to try to identify initiatives that would send a clear message to everybody in Iraq—Sunnis and Shiites alike—that now was the time for us to come together. The flow of initiatives the staff thought up was extensive and, in some cases, remarkable. "Why don't we have Saddam issue a surrender declaration?" someone suggested. "That might help us with the Baathists and former regime elements. Let's have him say that the fighting should stop for the sake of the country."

When this suggestion was brought to Ambassador Bremer, however, he responded, "We aren't going to do that. That will lend legitimacy to Saddam."

"Well, heck," I remember thinking, "he *was* the president of Iraq. Even Rumsfeld visited him back in 1983." GEN Abizaid was lukewarm to the suggestion, as well. And I found out later that this idea became a big joke among the CPA staff.

We also came up with a proposal to offer amnesty to insurgents in order to stop the violence and reintegrate them into Iraqi society. We pushed this idea very hard. But there was no appetite among the CPA, the Washington establishment, or the Iraqi Governing Council for even a mention of the word "amnesty." "We can't give amnesty to anybody who has attacked or killed Americans or members of the coalition," Bremer told me. "Then who can you offer amnesty to?" I replied. "How are we going to end this thing?" That concept also went nowhere.

We continued to suggest more initiatives to gain some semblance of momentum. Let's embrace the Shia tribal militias. Let's engage the insurgents in dialogue. Let's get the regional reconstruction packages funded so that we can make an economic impact on the country. Let's have a major meeting with the Sunnis, similar to what they did in Afghanistan, where we attempt to find a suitable national leader. All these ideas, and more, came from the CJTF-7 and CENTCOM staff.

In addition, every division undertook a major effort to advance political, economic, and security initiatives inside its own sector. The proposals were wide-ranging and specific—down to individual schools, factories, government buildings, job programs, neighborhood councils, and the like. Each package constituted a joint effort by the division commander and the CPA regional administrator. But we could not get Ambassador Bremer or anybody else in the CPA to take our proposals seriously. They were just too busy focusing on the transfer of sovereignty and the major reconstruction projects that would be funded by the $18 billion supplemental.

At this point, we were beginning to see the cascading effect of the CPA's complete shift in focus after the November 15 announcement. The initiatives we had suggested were long-term solutions that required significant amounts of time and work. But the CPA was interested only in performing short-term tasks that related directly to their major goals of writing a constitution, setting up elections, and creating a transitional government. They were not interested in doing anything that might extend past July 1, 2004, because on that date they would be leaving the country. Another more practical reason the CPA would not even consider taking on any new initiatives was the simple fact that it did not have the people to perform the work. Ambassador Bremer had been severely understaffed right from the very beginning and, as far as I knew, very little had been done in Washington to fix the CPA's manning problem.

In December, GEN Peter Pace, Vice Chairman of the Joint Chiefs of Staff, had visited Iraq to survey the overall situation. When I met with him in Baghdad, he had asked what he could do to help. And I believe he was surprised to hear me request that he make it a priority to help staff the CPA. "Sir, they just have too much to accomplish for the transfer of sovereignty, and not enough people to make it happen," I said. "We cannot allow Bremer's organization to stay undermanned. If we do, we'll never address the political and economic problems in Iraq, and come July 1 the military will be left holding the bag." GEN Pace responded by saying that he would work that piece of the equation. Afterward, some effort was made by the Pentagon to help out Bremer, but it really didn't amount to much. And the CPA never did achieve the proper levels of staff manning.

GEN Pace was just one of a number of people from Washington who flowed into the country to make sure the July 1 deadline would be met. From the Department of State, Ambassador Frank Ricciardone and retired LTG Mick Kicklighter made several trips to Iraq in an effort to set up the new American embassy. Not only would it assume most of the CPA duties, but the embassy relationship with Iraq would be traditional in nature—which meant that the United States' capacity to influence and execute the political and economic actions necessary to rebuild Iraq would diminish considerably. On July 1, 2004, the United States would revert back to an advisory role, with the same authorities it brought to any other sovereign nation.

As the embassy process began to take shape, the CJTF-7 staff became concerned about what appeared to be a complete disregard for the political and economic coalition after the transfer of sovereignty. In fact, there was tremendous bitterness and discord among some of the nations who had military forces in Iraq. Most of them had not been consulted when the U.S. government made the decision to speed up the transfer of sovereignty. How was the coalition effort going to be sustained in Iraq after July 1, 2004, they wondered?

It was a valid question—and one that needed to be addressed. So I took up the issue with Ricciardone and Kicklighter. "I've always had responsibility for the military aspect of the coalition," I said, "but somebody needs to pick up responsibility for sustaining the political and economic elements after the transfer of sovereignty. We should set up a coordination committee that embraces all of the nations on our team. We can't just disregard them."

"That's not our job," they responded.

"Well, whose job is it?" I asked.

"We don't know. Our mission is to form the embassy, and that's all."

So I went to Ambassador Bremer and raised the issue with him. "That's not my job," he said. "That's a Department of State issue."

A bit later, I had a long discussion with Deputy Secretary of State Richard Armitage when he came back into the country to check on the status of operations. "Mr. Secretary, you do, of course, realize that we have an incompetent line of political and economic operations in this country, don't you?"

"Yes, you are essentially correct, General," he responded. "But that's not your responsibility. It's Bremer's responsibility."

Even when Secretary of State Colin Powell came into the country for a status update, he seemed uninterested in anything except progress with the embassy. In my meetings with him, he did not concern himself with Bremer's work at the CPA, which is where the bulk of his responsibility rested. That was a boundary he would not cross.

The Department of State was reluctant to deal with the CPA, which was now under the control of the National Security Council. The CPA was reluctant to deal with the Department of State or the Department of Defense. There was no indication that the NSC was making any effort to synchronize all government agencies to achieve unity of effort in Iraq. It was either incapable, incompetent, or unauthorized to perform the task. So I began to wonder why the administration ever put the CPA under the NSC in the first place.

The absurdity of the U.S. government's bureaucracy really came to the forefront in February 2004 when the congressional supplement, which had been passed back in November, finally became available. At this point, after the long wait to receive funding for critical projects, the CPA was put in the position of having to spend $18 billion in a little over four months—by the July 1 deadline for transfer of sovereignty. Worse yet, it controlled the entire supplement, but did not have the contracting capacity or the proper mechanisms set up to identify projects, prioritize efforts, allocate funds, or monitor execution of contracts.

I found out very quickly that the CPA was solely focused on spending the supplement on multimillion-dollar high-impact projects, such as power plants, major sewer projects, and the like. To coordinate the efforts, the CPA set up IRMO, the Iraqi Reconstruction Management Office, which assumed responsibility for all reconstruction matters. Manned mostly by U.S. people with some coalition help, IRMO's mission was to plan, execute, and manage the reconstruction initiatives of the country. Its goal, therefore, was to distribute the entire $18 billion before the transfer of sovereignty. It was at this time that a significant contract management capability was deployed into Iraq. Despite that fact, the conditions were ripe for fraud, waste, and abuse across the entire country's reconstruction effort.

ON MARCH 8, 2004, two important events occurred. First, the Iraqi Governing Council approved a new interim constitution for the nation. Essentially, the United States had written the document for them, modeled, of course, after the U.S. Constitution. Passage had been delayed for about a week, because five Shia members of the Council refused to sign unless certain changes were made. Once a compromise was reached, a signing ceremony was held where a few explosions could be heard in the background from insurgent attacks that had been launched in protest. These accompanying attacks were a clear indication that, in order to slow the march toward sovereignty, the insurgency was targeting anyone who cooperated with the coalition.

Also on that day, Secretary Rumsfeld finally gave responsibility for building and training the Iraqi police and security forces to CJTF-7. This decision had been a long time coming. Previously, Rumsfeld would not accept the judgment that GEN Abizaid and I were providing and he had continuously refused to order Bremer to allow the U.S. military to be involved. We first requested the move back in September 2003, but the Secretary simply told us to work it out on the ground. Then, in November, during discussions about the transfer of sovereignty, GEN Abizaid formally requested the change, which, in turn, spurred Rumsfeld to send in a team led by MG Karl Eichenberry to assess the entire security line of operation (including the ICDC, the police, and the new Iraqi Army). Eichenberry filed his report in mid-February, which basically found that the overall effort was fractured, dysfunctional, and needed to be consolidated.

Unfortunately, the stage was now set for a tug-of-war within the Interagency (Department of State, National Security Council, Joint Chiefs of Staff, and Department of Defense) to pay for equipment and the building of Iraqi police and security forces. While Bremer balked at providing the necessary funding, final decisions languished in the NSC. Compounding the situation, the Department of State—which was set to take over from the CPA on July 1—didn't want the money spent, because it would impact their own future budget plans. So State used its position on the National Security Council to question, stall, and veto our various funding requests. When Deputy NSC Advisor Stephen J. Hadley came to Iraq and told me what was going on, I expressed my frustration with the entire process. "This is ludicrous," I said. "What say does the Department of State have in the things I'm doing here?"

"Well, they question the need for special police capacity," he replied. "You will need to state your position in order to break the impasse." he said.

"Sir, I've stated my position over and over again," I said. "We have to do this. It's vital. If we're going to stand up the Iraqis by July 1, then there is no other option."

At various times, the Department of Defense inflated the numbers of effective Iraqi forces (including the ICDC, the Army, the regular police, and border security) that were operating alongside the coalition. The enduring challenge was building capable and effective Iraqi forces rather than simply adding numbers. Ambassador Bremer immediately argued that the estimates were far too high, which resulted in questions from Congress. The continuing complaints about our inability to build the Iraqi Army, to adequately train police, and to recruit the proper number of troops were eventually presented by the national media as a major controversy. Secretary Rumsfeld then issued his March 8, 2004, order finally giving responsibility to CJTF-7.

When Rumsfeld's order came down, Paul Eaton and the new National Security Advisor, David Gompert, became quite upset, and Ambassador Bremer rejected the edict outright. It took some time before all three came around and accepted the reality of the situation. On the other hand, the CPA advisor to the minister of the interior immediately told me that he felt it was a good decision. "Boy, am I glad we're working with you all now," he said. "We can't get a damn thing done over at CPA. Maybe now, we'll get this stuff moving."

When Deputy Secretary Paul Wolfowitz called to give me the word of Rumsfeld's decision to consolidate the building of security capacity, he issued a warning. "Well, you got what you wanted," he said. "But be careful, because in two weeks, we'll be asking you why you're failing."

Sure enough, a couple of weeks later, the Pentagon inquired why we weren't progressing with supplying and arming the new Iraqi forces. We had to remind them that the reason for the consolidation of this effort was that not only had no major contracts for materials been signed, but national requirements had not even been defined by the CPA. We were establishing equipment requirements, facility needs, and manning levels from scratch, even though we had been providing that information all along. In effect, CJTF-7 had to begin the process anew. It would take thirty to forty-five days to document the requirements and line up the contracts, and another forty-five to sixty days before we actually started seeing equipment arrive in the country. By then, we would be right on the brink of the transfer of sovereignty—provided, of course, that all the steps progressed smoothly.

In order to effectively manage the entire process of manning, training, and equipping the Iraq security and military forces, GEN Abizaid and I concurred that we needed to put a three-star general in charge. We recommended MG David Petraeus for the job. Petraeus, who had gone back to the United States during the normal force rotations, agreed to take on the challenge and was nominated by Secretary Rumsfeld and President Bush. I was very surprised, however, when the Bush administration turned the announcement into a major public relations event. At a formal press conference, the President had Petraeus sitting next to him to demonstrate that he was on top of the situation. After having ignored the issue for eight months, the administration was spinning the event to make it look like Petraeus was coming to Iraq to salvage a failed military effort. It was all political maneuvering ahead of the upcoming presidential election.

David Petraeus eventually arrived back in Baghdad in late May 2004, and performed very well. Petraeus benefited tremendously from the work that took place in March and April before he arrived. During those two months, the CJTF-7 staff, working with CPA personnel, defined all police equipment requirements, refined and consolidated military equipment needs for both the Army and the ICDC, put contracts in place, and had equipment on the way to Iraq.

Concurrent with the effort to build and train the new Iraqi security forces, we were also finally setting up a Ministry of Defense and creating the entire national vertical chain of command. But while Secretary Rumsfeld had wrested authority for us to build and train the forces, the CPA retained responsibility for selecting individual leaders from the battalion level on up—and Ambassador Bremer would not let that go. I found this to be quite ironic, because before November 15, 2003, Bremer had adamantly refused to work with John Abizaid and me to establish the higher levels of Iraqi military leadership. As a matter of fact, Abizaid had been told point-blank by Secretary Rumsfeld to drop the issue. Back then, Bremer had only been interested in building a slow-paced, lower-level Iraqi Army from the ground up. But now he was forced to take action because of the July 1 deadline, so the CPA went into a mad scramble to find leaders to run the Ministry of Defense, the Joint Chiefs, and the higher levels of the Iraqi Army. CJTF-7's involvement was limited to interviewing candidates and providing feedback to Bremer. Of course, once the new Iraqi military leadership took over after the transfer of sovereignty, they would have no processes, systems, or capacities to function properly, because the coalition had not adequately addressed those elements. This was another instance in which the administration's attitude was basically to hand the problem off to the Iraqis. They would have to fend for themselves. In April, CJTF-7 started a major effort to build the capacity of the military hierarchy. But it was way too late.

There were two other key issues that the CPA took on during this time frame that it had previously refused to address—de-Baathification and the problem of local militias. In early April, Ambassador Bremer met with Governing Council member Ahmad Chalabi, who had been placed in charge of the entire de-Baathification process back on November 4, 2003. Bremer's message was straightforward: Serious problems had emerged since the Iraqis took control, he said. Among other things, Bremer told Chalabi that schools in many parts of Iraq were without teachers, and that a blanket exception needed to be made so that children could get their teachers back. He also urged Chalabi to implement the review committee process that had been envisioned for granting exceptions for Baath party members. In these instances, Bremer was now trying to force Chalibi to do what he had neglected to do himself.

With regard to the tribal militias, Ambassador Bremer and his national security advisor came up with a plan to disarm, demobilize, and reintegrate all the major militias in Iraq. But it all was just a concept with no real substance. The timelines they imposed, for instance, were absurd, because after July 1, the Iraqis were supposed to complete several major milestones within only two weeks, even though no resources had been assigned and no one deemed responsible for execution. As the CJTF-7 staff reviewed the plan, major questions arose. "How are you going to enforce this if the militias don't want to abide by it?" we asked. "For instance, how are you going to impose this on the 8,000 or 9,000 Kurds up in Kurdistan who will have no desire to participate?" "What are the resources required to execute this plan?" The only answers we received were: "Oh, that's an Iraqi problem," or "The Iraqis will be taking care of that." The CPA and national security advisor had no ability to disarm or reintegrate the tribal militias, nor were they truly committed to making it happen. They offered no plan to execute, no resources, and no funding. They only issued a theoretical piece of paper and declared success.

CJTF-7 was also undergoing major changes during this period. After a lot of machinations, we had finally received approval to establish a four-star command that would assume responsibility for the strategic- and operational-level functions that CFLCC had performed before it was shut down back in early June 2003. Initially, there had been serious thought given to actually bringing McKiernan and his entire headquarters back to Baghdad, but it was finally determined that he could not abandon the rest of CFLCC's theater mission. Rather, the decision was made to build the new command on a piecemeal basis by bringing in individual expertise from across the armed services. Although not as efficient as building around an established headquarters, the command would, at least, provide the capacity that had been so desperately needed during our first year of operations. According to our timeline, the four-star headquarters would be up and running by May 15, 2004.

On the international level, the United Nations became involved in the transition of sovereignty at the request of the Iraqi Governing Council. Selected as special UN envoy to Iraq was Lakhdar Brahimi, a Sunni Muslim from Algeria who came into the country for the express purpose of advising on the selection of future government leaders, such as president and prime minister. Brahimi, however, immediately butted heads with Paul Bremer, whom he later referred to as "the dictator of Iraq." Clearly, the U.S. government was not happy with the United Nations' involvement in the Iraqi transfer to sovereignty at this stage in the planning process. In fact, the NSC was so concerned about protecting U.S. interests in the matter that it dispatched Ambassador Robert D. Blackwill back to the country as a kind of "watchdog."

During several meetings, Blackwill forthrightly stated that he had specific orders from Washington. In essence, his mission was threefold: (1) monitor UN efforts and make certain they were acceptable to the NSC; (2) make sure that no actions were taken by CJTF-7, the CPA, or the United Nations that might jeopardize the July 1 transfer of sovereignty; and (3) prevent any military strikes, political activity, or comments in the press that might have a negative impact on opinion polls back in the United States.

Ambassador Blackwill sat in on nearly all meetings between the CPA and CJTF-7. Sometimes, if he was concerned about an issue, he would state something to the effect of "this is going to be bad," or "you can't do that, because it won't look good on CNN." Blackwill's presence in our meetings reminded me of the time I was in Kosovo and a Russian commissar sat in on every discussion I had with the Russian commander who reported to me. Only this time, I was the military officer who was being monitored.

Blackwill's candid statements on why he was in Iraq, coupled with Bremer's newfound enthusiasm for such issues as fixing de-Baathification, building a Ministry of Defense, and dealing with tribal militias, left no doubt as to the reason for the quick transfer of sovereignty. It was now crystal clear that a major success had to occur in Iraq before the presidential elections. Critical decisions affecting Iraq would be tied directly to ensuring the success of President Bush's reelection campaign.

THE FIRST THREE MONTHS of 2004 saw a general erosion of the relationship between the CPA and CJTF-7. While Ambassador Bremer and I maintained a personable enough relationship, there were multiple contentious issues, both military and political, that kept pounding at us. The Joint Task Force's constant pushing of initiatives (amnesty, surrender declaration, engaging the insurgents, etc.) added considerably to the tension. We also were squabbling over the role of the military after July 1, 2004, specifically in the preparation of a Status of Forces Agreement (SOFA). Ambassador Bremer had been told by Paul Wolfowitz not to negotiate the agreement with the interim governing council. "That's not the CPA's business," he had said. "That's DoD's business." This particular issue was very complicated in nature, because there are different levels of SOFAs between nations, with varying lengths and varying amounts of details. Wolfowitz wanted to handle the negotiations from Washington, but Bremer forged ahead anyway, telling us all the while that he was not doing so. Moreover, tensions between CJTF-7 and CPA were considerably heightened by the Abu Ghraib incident, which at this time was still an internal matter. However, Ambassador Bremer and the CPA, rightfully so, saw it as a real danger to the process of transferring sovereignty.

In early February, Bremer and I met with the International Committee of the Red Cross (ICRC) at CPA headquarters. When initially informed of the meeting, I assumed it was just a courtesy call. However, once I arrived, I found that the ICRC was actually presenting a key paper to Bremer that normally would have gone directly to the military unit that was the subject of the report. The underlying reason for this meeting was to secure a significant change in the established procedures for Red Cross reporting. CPA Deputy Dick Jones, who orchestrated the effort, wanted the reports to go outside the military chain of command and directly to Ambassador Bremer so he could monitor us. He also wanted to set up a direct reporting chain to the State Department. This proposed change became problematic when Dick Jones sent a message to the State Department and other agencies in Washington that assumed all allegations made in an ICRC memorandum were true. At that point, I confronted Jones directly to put a halt to this counterproductive procedure.

In that meeting, the ICRC representatives pointed out that their report mentioned a number of prisoner abuse allegations at Abu Ghraib (as well as some that dated back to the ground war). After the meeting, COL Warren and I went through the report paragraph by paragraph and determined that all of the ICRC concerns were going to be addressed in the investigation already being conducted by MG Tony Taguba. So we decided not to initiate a separate investigation, but instead forwarded the ICRC report to Taguba so he could embed it in his ongoing work.

In February, just before he presented his final investigative report to CENTCOM, MG Taguba arrived in Baghdad to give me a back briefing on his findings. It was around this time that I first saw the abuse photographs contained on the CD, and they made me sick. I just couldn't understand how our young soldiers could have done such things to the prisoners. Although Taguba's investigation was concerned only with the military police aspect of the incident, one of the first things he told me was that some military intelligence personnel were clearly involved in the abuses. In fact, he found that, contrary to MG Ryder's previous report, MI interrogators had requested the MP guards set physical and mental conditions for favorable interrogations. Because of regulations governing the conduct of military intelligence personnel, I immediately requested a new and separate investigation from CENTCOM. The Army subsequently assigned MG George Fay to lead the effort.

Overall, the substance of the Taguba report was quite damning. "Numerous incidents of sadistic, blatant, and wanton criminal abuses were inflicted on several detainees at Abu Ghraib prison," he wrote. In listing the specific abuses that were perpetrated by the MPs, he basically confirmed everything I had already been told, but the details seemed much more horrific in some cases—especially when it involved beatings, use of unmuzzled dogs, and sexually deviant behavior. I was extremely disappointed that American soldiers had displayed such a lack of discipline and completely disregarded the dignity of other human beings.

MG Taguba confirmed several other facts that had contributed to this catastrophic failure. Specifically, he mentioned:

- "Prior to deployment to Iraq, [the MPs] received no training in detention/internee operations . . . [and had] very little instruction or training on the applicable rules of the Geneva Conventions."

- "There is a general lack of knowledge, implementation, and emphasis of basic legal, regulatory, doctrinal, and command requirements within the 800th MP Brigade and its subordinate units."
- "The Abu Ghraib and Camp Bucca detention facilities are significantly over their intended maximum capacity while the guard force is undermanned and under resourced."

Taguba also noted in his report that he had conducted a four-hour interview with BG Janis Karpinski, during much of which she was extremely emotional. "I found particularly disturbing," he wrote, "her complete unwillingness to either understand or accept that many of the problems inherent in the 800th MP Brigade were caused or exacerbated by poor leadership and the refusal of her command to both establish and enforce basic standards and principles among its soldiers."

One of MG Taguba's recommendations was that Karpinski be relieved from command and given a written reprimand. He also recommended that COL Tom Pappas (whom I had placed in charge of the defense of Abu Ghraib back in November 2003) be both reprimanded and formally investigated for his possible involvement in the abuses. MG Taguba found that there was "clear friction and lack of effective communication" between COL Pappas and BG Karpinski, that this "ambiguous command relationship was exacerbated" by my placing Pappas in charge at Abu Ghraib, and that my action was "not doctrinally sound due to the different missions and agendas assigned" to the MP and MI specialties. I was quite surprised at these findings, because neither Pappas nor Karpinski had ever mentioned any kind of friction between them. Furthermore, there was no basis to the statement that my action was not doctrinally sound. The issue of command relationships had been debated extensively over the years, but had never been resolved by the Army.

Subsequent to the Taguba report, on March 12, 2004, MG Fay, who was in Iraq conducting the new investigation I had requested, came into my office for a brief discussion. "Sir, I'm getting some indications that you might have known something," he said. "So I'm going to have to question you."

"Okay, stop right there," I replied, thinking back to my days in the Army Inspector General's office. "This investigation now has to move to a higher level. You currently do not have the authority to interview me in this capacity. I've got to go back to CENTCOM and request a higher-ranking investigating officer."

As soon as MG Fay left, I sat down and wrote a memorandum to GEN Abizaid informing him that I was now the subject of an investigation, that it was beyond the authority of a major general, and that he or the Army should assign a higher-ranking officer. GEN Paul Kern (a four-star) was subsequently designated to lead the effort, and, in turn, he appointed LTG Anthony Jones to work with MG Fay in a combined investigation into the military intelligence aspects of the Abu Ghraib abuses, including my possible involvement.

Around this time, I also reviewed the consolidated report I had previously asked Marc Warren to generate involving other abuses that had occurred in Iraq. There had been a clear and steady series of incidents over the previous eight or nine months indicating that soldiers had been mishandling prisoners. The 3rd Armored Cavalry had a significant number, as did the 1st Armored Division. The good news was that the system was working, because the commanders had taken immediate and appropriate actions. In one case, an allegation had been made against a lieutenant colonel who had fired his pistol and threatened to kill an Iraqi he was questioning. "Marc, what is Ray Odierno [the division commander] doing about this?" I asked COL Warren.

"C'mon sir, you know it's not appropriate for you to be asking commanders to give you situation reports on judicial proceedings in which they're involved," Warren replied. "That constitutes command influence. You have to rely on the system and your commanders and lawyers to do the right thing in these cases." I knew that Marc was absolutely correct. In fact, that's what I had been taught over the course of my career. However, I felt a duty to ensure that we were doing the right thing in these cases, so I tasked him to figure out a way to keep me informed through his legal channels.

COL Warren's report, the ICRC report, and the Taguba investigation all reminded me of GEN Barry McCaffrey's admonition that there would always be 10 percent of the force that would cause problems, and that we needed to be aggressive in tempering their impact on the rest of our soldiers. In turn, that spurred me to speak about the issue in our regular commander's conferences. "When you receive reports that abuses may have been committed, then you have to take action regardless of who may be involved," I advised. "You must do the right thing, even if it means implicating a senior leader who may have done something wrong. That is part of your responsibility."

I also sent out a series of written communications reminding our soldiers to respect the rights of the individual, follow the laws of war, and treat all Iraqis with dignity and respect. The following memorandum, issued on March 2, 2004, is a representative example:

> MEMORANDUM FOR All Coalition Forces Personnel
> SUBJECT: Proper Conduct During Combat Operations
>
> 1. Purpose. This memorandum reemphasizes the responsibility of Coalition Forces to treat all persons with dignity and respect . . . [and] comply with the law of war.
>
> 2. Humane Treatment of the Iraqi People. Coalition Forces are committed to restoring the human rights of the Iraqi civilians and the rule of law. We must treat all civilians with humanity, dignity, and respect for their property and culture. Coalition Forces preserve human life by avoiding civilian casualties and rendering prompt medical attention to persons injured during combat operations. Use judgment and discretion in deciding whether to detain civilians. In all circumstances, treat those who are not taking an active part in hostilities, including prisoners and detain-

ees, humanely. Remain particularly aware of heightened sensitivity in the Iraqi and Islamic cultures concerning the treatment of females. Whenever possible, females will be searched in a nonpublic location; unless absolutely necessary, male soldiers will not search females.

3. Force Protection and the Lawful Use of Force. We are conducting combat operations in a complex, dangerous environment. Coalition Forces must remain bold and aggressive, yet disciplined, in their use of force. When in contact with the enemy, use only that force necessary to accomplish the mission while minimizing unintended damage. Our posture must be strong and determined, while remaining firmly in control of the destructive power of our weapons. You have learned the principles of the law of war throughout your military careers. As professional soldiers, you must follow them, and comply with the rules of engagement. Before engaging any target, you must be reasonably certain that it is a legitimate military target. Self-defense is always authorized . . .

5. A summary of "Rules of Proper Conduct During Combat Operations" is enclosed. This memorandum and the Rules of Proper Conduct will be distributed down to the platoon level. Leaders will ensure that all CJTF-7 personnel are trained on the Rules of Proper Conduct. Additionally, leaders will ensure that all CJTF-7 personnel receive refresher training on the rules of engagement, which includes training on the disciplined use of force . . .

6. Conclusion. Respect for others, humane treatment of all persons, and adherence to the law of war and rules of engagement is a matter of discipline and values. It is what separates us from our enemies. I expect all leaders to reinforce this message.

RICARDO S. SANCHEZ
Lieutenant General, USA
Commanding

CHAPTER 18

The Shia Rebellion

By March 2004, concern had reached a peak in Washington about influence on the Iraq Shia population by Muqtada al-Sadr, whose militia, the Mahdi Army, had grown to between 5,000 and 10,000 in number. Both the National Security Council and the Department of Defense held multiple discussions and came up with their own plans about how to mitigate his impact and possibly kill him. Paul Wolfowitz, one of the government's point men in the effort, called and we discussed the strategy.

"We're calling it the 'boa constrictor' plan," he said. "We want to eat away at al-Sadr's support base across the country. His real strength, we believe, comes from his lieutenants. So our idea is to gradually take them out and then get al-Sadr himself. What do you think?"

"Well, we can do that," I replied. "We know who his lieutenants are, and whenever we have the opportunity, our forces will be more than glad to launch these kinds of operations. Of course, there will be a reaction from al-Sadr, so we'll have to obtain approval from Washington. But we can handle it."

"I don't think obtaining approval will be a problem," said Wolfowitz.

Soon we were developing plans to identify, track, and capture al-Sadr's key lieutenants, beginning with Mustafa al-Yaqoubi, whom we knew could be taken without too much trouble. Our idea was to grab him in a predawn raid at his home in Najaf and then move him to an Iraqi prison up north near Mosul.

Also during this time frame, the weekly newspaper *Hawza*, which was operated by Muqtada al-Sadr's organization, ran a series of articles suggesting that violence against coalition forces was more than appropriate. The paper (which had a circulation of 50,000) also took a shot at Ambassador Bremer by stating that he was following in the footsteps of Saddam Hussein. I'm not sure what set Bremer off, but he became very upset and started talking about sedition and inciting violence. "We've got to shut that damned newspaper down," he said. "I'm not going to tolerate this."

On March 27, 2004, Bremer gave the order to close *Hawza* for sixty days. After coordinating with CENTCOM, CJTF-7 executed the order. The brigade commander went in, told all the people to go home, and then chained and locked the doors. There was no resistance and no violence. Over the next few days, however, thousands of al-Sadr followers staged angry but nonviolent demonstrations in the streets of Sadr City. "Where is democracy now?" they shouted. "What about freedom of the press?"

At virtually the same moment these demonstrations were taking place, on March 31, 2004, a two-vehicle convoy of private security company Blackwater USA employees was ambushed while traveling through the Sunni stronghold of Fallujah. A group of masked gunmen opened fire with assault weapons and grenades and killed all four of the Americans. A mob then mutilated, burned, and dragged the bodies through the streets of the city, finally hanging them from a bridge that crossed the Euphrates River. As horrible as this was, such an incident was not considered a major tactical setback, especially because the Blackwater contractors had been previously warned about going through downtown Fallujah and should not have been there in the first place. However, because the gruesome images of charred corpses hanging from the bridge were broadcast back in the United States, the resulting media frenzy and emotional firestorm of controversy elevated the Blackwater incident to top priority at the White House. When Democrats in Congress took advantage of the situation to attack the administration's war policy, the White House had to react.

There was consensus across the board from civilian and military leaders (including me) that we had to respond with force. Our reasons, however, were very different. My view was that the Republicans were concerned about being only seven months away from the presidential election. Were they going to look strong or were they going to look weak? A forceful response was a way for the administration to show resolve. Military leaders (including me) believed that we had to go on the offensive to prove to the insurgents that we would stand up against these attacks on Americans. Force seemed to be one of the few things the Iraqi insurgents clearly understood. The events also opened up an opportunity for us to eliminate some of the threats posed by Fallujah, which had been used as an IED-production sanctuary and was a known safe haven for insurgents.

In the early days of April, there was a flurry of communications between Washington and Baghdad—telephone calls, e-mails, and video teleconferences (VTCs)—all to sort out the coalition's response to Blackwater. Some of the discussions were, at times, both contentious and lively. Secretary Rumsfeld, for instance, was particularly forceful in his demand for action. "We've got to pound these guys," he said. "This is also a good opportunity for us to push the Sunnis on the Governing Council to step forward and condemn this attack, and we'll remember those who do not. It's time for them to choose. They are either with us or against us."

While there was a steady drumbeat from Washington to take swift action, there was also caution on the part of the military with regard to timing. The Marines, for example, were reluctant to launch too quickly. They had recently transitioned into Fallujah, taking over from the 82nd Airborne, and were concerned about situational awareness and composition of forces. The Marines were also settling into the western part of Iraq, and peacefully engaging with the Sunnis in what they called a "kid glove" approach, which they had used successfully in the south with the Shia population. Their overall strategy was to practice the same philosophy in Anbar province, which included Fallujah. It was a good idea and I supported them.

Despite our concerns, Secretary Rumsfeld directed CENTCOM and CJTF-7 to begin planning for an immediate offensive into Fallujah. Accordingly, we developed our attack plan, determined the forces necessary, looked at the timing, and weighed the risks. In a VTC with Rumsfeld and Bremer, GEN Abizaid presented the overall plan and stated very frankly that CJTF-7 and CENTCOM agreed with the Marines. "The timing is not right and they haven't had time to implement their engagement program," said Abizaid. "We should wait."

"No, we've got to attack," replied Rumsfeld, obviously reflecting the mood of the NSC and the White House. "And we must do more than just get the perpetrators of this Blackwater incident. We need to make sure that Iraqis in other cities receive our message."

Our final step in the decision-making process was to review our recommendations and strategy for the President and the National Security Council. In that VTC, we presented our plan for Operation Vigilant Resolve. Our missions, as approved by the White House and the NSC (which were also involved in defining them), were: (1) eliminate Fallujah as a safe haven for Sunni insurgents; (2) eliminate all weapons caches from the city; (3) establish law and order for long-term stability and security; and (4) capture or kill the perpetrators of the Blackwater ambush. The Marine Expeditionary Force would lead the effort and be joined by the ICDC and elements of the new Iraqi Army. Attacks would be powerful, precise, and sustained. Our best estimate, we told the President, was three to four weeks of intense fighting.

We also estimated impacts on the population, devastation in the city, and the monetary costs of rebuilding. GEN Abizaid further made it clear that we preferred not to launch the attack right now. He pointed out that the Marines would like to have time to implement their "kid glove" strategy with the Sunnis, and we supported their recommendation. President Bush stated that he appreciated our caution, but then ordered us to attack.

We acknowledged the President's decision, but then reaffirmed that the Fallujah offensive was going to be a pretty ugly operation, with a lot of collateral damage—in both infrastructure and the inevitable civilian casualties. We also expressed our concern regarding the Arabic television network Al Jazeera, which was sure to broadcast live reports on the battle. One reporter was already in there and there was no way we were going to be able to get him out. In turn, that would create a major strategic communication problem for us in the Arab world. "If we're going to proceed, we must be prepared to counter Al Jazeera with a coordinated strategic communications plan," we said. Such a plan would require the development of a strategy at the interagency level for communicating to the world, the nation, and the Iraqis our key objectives. Part of this effort would require a near-real-time capability to counter the inevitable disinformation to be released by Al Jazeera.

All heads around the table nodded affirmatively. "Yes, we understand," President Bush replied. "We know it's going to be ugly, but we are committed."

"Very well, Mr. President. Then Operation Vigilant Resolve is a go."

As soon as I got out of that video teleconference, I called Jim Conway, commanding general of the 1st Marine Expeditionary Force. "Jim, the decision has been made to execute Vigilant Resolve. We communicated your concerns to the President," I said. "But we are launching the offensive anyway."

"Okay, General," he replied. "I don't like it, but we're prepared to execute."

. . . .

CONCURRENT WITH OUR DISCUSSIONS about Fallujah, we were also refining our analysis and planning for Wolfowitz's "boa constrictor" strategy in regard to Muqtada al-Sadr and his lieutenants. After completing our plans and briefings, GEN Abizaid and I had a long conversation about the operation and its possible ramifications. "Ric, what do you think about striking on two different fronts at this time?" Abizaid asked.

"Well, sir, it can get tough on us," I replied. "But we've been looking at this guy for six months now, and every bit of intelligence we have says al-Sadr will respond with some demonstrations and rioting. There will be violence, but I've discussed it with my subordinate commanders, and they all believe we can handle it. We expect some major demonstrations, but not a general call to arms. Also, I'm not sure there will ever be a better time to strike at al-Sadr, because we're right in the middle of transitioning U.S. forces. We have an overlap and, at our peak, we'll have a maximum number of 170,000 troops on the ground."

"Okay, Ric," Abizaid replied. "I'm not sure we should be going on two different fronts in this environment. But given what you said, I'll support you." GEN Abizaid received approval from Washington (all the way up to the level of the President), and then issued the order for CJTF-7 to launch against al-Sadr's lieutenant. In retrospect, however, John Abizaid's instinct was right on the mark. His red star cluster was going off and I should have listened more carefully to him.

Four days after the Blackwater incident, on April 3, 2004, Special Forces executed the predawn raid in Najaf at the home of al-Sadr lieutenant Mustafa al-Yaqoubi, arrested him, and spirited him off to a prison in northern Iraq. Everything went according to plan. But with this action coming on the heels of *Hawza*'s shutdown, Muqtada al-Sadr immediately rose up and put out a call to arms. "I and my followers of the believers have come under attack from the occupiers, imperialism, and the appointees," he said publicly. "Be on the utmost readiness, and strike them where you meet them."

All hell broke loose over the next two days, as al-Sadr put out more direct calls for his followers to attack coalition troops and government buildings across the country. The Mahdi Army came out very aggressively in Sadr City by taking control of five police stations in one night. On Sunday, April 4, 2004, a platoon of the 1st Cavalry that had just taken over their sector from 1st Armored Division was ambushed on a routine patrol and suffered heavy casualties. In fact, the ambush turned into a huge battle in which we had to deploy major reinforcements. In the After Action Report, we learned that the soldiers in this new platoon were totally unaware that the operation to seize al-Sadr's lieutenant had taken place or that it had the potential to create violence in their sector. There had been a major breakdown in communications somewhere between the division commanders and the soldiers on the ground. As a result, they went into a volatile area at the wrong time and were decimated (suffering eight dead and many more wounded). The incident reinforced the tremendous vulnerability of units in transition—and at this particular point in time, we were transitioning troops all across Iraq.

Al-Sadr's violent attacks did not stop in the Sadr City area of Baghdad. His forces extended their operations across southern Iraq and seized control of four provincial capitals—Najaf, Al Kut, Nasiriyah, and Basra. In Nasiriyah, they moved to control key roads and engage Italian coalition forces. In Basra, they aggressively took over government buildings and pushed back the English and Dutch troops stationed in the city. And in Najaf, the Mahdi Army staged a major attack on the ICDC compound and killed many of their own people. Muqtada al-Sadr stationed himself in the Great Mosque in the nearby holy city of Kufa, while hundreds of his militia drove out the Iraqi security forces. Al-Sadr had routinely delivered his political sermons during Friday prayers at the Great Mosque there, and he was shortly to tell his followers, "Every person has to take a stand, either with us or against us. Neutrality doesn't exist between us and the Americans."

Overall, the fighting was intense and bloody. Al-Sadr's people staged coordinated and synchronized attacks against high-value targets all over Iraq. During the first few days of this Shia rebellion, it became painfully clear that our intelligence assessments concerning Muqtada al-Sadr's resolve and capabilities were terribly wrong. We had underestimated the enemy and were now paying the price for that failure.

Serving to further complicate events was the fact that we really had no offensive military capability in southern Iraq, because the multinational coalition forces had too many restrictive rules of engagement. So when Muqtada al-Sadr's supporters launched their attacks, a number of the coalition forces deserted their posts. The Ukrainians, for example, left the bridges they were supposed to protect in Al Kut, withdrew back to their compound, and assumed a defensive position. That left some of our Special Forces and CPA personnel at great risk, so we depended on close air support to protect them for a period of time. The overall lack of offensive military capability in the South was precisely why al-Sadr's supporters were able to seize control of the four provincial capitals.

On April 4, 2004, right in the middle of all this turmoil, I was at my command post in the Green Zone when we began receiving reports over satellite radio from a young Marine Corps Reserve major at the CPA building in Najaf. "There are hundreds of Iraqis attacking the compound," he said. "With each wave, they get more aggressive. The Spanish forces have abandoned their posts and left us to fend for ourselves. They are a bunch of cowards. Some of the Iraqi Civil Defense Corps have been killed and their compound has been overrun. We need help. Repeat. We need help."

Ambassador Bremer had received a call from his regional coordinator inside the building and he, too, called me to express his concern about what was happening in Najaf. The situation sounded so dire that I immediately ordered close air support (Air

Force jet fighters) and the deployment of attack helicopters from Baghdad to Najaf. But when the pilots flew over the area, they radioed back that they could see no enemy activity on the ground below them and that there was nothing to shoot at. Thinking that the fighting had stopped, we radioed back to the young major, but again he responded frantically. "We are still being attacked," he said. "Fighting everywhere. This may be the last radio call we can make before we get overrun. Send help."

The information was too conflicting, and I wondered what the hell was really going on. "What reports are we getting from the coalition division commander's headquarters?" I asked my executive officer.

"Sir, they're reporting enemy contact but it's not critical," he replied. "However, the Polish commander is in Babylon and it has been difficult getting clarity on the ground situation."

We could not allow a major defeat to take place at the CPA compound in Najaf. However, sending U.S. ground forces from another area to Najaf was not an option, because they would never arrive in time to assist in the battle. So I turned back to my executive officer. "Get my helicopter ready," I said. But he had anticipated my move.

"Sir, it's already sitting on the pad revving up," he quickly responded.

"Okay, let's go to Najaf."

When we got into the helicopter, I asked the pilots how long it would take to get there. "Middle of the afternoon, sir. Good weather. Should take about twenty to twenty-five minutes," came the response.

"Well, go as fast as you can," I said. "This sounds like a bad situation and we need to get there ASAP."

During the flight, we monitored communications coming from the compound and the described events continued to get worse and worse. "Being attacked from the South. We're about to be overrun and killed," said the major. "If you lose our signal, you'll know what happened."

As we approached Najaf, the pilots asked where I wanted to go. "I want to go into the compound," I said.

"But sir, with these reports, that'll be right in the middle of a firefight."

"I don't care! Get in there! Figure out how you're going to drop me off!"

Our executive officer radioed the major that Victory Six, my helicopter, was inbound. The response came back: "Hell, we can't have the corps commander in here in the middle of a fight."

"Well, get ready. He'll be there in two minutes."

Our pilots decided to approach from the north. They skimmed along close to the ground and then, all of a sudden, popped up above the compound walls, and landed in the courtyard. I jumped out with a couple of men and went into the CPA building as the helicopter took off. I heard a couple of shots in the background, but nothing that sounded like a major firefight.

Once in the building, I quickly found the young major reporting on his radio. "What is your name?" I asked. He identified himself.

"Where is the main attack on the compound?" I asked.

"Sir, it's right outside the front gate."

"How are you getting your reports?"

"Well, these Blackwater and CPA guys are telling me what's happening."

"Have you looked out there yourself to see what's going on?"

"Well, no, sir. I'm just reporting what I'm being told."

"Okay, Major, how do I get to the roof of this building," I asked.

"Oh, you don't want to go up there, sir. We just had people shot on the roof. We're taking sniper fire."

My men and I climbed two stories up and opened the door to the roof. The first things I saw were a couple of used plastic IV bottles and a pool of blood on the floor. Then I noticed two soldiers huddled behind the roof's wall. "I'm going to go talk to that soldier right there," I said pointing to a young sergeant. "Oh, no,

General, don't do that," said the major. But I ran across the roof and dropped down next to the sergeant.

When he saw the three stars on my helmet, he said, "Jesus, sir. What are you doing here?"

"Tell me what is happening, Sergeant."

"Aw, sir, we've got this damned sniper up on the top of the hospital and we can't get him."

"I've been getting reports that there are hundreds of Iraqis attacking."

"Well, sir, there was a mob out there a long time ago. But they're gone."

"Are you sure?"

"Oh, yes, sir. We're just trying to get this sniper now. Two of my soldiers got hit. But they're not critically wounded, sir. They'll be okay."

"That's good. Do you have anybody helping you?"

"Yes, sir. There is another unit moving up through the hospital to flank the sniper."

"Okay, Sergeant, looks like you've got the situation well under control," I said. "Why don't you fire a couple of shots at that guy to give me some cover so I can get back to the door." So he fired five or six shots in that direction and I ran back across the open roof and went into the building.

As soon as I got back downstairs, I went across the compound to see the Spanish brigade commander and his soldiers. "Sir, the base was attacked some time ago, but we repelled the enemy. However, the ICDC compound [about a mile away] was overrun and is now controlled by the enemy. One of our platoons got into a pretty good fight there, ran out of ammunition, and pulled back. We have dispersed our forces around the compound and are in good defensive shape."

"I received a report that you had deserted your posts and left the Americans to fend for themselves," I said.

"Not true, sir," replied the commander. "Those Blackwater and CPA guys wanted us to put all of our troops back and surround their building. But we didn't need to do that, because there was never any threat of being overrun. Besides, it was better for us to protect the entire compound rather than just one building."

"Very well, Colonel," I replied. "Thank you very much for the update. Keep up the good work."

Before flying back to Baghdad, I sat down with the young Marine major who had been doing all the broadcasting and chewed him out. "Major, nothing you were telling us before I got here was true," I said. "You were not being attacked by hundreds of Iraqis. The Spanish are not cowards. And you were in no danger of being overrun. These civilians were not providing you with accurate information."

"Oh, sir, I didn't realize," he responded.

"Major, *that's what we do*! If you are the senior man on the ground, and you are reporting back to headquarters, you better damn well have good situational awareness. That's what we count on you for. Do you understand?"

"Yes, sir. I understand. It won't happen again, sir."

"I'm sure it won't."

When I got back to the Green Zone later that afternoon, I went in to see Ambassador Bremer and told him the whole story. "There was a mob," I said. "The compound did get attacked, but the enemy was repelled. We took some casualties. But we were never in any danger of being overrun, losing your regional administrator, or having all your people killed. In fact, we were far from it."

"Well, that's not what my guys were saying," Bremer replied.

"I know. But I flew down there to see for myself. They were clearly exaggerating."

Although the Ambassador didn't want to believe it, what had really happened was that the CPA personnel had panicked and the Blackwater civilians were aggravating the situation by having the young major relay bogus information. The CPA, including Bremer, didn't trust the Spanish, so they had exaggerated the reporting in order to get CJTF-7 to deploy American troops.

During my unscheduled trip down to Najaf, the CJTF-7 staff continued crisis action planning to quell the violence and reestablish control in southern Iraq. There was no question we were going to have to attack the Mahdi Army. We could not allow al-Sadr to remain in control of key government buildings in Basra, Al Kut, Nasiriyah, and Najaf. The question was how to do it, given the facts that we were getting ready for a major operation in Fallujah and we had no offensive capability in the south (except with the UK and Italian forces). We were obviously going to have to move American combat forces down there due to the lack of offensive capabilities with the international troops. But which ones? The Marines were locked into Fallujah with some Army and Air Force support. The 1st Armored and 1st Cavalry Divisions were both half in and half out of Baghdad. First Armored, which was leaving the country, already had troops in Kuwait and back in Germany. And it was the same for 1st Cavalry, which was on the way into Iraq. So which division should we send down south? If we deployed 1st Cavalry we'd have a unit with no situational awareness going on the offensive. And if we moved 1st Armored, we'd be removing from Baghdad the unit with the best situational awareness in an area where there was major instability. By taking troops from one place, we'd be leaving that area open to further danger. Based on an extensive risk assessment and the analysis of the options available, I ordered two operational-level unit movements.

First, I called MG Marty Dempsey and informed him that I was working with Abizaid to keep the 1st Armored Division in the country for another ninety days. Their mission would be to disengage from Baghdad, bring back the elements already in Kuwait and Germany, and lead the main effort in reestablishing control of the south. It was my former division, so I didn't hesitate to give the order. The impact of this decision was not as great as it might have been on another organization. The stage had been set for such a possibility the previous year when the soldiers and their families were told to be prepared to stay beyond the one-year mark, if that's what it took. In addition, GEN B. B. Bell's headquarters and the 1st Armored Division rear detachments all worked hard to make sure that families back in Germany understood the situation. Everybody at 1st Armored handled the news like true professionals, and I was supremely proud of my former organization.

The second major move we made was to order the Stryker Brigade down from Mosul. This unit was comprised of Strykers, which were highly mobile all-wheel-drive eight-wheeled combat vehicles originally designed for swift and maximum maneuverability on the battlefield. Our plans were for them to initially move into the Diyala province, then Baghdad, and, finally, all the way down south to engage in operations with the 1st Armored Division. Within a matter of hours, the Stryker unit was on the move and available for employment in Diyala. It was the first time a Stryker formation would conduct an operational maneuver of over 300 kilometers and perform exactly as designed.

One of the final steps in our attack planning process was to declare Muqtada al-Sadr's Mahdi Army a hostile force. For that, we had to work a justification process through Bremer, Abizaid, Rumsfeld, the Joint Staff, and the White House. But everybody came on line and agreed to the move. Now, rather than having to wait for the Mahdi Army to attack us, our military units had the ability to engage as soon as they were identified. And it was pretty easy to spot them in their black "pajama" uniforms with black headbands, carrying AK-47s.

Once we had completed all of our planning, CJTF-7 held a formal briefing for GEN Abizaid to present the entire battle strategy and to seek approval for the retention of 1st Armored Division and other combat forces to fight down south (which he quickly obtained from the Secretary of Defense). GEN Abizaid also worked the details of the mission statement through his channels to the Department of Defense, as did Ambassador Bremer through the National Security Council. During this process, the White House got deeply involved in the specific wording of our missions, especially when it came to killing or capturing

Muqtada al-Sadr. The final decision rested directly with the President of the United States. Once finalized, our specific missions were: (1) capture or kill Muqtada al-Sadr; (2) defeat the Mahdi Army; (3) restore stability in the southern provinces; and (4) assist CPA in reestablishing civil authority and security.

There was no VTC with the President discussing our movement to the south, as there had been for Fallujah. However, a decision was made relatively quickly, and as soon as Abizaid received approval, he called me. "Ric, you've got the authority to engage the Mahdi Army and to capture or kill Muqtada al-Sadr," he said. "Execute as planned."

As we engaged in combat on two fronts over the next couple of weeks, operations at CJTF-7 became extraordinarily complex and challenging. We not only had to deal with all the military aspects, we also had to interact with the CPA, the Iraqi Governing Council, and the chain of command back in Washington. Moreover, there were several major transitions going on in Iraq all at the same time, which considerably added to the pressure. The entire coalition force was either moving in or out of the country or, as in the case of 1st Armored, caught in stride. It was part of our normal rotation of forces, but it led to a huge drop in situational awareness across the country, because there was just no way to quickly replace the experience and instinct that had been built up over the course of a year. CJTF-7 was also planning for the transition to a new four-star headquarters, which had to be completed before the transfer of sovereignty.

The one thing in Iraq that could not be put on hold was the timeline for the transfer of sovereignty. We were only ninety days away from the July 1 deadline and we simply had to continue all of the programs involved, including dealing with the vast array of political initiatives and dynamics. The United Nations was in the country. The new embassy was being set up. The individual nations of the coalition were being ignored. The Iraqi Governing Council was upset with the timing of the moves to fight in both Fallujah and the south. The CPA was strapped for help.

On April 6, 2004, I joined Ambassador Bremer for a meeting with a newly formed subcommittee of the Governing Council, which was set up to coordinate and synchronize security operations with the coalition forces. When its members began questioning the need for action against al-Sadr, Bremer was adamant in his response. "Muqtada al-Sadr and his Mahdi Army have declared themselves enemies of the Iraqi people," he said. "This challenge must be addressed swiftly, with no ambiguity, and without leniency. Failure to respond will send a clear signal that violence is effective against the coalition." Bremer's statement essentially summed up the Bush administration's position on the entire matter.

Before launching our offensive to regain control of government buildings in the four southern provincial capitals, CJTF-7 conducted an overall assessment of our enemies on the ground. First, we had the former members of Saddam Hussein's regime, who were leading the main insurgency. While their attacks against coalition forces were still strong and consistent, they had also started attacking al-Sadr's Shia supporters. Essentially, they were fanning the flames of Shia anger and trying to blame it on coalition forces. Second, of course, we were fighting the Shia extremists associated with Muqtada al-Sadr. Third, there were the Sunni extremists (led by Abu Musab al-Zarqawi) not associated formally with Saddam's former Baathist regime. This enemy was located primarily in Fallujah and in other areas of the Sunni Triangle. Finally, there was a small but steady stream of foreign fighters who had entered the country. At this time, however, we had no reliable intelligence that they had any formal linkage to the al-Qaeda terrorist organization. These enemy intelligence assessments were ongoing and continuously passed back through the chain of command to Washington, including the Pentagon, the National Security Council, and the White House.

At a time when our military forces were preparing to fight a two-front war in Fallujah and southern Iraq, I believe the Bush administration was engaged in a two-front war of another kind. One front was the war in Iraq, which, although important, was given secondary priority. Their main concern was the war to retain power at home. This front was opened in the fall of 2003 when the decision was made to transfer sovereignty by July 1, 2004. And as events in the spring of 2004 unfolded, it became abundantly clear that there was nothing that would keep the Bush administration from winning the upcoming presidential election. Nothing.

CHAPTER 19

Fallujah and the Onset of Civil War

My Humvee drove up the narrow road leading into the cemetery—past the grave markers, the headstones, and the tombs—until we reached a small structure that reminded me of the first house I had lived in as a kid back in Rio Grande City. It was only one room, about fifteen feet wide by twenty-five feet long. It had no doors, no windows, and, even worse than my first house, it had no roof. It was just a shell of a structure that the Marines had chosen as the site for our meeting. As I walked in, I was greeted by MAJ GEN James N. Mattis, commander of the 1st Marine Division. Also present were the regimental commander, the battalion commander, several staff officers, and a couple of other Marines.

"How are you guys doing?" I asked.

"Very well, General Sanchez"; "Good, sir"; "Couldn't be better, General," came the replies.

The early afternoon sun beat down on us as the Marines pointed to their maps and gave me a brief overview of their operational plan for the launch into Fallujah, which was set to begin in a few hours. We could also hear a number of enemy mortar rounds explode on the outskirts of the cemetery, a block or so from the nearest Sunni neighborhood.

"This is a terrific plan," I said, after they had completed their presentation. "Is there anything else you need in the way of supplies, support, or troops? We have things going on elsewhere, but if necessary, we can move additional forces in to help you."

"Thank you, sir, but we have sufficient forces to accomplish the mission. It's going to be a tough fight, but we'll be okay, and I believe we have everything we need."

"Very well," I replied. "But I want you all to know that we've got other offensive operations supporting you on this operation. If there's anything you need, anything at all, let us know immediately and you'll have it."

"Yes, sir, General Sanchez." "Thank you, sir."

"Good luck, men. And God speed."

It was April 6, 2004, and there was no question in anybody's mind about the significance of this mission. All the plans were in place and all the approvals had been given (all the way up to the President of the United States). Now it was time to execute. My meeting with the Marines lasted only fifteen or twenty minutes. But I felt it was important for me to be right there with them before they attacked. I not only wanted to get a good idea of how they were going to conduct their offensive, but I needed to confirm they were properly resourced down to the regimental level. It was also important for me to make sure these warriors knew that I cared about them and would be there for them in the toughest of times. It was all part of being in the box, of leading under fire.

Over the previous couple of days, we had cordoned off Fallujah by blockading all main roads leading in or out. In response to these and other preliminary combat operations, nearly a third of the civilian population fled the city. So when more than 2,000 Marines launched the attack, everybody in Fallujah knew what was coming. The fighting was house-to-house and door-to-door. It was fierce, bloody, and extraordinarily aggressive. Close air support was an integral part of the offensive, including Air Force AC-130 Spectre gunships and fighter aircraft. Hellfire missiles were shot from Super Cobra helicopters and pinpoint air strikes were carried out. The destruction was massive in certain parts of the city and, in addition to inflicting heavy enemy losses, a lot of civilians were injured.

For the first time, we incorporated Iraqis into major combat operations as backup for the Marines. The two Iraqi Army battalions seemed to be properly trained and ready for battle. However, their buses were attacked upon leaving the compound and they refused to continue their movement to Fallujah. Their participation was a flat-out failure. The ICDC units fared a little better, but not much. One battalion, a Special Forces–trained group of mostly Kurds from the north, performed extremely well. However, having Kurds attack Sunnis created another political controversy that we certainly did not need. Shia and Sunni members of the ICDC from other areas, for the most part, would not even come into Fallujah. "There is no way we are going to fight against our brothers," they said. "We committed only to protecting our region."

On the second day of the offensive (April 7, 2004), the prominent Sunni religious leader Shaykh Abdul Qadr al-Ani demanded a unilateral withdrawal of coalition forces from Fallujah to the outskirts of town. That would provide time for the dead to be collected and buried, the wounded to be taken to hospitals, and for humanitarian aid to enter the city. Afterward, al-Ani advocated a dialogue to resolve political and military issues. The CPA, however, immediately shot back a communication that a unilateral withdrawal was not possible, and that the Sunnis should lay down their arms and surrender to the coalition forces.

That afternoon, Ambassador Bremer and I sat in on a VTC with President Bush and the NSC. After I provided an update on Fallujah and the situation in the south, Bremer gave an upbeat assessment of the Iraqi Governing Council's support. "They've been quite good," he said. "They've issued another positive statement. But [leading Shia cleric, the Grand Ayatollah Ali al-Sistani] has also issued three statements. He doesn't believe we're serious about taking out al-Sadr. He has a 'we've-heard-you-guys-before-and-you-didn't-do-anything' kind of attitude."

Then Secretary of State Colin Powell spoke up. "There's a broader and deeper problem here," he said. "The masses are willing to come out and pile on when there is a Humvee burning. We've got to smash somebody's ass quickly. There has to be a total victory somewhere. We must have a brute demonstration of power. This is broader than anything we have seen."

"Right, we have to strike wherever we find them," agreed President Bush. "The Mahdi Army is a hostile force. We can't allow one man [meaning Muqtada al-Sadr] to change the course of the country. It is absolutely vital that we have robust offensive operations everywhere down south. At the end of this campaign, al-Sadr must be gone. At a minimum, he will be arrested. It is essential he be wiped out."

"What do we call this, anyway?" asked Secretary Rumsfeld. "Is this high intensity, low intensity? What?"

Before anyone could answer Rumsfeld's question, President Bush launched into what I considered a kind of confused pep talk regarding both Fallujah and our upcoming southern campaign. "Kick ass!" he said, echoing Colin Powell's tough talk. "If somebody tries to stop the march to democracy, we will seek them out and kill them! We must be tougher than hell! This Vietnam stuff, this is not even close. It is a mind-set. We can't send that message. It's an excuse to prepare us for withdrawal.

"There is a series of moments and this is one of them. Our will is being tested, but we are resolute. We have a better way. Stay strong! Stay the course! Kill them! Be confident! Prevail! We are going to wipe them out! We are not blinking!"

To say that the Fallujah offensive angered the Sunni Muslims of Iraq would be a gross understatement. Up to this point, many had still been on the fence and were working with us to create a more stable government. But when the images of destruction were broadcast on Al Jazeera, most Sunnis felt Fallujah was an attack on their very existence. It appeared that U.S.-led coalition forces would spare no effort to eliminate them. After all, not only were we using all our combat power, but other Iraqis, both Kurds and Shiites, were part of the attack. Now they felt there was no hope for a better Iraq, and no choice but to fight back.

When tribal leaders put out a call to arms, Sunnis everywhere responded, and the Sunni Triangle exploded in violence. There were synchronized and well-targeted actions in central Iraq along the Lower Euphrates River, in the Diyalah province, in Baghdad, north to Mosul, and south to Karbala. The fighting, again, was very heavy and very bloody. The Sunnis took aim at bridges and roads along the rivers. At one point, our logistics flow was so disrupted that civilians in the Green Zone began complaining about not having the full complement of Brown and Root menus in the mess hall. "Don't even consider this a problem," I told the staff when the complaint surfaced. "Give them MREs and they can survive. Right now we have to make sure we get fuel and ammunition to the troops. That's our top priority."

The new fighting, on top of what was already going on in the Shia areas of Sadr City and southern Iraq, plunged the entire country into violence at some level. Coalition forces were fighting Sunnis, Shiites, and Saddam's insurgents all at the same time. The insurgents were fanning the flames by attacking both Shia and Sunni tribes and blaming it on the coalition. Sunni and Shia tribes were fighting each other. Shia-on-Shia infighting was just below the surface between the Mahdi Army and SCIRI militias. The Badr Corps was itching to get involved. And in the area south of Baghdad where the Sunni Triangle overlapped with the northern portion of the Shia front, Sunnis and Shiites were actually fighting *together* against coalition forces. We were now in the middle of a civil war. And what's more, we had created these conditions ourselves.

The Sunni rebellion would not have been of such magnitude had it not been for the Arabic television network Al Jazeera. As soon as we launched into Fallujah, it started broadcasting images of the devastation. And I have no doubt the network was complicit with the enemy. Again and again, Al Jazeera's television cameras and reporters showed up just at the right time to record major attacks against coalition forces, which, of course, elicited strong responses. The intensity of the fighting, selective editing, and a reporter who consistently portrayed the side of the Sunnis incited resistance against the coalition. And it wasn't just the opposing forces and neutral elements that were motivated to act. We also began receiving significant amounts of political and emotional pressure from our own side.

It started with a "sky is falling" attitude from the civilians at the CPA, who not only were upset that their food menus had been impacted, but were truly concerned that the Green Zone was about to be overrun. There was also some loose talk among some coalition forces and CPA staff about the evacuation of noncombatants. However, from both an operational and strategic level, at no time did coalition military leadership ever feel that we were at risk of losing a battle. The panic was caused, for the most part, by the Al Jazeera television broadcasts, which Abizaid and I had warned the administration about. Washington had done nothing to coordinate strategic information activities that would counter Al Jazeera's impact.

Lakhdar Brahimi also reacted to the press coverage by threatening to halt the UN effort and leave the country if the fighting did not stop immediately. "The political process is in danger of becoming irrelevant, because there's a war going on," he told Ambassador Robert Blackwill. Groups of Sunni tribal leaders also came in and protested. "This is wrong and you must stop," they said. "You're killing our people." The Sunni members of the Iraqi Governing Council were absolutely outraged at the amount of civilian casualties. "These operations by the Americans are unacceptable and illegal," declared Adnan Pachachi.

All of these complaints sent Paul Bremer into a tailspin, especially when most of the Sunni council members threatened to leave the government. The Ambassador was forced into a continuous effort to prevent the council from completely collapsing. Eventually, he persuaded them to suspend their membership rather than terminating it. In addition, he got the Sunni members to commit to an effort to communicate with the enemy and try to achieve some sort of peaceful solution.

At one point, Bremer called me and asked, "What the hell are you guys doing on the ground?"

"Sir, our mission is clear and it hasn't changed. We're executing according to our plan," I responded.

"How much longer will it be before you complete the plan? Are you about to finish?"

"No, sir, we are not about to finish. Remember, we said it would take several weeks. This was the plan we briefed to you, the NSC, DoD, and the President."

"Well, this is bad. This is bad. We're going to have to make some adjustments."

In addition to Bremer, coalition member nations were putting tremendous pressure on us to stop the fighting. In the early days of the Fallujah offensive, it became very apparent that the U.S. government had not cleared the decision to launch with the political leadership of the coalition nations. And the leaders of these nations were upset about it. My British deputy commanding general had been involved in all of the internal planning, and all of our coalition commanders were full partners in the execution of our offensive plan. But their entire philosophy was very different from that of the Americans. When al-Sadr's militia overran Basra, for instance, the UK troops simply let them occupy the political buildings, and then tried to negotiate their withdrawal.

The British three-star general on the CJTF-7 staff participated in all our planning sessions and communicated our intentions to London on a daily basis. Consistently, he voiced his government's concern about our planned offensive, and I'm certain that lively discussions took place between the White House and 10 Downing Street. London believed that we were being far too heavy-handed, but President Bush still gave the order to launch.

Further adding to the angst among coalition nations was the

fact that the insurgents had instituted a new campaign to kidnap various foreign diplomats and either hold them for ransom or kill them outright. The Italians and Koreans were targeted first, and they voiced their concerns to us in emphatic terms.

So within a matter of twenty-four to thirty-six hours, the Bush administration began to receive pressure from all sides—from the coalition nations, from the UN effort in Iraq, from the Iraqi Governing Council, from Bremer and the CPA, and, most important, from the American people and the Democrats in Congress, who were viewing images of the fighting via CNN, MSNBC, and the various other national networks. E-mails began flying back and forth between Washington and Baghdad. After a series of phone calls, it became apparent that nearly all the civilian leaders were having doubts about continuing the offensive. While GEN Richard Myers, GEN Abizaid, and I reminded everybody that we were executing exactly as ordered and that everything was going according to plan, we still had several more weeks before we could achieve our mission. But we were fighting a losing battle. With the transfer of sovereignty at risk and the presidential election weighing on their minds, the White House and the NSC were already backing away from the decision to go into Fallujah.

On the afternoon of April 8, 2004, almost exactly forty-eight hours after I had met with the Marines in the Fallujah cemetery, Ambassador Bremer, GEN Abizaid, and I met in Bremer's office at the Republican Palace. "Ric, it's been decided that you've got to stop your offensive operations and withdraw from Fallujah immediately," said the Ambassador.

I had no doubts that this decision had been handed down by the National Security Council and the President. Bremer would never have dared to convey such guidance without the full force of the administration behind him. I looked over at GEN Abizaid and waited for him to respond first.

"I can understand the political situation," he said.

"Well, I do, too, sir," I replied. "But we're fighting toe-to-toe with the enemy. We're in the streets and in middle of neighborhoods. We can't stop now. If we don't finish the mission, we're going to have to come back and do it later."

"You need to withdraw your forces right now," Bremer reaffirmed. "Otherwise, you're going to screw up the political process."

"No, I can't do that," I replied. "Currently we control less than 50 percent of the city and we are in contact with the enemy across the entire front. The enemy is attacking and defending itself effectively. If we pull out under fire, it will be a strategic defeat for America. And you know that the first thing Al Jazeera will report is that the enemy caused the retreat."

"You've got to withdraw! The transfer of sovereignty is in danger!"

"I won't do it!"

Bremer and I were now shouting at each other. So I turned to Abizaid and said, "Look, sir, I am *not* going to issue that order. If you want that order issued, you will have to find another commander."

There was a long pause in the room before Abizaid finally responded. "I agree with Ric that we cannot withdraw under fire," he said, turning to Bremer. "If we pull back now, it will be very bad for us. However, we've got to consider the political process."

"I understand the political process is subject to collapse," I responded, "and that we did not anticipate the extent of the Sunni uprising. But that is irrelevant at this point. If you want us to really suffer a strategic defeat . . ."

"That's not what I want, Ric," replied Abizaid. "And it's not what I'm talking about. There's a bigger issue here."

"I understand the bigger issue, sir. But for God's sake, we can't just quit after only two days of a three- to four-week mission. The United States military does not do that!"

"Well, we're going to have to come up with something, otherwise the decision will be made for us. I can assure you that the White House is going to stop the offensive whether we issue the orders or not."

"All right, all right," I finally relented. "We can stop the offensive, but I'm not going to withdraw until we obtain the right separation of forces, and until we can do so under much more favorable circumstances."

"So you're suggesting a cease-fire?" asked Abizaid.

"Sir, I'm suggesting that we unilaterally cease offensive operations until we can achieve the separation that will allow us to withdraw, but not under fire."

"I can agree with that," Abizaid replied.

"That's okay with me," said Bremer. "Now we just have to figure out how to announce it."

As soon as we walked out of Bremer's office, Abizaid looked at me and said, "Boy, Ric, I've never seen you that fired up, before."

"Well, we've got Marines out there fighting and dying," I said.

As soon as I got back to my office, I placed a call to LT GEN Jim Conway, commander of the 1st Marine Expeditionary Force. "Jim, we're going to cease offensive operations." I said.

"What?" he said. "What the hell are we doing? We're right on the verge of breaking this thing wide open."

"Look, it's political and we really don't have a choice. The Iraqi political process is about to collapse and the transfer of sovereignty is at risk. The order will come down to you immediately, and you will have about eight to twelve hours to implement it. Do what you need to do until then."

"Yes, sir," he responded. "But I must retain the authority to conduct defensive operations."

"You have that authority. Make sure you conduct very robust counterattacks if your positions are attacked," I replied.

At noon the next day, April 9, 2004, Ambassador Bremer formally announced that coalition forces would unilaterally cease offensive operations in Fallujah. He stated that the CPA wanted to facilitate negotiations between the Iraqi Governing Council and city spokespersons, and use the time to allow humanitarian supplies to be delivered to residents. Amid media speculation of incompetence, however, Fallujah came to stand for a failure to achieve clearly stated and defined missions. It was really a strategic defeat for the United States and a moral victory for the insurgents.

My final orders to the Marines were to halt the offensive, but to fight back when attacked and eliminate any pockets of resistance associated with the attacking forces. Of course, I knew that when you give the Marines an order to do such limited attacks, they might do so with an entire regiment.

The administration wanted us to cut and run. But from the warrior's perspective, we did not withdraw our forces under fire. We held our ground. In other words, we cut, but we did not run.

ON THE AFTERNOON OF April 9, 2004, only a few hours after Bremer made his announcement regarding Fallujah, he and I sat in on a VTC with GEN Abizaid and Secretary Rumsfeld, which was held in preparation before a broader conference with the President.

When Abizaid and I cautioned Rumsfeld on the need to be careful about launching a major attack against Muqtada al-Sadr, Bremer expressed concern that a strike might lead to instability in the country, which could impact the transfer of sovereignty on July 1. "Well, why do we have to transfer sovereignty on the first of July?" asked Rumsfeld. "If we have to conduct some military operations in order to achieve stability, let's do that, and just delay the transfer of sovereignty."

Secretary Rumsfeld's statement was so spontaneous that it made me wonder if he was out of sync with the NSC and the White House. Had he not been participating in the in-depth discussions regarding the decision to transfer sovereignty in the first place? The Secretary did not repeat his questions when we went into our VTC with the President and members of the NSC.

GEN Abizaid began with a briefing about the status of operations in Fallujah. "At about 1200 local time today, we suspended our operations," he said. "We're on hold due, in part, to a request from the Governing Council."

"There's a real threat here," said Ambassador Bremer. "The top Sunnis threatened to leave last night."

"What is the mood that causes Governing Council members to speak out?" asked President Bush.

"It's huge pressure from TV coverage," Bremer replied.

"Al Jazeera overwhelms us in the press," agreed Abizaid.

"A lot of people in Iraq are committed to democracy," said the President. "The IGC is not learning from them."

"Governing Council members are talking to opposition leaders in Fallujah," said Ambassador Bremer. "They welcome the unilateral cessation of offensive operations, but are retaining a right to self-defense. However, we're not likely to get the enemy to cease their attacks."

"Our objectives are the destruction of the insurgent forces and to bring to justice those responsible for the contractors' murders and mutilations in Fallujah," said Bush.

"Yes, Mr. President," replied Bremer. "Our demands are to turn over the attackers of the Blackwater contractors; turn over the mutilators; turn over foreign fighters; anticoalition forces are to lay down their arms and stop all operations against coalition forces; local leaders must remain in the city, produce the Al Jazeera reporter who has been inciting violence, and bring him out of Fallujah."

At that point, I reiterated that we had agreed not to conduct offensive operations, but that it was definitely not right to characterize this as a cease-fire. "There is still going to be a lot of fighting," I cautioned the group. "And Al Jazeera will continue to report the fighting."

GEN Abizaid then took the opportunity to persuade the President to resume the assault on Fallujah. "From a tactical point of view, we must continue the attack," he said. "Our plan would be to trap the enemy in the southwest portion of the city. It will take three to four more days to sweep through the city. This is a great military opportunity. However, we appreciate the political sensitivities."

Chairman of the Joint Chiefs GEN Richard Myers then jumped in to support Abizaid's recommendation. "The Marines are very confident and wish to continue," he said. "But we don't want to delay beyond twenty-four hours. Give us at least four days and we can finish this job."

I added a few words confirming that the military was definitely able to continue the mission and that we needed to finish the offensive. However, there was no response to our recommendations, and after a pause, GEN Abizaid asked me to give a brief update of the situation in southern Iraq.

Over the next several minutes, I meticulously gave the status of each city—where we were in control and where al-Sadr's forces were in control; where the ICDC and Iraqi police had performed well and where they had abandoned their stations; and how the coalition forces were responding or not responding. Then I informed the President that we were going to launch the major offensive on April 10, the next day. "We must be prepared," I said, "because Najaf might turn into a Fallujah-like situation."

At that point, Condoleezza Rice jumped into the conversation. "We have a two-front political problem here," she said. "We must keep that in mind. How do we deal with the potential resignation of Governing Council members? If that happens, we could lose the chance for the IGC to be the center of the future government. That could be quite a serious problem. It is very important for us to stay with the June 30 date [transfer of sovereignty]."

Ambassador Bremer agreed with Rice's concerns about the fate of the Governing Council. "If it all falls apart, we become the ultimate occupying power with no endgame," he said. "We'll have nobody to transfer power to on July 1. We have a huge problem if it falls apart."

"Jerry ought to be screwing his head into what his course of action will be if the Governing Council walks away," interrupted Secretary Rumsfeld, as if Bremer wasn't even present. "It's very important to stay with the June 30 date."

"Do these people even want us here?" asked President Bush. "What we have seen is erosion of them wanting us here. Can you find anybody to thank us for giving them democracy and freedom?"

"Mr. President, part of the problem is that there is no strong Iraqi leader," said GEN Abizaid. "The chain of command is also very vague. And they don't want to fight for us [Americans]."

"The Governing Council did issue a couple of good statements," Bremer reassured the President. "They're all advocating tough actions against al-Sadr, but only in private. Sistani, however, is dead set against any military operations in Najaf and Karbala. He is desperate to avoid another Fallujah. His message is that we ought to be looking for a peaceful solution."

At that point, Colin Powell offered up some advice. "We must push ourselves as hard as we can to create this interim government," he said. "We must put an Iraqi face on it as fast as we can. We must show strength and not weakness, but keep an eye on the political situation. I think we should give [Fallujah] another eighteen hours."

"Well, if we don't have a government to hand over sovereignty to," said the President, "we won't keep our coalition partners. As much as we want al-Sadr, it may unleash forces we can't control in the country."

"I have a follow-on briefing this weekend," said Secretary Rumsfeld. "What high-level actions must be made in the media to reassure the public?"

WHEN THE LEAD ELEMENTS of the 1st Armored Division reached the outskirts of Najaf, MG Marty Dempsey and I finalized our strategy for regaining control there. Most important, I directed Marty to create exclusion zones to isolate various sensitive religious sites, and then set up a series of checkpoints for enforcement. Shortly after I approved Marty's operational plan, Ambassador Bremer asked me to come down to his office for a meeting. When I walked in, Dick Jones (the deputy CPA administrator), the British ambassador assigned to his staff, and the CPA director of operations were all huddled over a map on the table. "Ric, we need you to show us where your checkpoints are and how you're going to execute the tactical plan to control Najaf," said Bremer.

I was stunned. "Are you serious?" I asked.

"Yes, we'd like to know what the tactical plan is," he replied.

"Well, I'm not going to do it," I said. "I guarantee you that we have a tactical plan. I am comfortable with it and have reviewed it with the division commander. I know he can execute the orders he's been assigned."

"Well, we need to know . . ."

"Stop right there, sir. I am not going to give you the details of our tactical plan. You should not be worried about whether I can perform a military mission. I guarantee you we can. I know what I'm supposed to be doing. What I'm not sure of is what *you're* supposed to be doing. In Najaf, there is nothing in place to ensure that either the political or economic pieces of the equation are met. And that is the CPA's responsibility. We are not going to succeed in Iraq unless we bring all the pieces together."

There was a long pause in the room before Bremer finally said, "Well, okay, Ric. Let me talk to my guys." A few days after that meeting, the CPA actually did propose and fund several initiatives in Najaf, including reconstituting the local government, infrastructure, police, and schools. It was the first time that we actually had a synchronized political, economic, and military plan for an area in Iraq. And it was a good one.

On April 10, 2004, coalition forces began the offensive to reestablish control of the southern Shia provinces. Our main focus was on Najaf, not only because we knew it would be the scene of the toughest fighting, but because a key component of our mission was to kill or capture Muqtada al-Sadr, and he had concentrated the majority of his forces there. Up to this point, we were constrained from conducting any offensive operations due to the millions of Shia pilgrims who had come to the holy sites in Najaf, Kufa, and Karbala in observance of the Muslim holiday of Arba'een. Saddam Hussein had banned the observance when he was president of Iraq.

From a military standpoint, Arba'een was a Godsend for us, because we didn't have the forces to conduct a major offensive operation in Najaf. So our strategy was to tackle the smaller hot spots in the south where we did not need major forces. In Basra, the British were able to talk the Shiites into moving out of the occupied government sites; the Italians soon regained control in Nasiriyah; and U.S. military elements had little trouble reestablishing control in Al Kut. During that time, we were able to move the 1st Armored Division and the Strykers down from the north. Once there, 1st Armored followed through on its plan and set up a series of checkpoints designed to isolate movements of the Mahdi Army and prevent reinforcements and egress. The exclusion zones also worked well in helping to gradually choke down the city as the flow of pilgrims left Najaf. One of our first battles occurred at a river bridge leading into Kufa, which was occupied by major elements of al-Sadr's forces. The Mahdi Army put up a fierce fight to prevent us from entering the area, but we eventually secured the bridge.

When word of our offensive reached Sadr City, calls went out for people to head to their local mosques, obtain weapons, and join the fight against the coalition. Iran issued statements of support and promises of material assistance, and former members of Saddam's regime (those currently involved in the insurgency), along with other Sunni extremists, began cooperating with the Mahdi Army. Casualties mounted rapidly in Sadr City and across the south as fighting became fierce.

During the middle of all this action in southern Iraq, Ambassador Bremer and I participated in another presidential-level VTC from Baghdad, but the majority of the focus was on Fallujah and resolving the political situation. GEN Abizaid began the conference with a military update and again lobbied to restart the offensive. "The Marines are cordoning Fallujah and killing anyone trying to maneuver against them," he said. "They have isolated a pocket of 500 to 800 fighters, probably mostly ex-Baathists. We can conduct offensive operations very quickly. Once again, Mr. President, we reemphasize that we only need three or four more days of precision attacks to secure the city."

While President Bush and Colin Powell remained mostly silent, Vice President Dick Cheney and Secretary Rumsfeld did most of the talking. That was unusual, because Cheney had been present at all the other VTCs and rarely said a word. "We had to stop military operations in Fallujah because of the Governing Council almost fracturing," he said. "That's a big stick to wield over us. What's the likelihood of them staying together in the future?"

"We need to be ready if they resign," said Rumsfeld. "We must think the unthinkable during crises in order to plan contingencies."

At that point, Ambassador Bremer tried to reassure everybody that he had things under control. "We have a command structure for the Sunnis in place," he said. "But the paper on Governing Council operations is very sensitive."

Rumsfeld instantly turned to Rice and said, "Don't even send Bremer's paper over if it is that sensitive. The hell with him. I will send you ours." Once again, Rumsfeld made that statement as if Ambassador Bremer was not even part of the VTC.

"We need Iraqis to stand up and make the case that they have an enormous opportunity and they're letting it erode," continued the Secretary. "The security situation in the Middle Eastern Gulf states will be pretty lousy if Iraq breaks up. Should someone on the political side go out to the region?"

After walking out of the conference, I remember thinking to myself that the administration, for the first time, was talking about working regional diplomacy. We were already well into a crisis and the suggestion came from the Secretary of Defense, of all people, rather than the President, the Vice President, the Secretary of State, the Director of the NSC, or the head of the CPA.

Overall, the main offensive lasted five days, from April 10 to April 15, 2004. Even though coalition forces were able to regain control of government buildings and other key sites in Najaf and Karbala, Shia fighters did not easily relinquish their positions. And afterward, violence and local skirmishes continued at a relatively high rate.

At this point, coalition operations turned toward setting up the capture of Muqtada al-Sadr. We knew where he lived, had documented his patterns of travel, and staked out the eight- to ten-mile route he took from his home to the Kufa mosque. After laying out a plan in excruciating detail, we decided to launch the operation while he was traveling so that we could spare any collateral damage to the Kufa mosque or nearby civilians. Once al-Sadr's exact location was confirmed to be inside our cordoned area, MG Marty Dempsey called and informed me that they were ready to go.

I then went back to Ambassador Bremer and told him that it was time to take out Muqtada al-Sadr. I briefed him on the overall plan and told him that, although we would try to capture al-Sadr, we might end up in a firefight that would result in his death. I also warned Bremer that the aftermath of this operation would be very, very ugly. Currently, we were back in control of most of the country, and the operation might create new instability. In the long run, however, it was the right thing to do, because we were sure we would be able to gradually reduce any remaining Mahdi Army resistance. Ambassador Bremer thanked me for the update and then placed a call to Washington, because the final order to execute had to be approved at the level of the President.

Once again, however, we were under tremendous political pressures. The Iraqi Governing Council, especially, was very weak-kneed about launching an operation against al-Sadr. Very shortly after I briefed him, Ambassador Bremer said to me, "Your guidance is the following: Do not create any condition where it will even remotely bring the possibility of having an encounter with Muqtada al-Sadr. Such an operation will endanger the transfer of sovereignty."

"So what are we going to do now?" I asked.

"That's your guidance, General."

"But we're walking away from the stated mission."

"Your guidance is not to execute the operation."

"How are we ever going to achieve any sort of stability in this country if we keep walking away from our missions?"

"I don't know," replied Bremer. "Maybe we can get al-Sadr to turn himself in to Iraqi authorities, or maintain stability without arresting him. We've got the Governing Council working on it." In effect, Bremer was telling me that we were going to defer on the problem and hand it off to the Iraqis.

So I immediately communicated the decision to Marty Dempsey, who was incredulous. "But, sir, we're ready!" he protested. "We've got everything lined up. All we need is an order to execute!"

"I'm sorry, Marty, but you're not going to get one," I responded. "The politics are just too heavy on this thing."

Over the course of the next week, while 1st Armored maintained the cordons around Kufa, I went back to Bremer several more times and asked him to secure approval to take out al-Sadr, but he just sat behind his desk and said, "No." I appealed to GEN Abizaid, but he also was resigned to the situation. "Ric, it's the political decision and we have to live with it," he said. "The Iraqis will have to figure out what to do with al-Sadr."

Shortly after that conversation with Abizaid, I received a note from the Department of the Army's General Officer Management Branch informing me that my nomination for command of SOUTHCOM (and promotion to four-star general) had been sent to the Secretary of Defense for subsequent submission to the President. The only thing that crossed my mind at that moment was that Secretary Rumsfeld had made good on his word. But there were so many other things going on—and I was so tired—that I simply set the note aside and got on with my business. There was work to be done in Najaf, Sadr City, and all across the south now that the decision had been made not to take out Muqtada al-Sadr. For instance, we had to create a new strategic plan to contain the Mahdi Army.

During the third week of April, I was approached by the top leaders of SCIRI, al-Sadr's rival Shia political party. They were more than ready, they said, to use their own militia, the Badr Corps, to contribute to security in Najaf, retake the Kufa mosque, and kill Muqtada al-Sadr. Of course, I declined the offer, because I had been ordered to avoid anything that would give rise to a conflict with al-Sadr, and because it was our policy not to allow any militias to stand up and assume security responsibilities. More important, we were not about to allow the Iran-backed Badr Corps a free hand to wage war on al-Sadr's Shia party. Doing so would have significantly escalated the already raging Shia-on-Shia conflict across the south. Besides, SCIRI wasn't fooling anybody. We knew it had a vested interest in taking out al-Sadr and controlling the Great Mosque of Kufa, Najaf, and Karbala.

In addition to the SCIRI leaders, a prominent Shia member of the Iraqi Governing Council approached me on multiple occasions to do something about Muqtada al-Sadr. "General, you've got to kill him," he said. "That is what is best for Iraq. And you've got to do it before the transfer of sovereignty."

"Well, if we're going to launch such an operation, you'll have to stand up publicly and advocate it," I said. "You'll also have to persuade the rest of the Governing Council and speak to the Iraqi people about it. Are you willing to do that?"

"Oh, no, General. I can't do that."

"Well, then, we will not be launching an attack on al-Sadr before the transfer," I replied.

AS SOON AS WE stopped our major offensive operations in Fallujah and in the southern provinces, the overbearing political pressure dissipated, and circumstances reached a steady enough state for us to continue the march toward transfer of sovereignty. Lakhdar Brahimi resumed his work, coalition member-nation kidnappings tapered off, and the CPA began an all-consuming effort to achieve its missions. The Governing Council conducted negotiations with Muqtada al-Sadr, which led to a gradual withdrawal from any remaining occupied facilities in the southern provinces. Eventually, and for the time being, members of the Mahdi Army merged back into the general population.

In the meantime, the CJTF-7 staff began a series of initiatives designed to stem the tide of a dramatic increase in surprise attacks on both private citizens and coalition forces. With the approval of GEN Abizaid, we held meetings with many Sunni tribal leaders and began communicating with leaders of the insurgency. However, as soon as the neoconservatives in Washington found out about it, they quashed our efforts. "How in hell can you be talking to people who killed Americans?" I was asked. "They have blood on their hands. We will never negotiate with those who have killed Americans."

This was the second time a major effort at opening dialogue with the insurgents was halted by Washington. The first had been during the lull in violent activity after the capture of Saddam Hussein. This time, it happened amid increased violence in the wake of Najaf and Fallujah. They wouldn't let us negotiate with the enemy during quiet times *or* violent times. America had conducted such negotiations in every other war in which we'd been involved. But not in Iraq—at least, not now. It was almost as if Washington did not want to stabilize Iraq. Everything became the Iraqi people's problem.

By the end of April, it appeared that we had finally resolved the situation in Fallujah. Ever since the cessation of offensive operations, coalition forces had remained in place and fended off a series of attacks by insurgents. However, we were still stuck there without an easy way out. After all, no one had anticipated that we would halt our offensive without completing the mission, and we certainly weren't going to turn it over to the Sunni insurgents who still controlled three-quarters of the metropolitan area.

A potential deal was brokered by LT GEN Jim Conway, whose idea was to turn security over to the Fallujah Brigade, a force comprised of Sunnis and led by former members of Saddam's military that had been organized by the CIA. When the suggestion was presented to GEN Abizaid and me, our initial reaction was that it was as good an alternative as any we had heard. We were trying to involve the Sunnis anyway. Why not give it a shot? However, nearly everybody else in both Washington and London questioned the move. In Iraq, some members of the Governing Council were against it, and Ambassador Bremer kept saying, "This is going to be a failure. This is going to be a failure."

"Okay, if somebody has a better idea, tell us what it is," we responded. But no one could offer a viable alternative that would allow us to disengage from the city. So Conway was given the go-ahead to set it up. In return for being armed with U.S. weapons and equipment, the Fallujah Brigade leaders agreed on certain timelines to pull back their forces, stop insurgent attacks, turn in major weapons, and turn over the culprits who had ambushed the Blackwater convoy. With those promises secured, we announced the program through the media via a news release and press conference. However, we quickly realized that the Iraqi general that had been placed in charge of the Fallujah Brigade was a former member of the Republican Guard and, to make matters worse, he looked just like Saddam Hussein. Shia leaders instantly complained that we were reestablishing Saddam's army and putting it back in charge. That particular general was immediately replaced with someone else, but most Iraqis continued to believe that coalition forces were going to reestablish the old regime.

With time, the Fallujah Brigade was completely eliminated, having been deemed a flat failure, because its leadership never kept any of its promises. The whole episode did, however, promote enough stability in the region for us to disengage from Fallujah and become comfortable enough to proceed with plans for the transfer of sovereignty. Unfortunately, we still had not made any major progress toward any of our originally stated objectives, and all levels of leadership knew it was just a matter of time until we had to go back in there and eliminate the insurgent safe haven. *[On November 6, 2004 (four days after the U.S. presidential election), U.S. Marines launched the second battle of Fallujah and were victorious after experiencing the heaviest urban combat since Vietnam. The battle lasted six weeks.]*

On April 28, 2004, just at the time we were finalizing plans for the Fallujah Brigade, the CBS television program *60 Minutes II* ran a prime-time story documenting the prisoner abuses at Abu Ghraib. For the first time, America and the world saw some of the scandalous pictures, including naked prisoners stacked in a pyramid, a prisoner standing on a box with his head covered and electrical wires attached to his hands, and American soldiers posing with prisoners while laughing, pointing, and giving a thumbs-up sign. The program also recounted a telephone interview with SSG Frederick, then under arrest and facing court-martial, who said he would plead not guilty. "We had no support . . . and I kept asking my chain of command for . . . rules and regulations," they quoted Frederick as having said. In addition, his lawyer blamed higher levels of command for creating the conditions that allowed the abuses to happen.

The *60 Minutes II* program pointed out that Abu Ghraib prison was "the centerpiece of Saddam's empire of fear" that included "torture beyond imagining" and "executions without reason." The obvious implication was that U.S. forces were continuing the tradition. Moreover, the entire story left the impression that the Abu Ghraib abuses occurred during formal interrogations of Iraqi prisoners.

THE LAST WEEK OF March and first two weeks of April 2004 were a strategic disaster for America's mission in Iraq. First of all, a lot of people were killed. In Fallujah, 40 Marines lost their lives and 600 Sunnis died, more than half of whom were considered to be civilians. And fighting with the Shiites in Sadr City and southern Iraq left 27 American soldiers dead and between 2,500 and 3,000 members of the Mahdi Army killed.

Second, the already tenuous coalition of nations began to fracture. Spain was the first to go. Shortly after Najaf was stabilized, Spanish military leaders informed me that they were standing down and would be leaving the country in a matter of days. Due to an erosion of national will, recent elections, and a new government fulfilling its campaign promises, Madrid had decided to pull the plug. While a couple of hundred operating troops were leaving, a few Spaniards at the staff level would stay a while longer to assist with the transition. We had to scramble to replace the Spanish, because they were the command element for the coalition brigade in Najaf. Nicaragua, one of the Latin American countries that had been reporting to the Spanish, also decided to make an early exit.

Our spring military actions had a serious long-term impact on the country. Closing Muqtada al-Sadr's newspaper *Hawza*, seizing his lieutenant Mustafa al-Yaqoubi, the subsequent action to recover ground seized by the Shiites, and the launch of our offensive into Fallujah led to a new level of violence that simply never let up. Oil terminals were blown up, lines of communications were cut off, government buildings were bombed, and dozens of key Baghdad infrastructure sites were targeted. At one point, the CJTF-7 staff put together a "worst-case scenario" to reinforce Baghdad that included moving in America's Strategic Reserve to stabilize the situation should the capital be threatened. There was no question about it: When we compared the levels of violence before and after Fallujah, a new bar had been set. The increase was simply unbelievable.

Our actions had undeniably ignited a civil war in Iraq and the ongoing insurgency had gained unprecedented strength. There was fighting in just about every province—from Mosul in the north, to Baghdad and the Sunni Triangle in the central part of the nation, all the way down to Basra in southern Iraq. And to top it all off, Al Jazeera coverage of the Fallujah offensive led to the rise of the Jordanian-born terrorist Abu Musab al-Zarqawi, who subsequently hooked up with Osama bin Laden. In turn, bin Laden's al-Qaeda terrorists swarmed into the country.

By stopping the attacks in Fallujah, and by not taking out Muqtada al-Sadr, we set the stage for increased violence ignited by the insurgency, a new civil war, and a major surge in al-Qaeda terrorist activity. The highest levels of the executive branch of the U.S. government gave us the order to attack in Fallujah. But when the hard fighting was shown on Al Jazeera and CNN, and pressure began to build from all sides, the Bush administration immediately backed away. Essentially, they ordered us to cut and run. And I found that guidance to be particularly troubling given how the administration characterized Americans who objected to our military involvement in Iraq.

On April 21, 2004—less than two weeks after the decision to stop offensive operations in Fallujah, and less than a week after the decision to stop all efforts to kill or capture Muqtada al-Sadr in Najaf—President Bush addressed the Newspaper Association of America in Washington, D.C. "We're not going to cut and run if I'm in the Oval Office," he stated, matter-of-factly. Five months later, at a campaign speech in Birmingham, Alabama, he blasted the Democrats for suggesting that America needed to withdraw from Iraq. "The party of FDR and the party of Harry Truman has become the party of cut and run," he charged.

In April 2004, we walked away from two important military missions. And for what? So the Bush administration could retain power. The timing and potential repercussions of our actions had not been thought through beforehand. Had things gone terribly wrong in Fallujah and Najaf, or had the transfer of sovereignty been jeopardized, there would have been no time to recover before the November presidential election. And that was a risk the Bush administration was not willing to take. They also could not afford for the American people to know that violence had increased everywhere, that terrorists were now moving into the country in larger numbers, and that low-level civil war had erupted in Iraq. Instead, they wanted to be able to say that the war was almost over, that sovereignty was smoothly being transferred to Iraq, and that the Iraqi people would soon live in a free and democratic state thanks to our efforts.

On April 13, 2004, shortly after the Fallujah cease-fire, President Bush addressed the nation in a prime-time press conference. He began with a brief statement about the situation in Iraq. In part, this is what he said:

> The violence we have seen is a power grab by these extreme and ruthless elements. It's not a civil war. It's not a popular uprising. Most of Iraq is relatively stable . . .
>
> One central commitment of [our] mission is the transfer of sovereignty back to the Iraqi people. We have set a deadline of June 30. It is important that we meet that deadline . . .
>
> The nation of Iraq is moving toward self-rule, and Iraqis and Americans will see evidence in the months to come. On June 30, when the flag of a free Iraq is raised, Iraqi officials will assume full responsibility for the ministries of government. On that day, the transitional administrative law, including a bill of rights that is unprecedented in the Arab world, will take full effect.

JUST BEFORE SPAIN'S TROOPS left Iraq, their commanding general came into my office to say goodbye. We had a long talk in Spanish and he explained that he felt very bad about the political direction he had been given, but that he had no choice in the matter. He said, "*Mi General, por favor entienda que nosotros somos soldados. Tenemos que obedecer nuestras ordenes.*"

I paused a moment and then got up and shook his hand in friendship. "General, we, too, are subject to civilian control," I said. "And believe me, I understand. We are soldiers. We must obey orders."

PART IV

ABU GHRAIB: AFTERMATH AND IMPACT

★ ★ ★

CHAPTER 20

The Perfect Storm

May 12, 2004, was a particularly important day in Baghdad, because Ambassador Bremer and I met with the Iraqi Governing Council to give them an update on the Abu Ghraib abuses. The two weeks after *60 Minutes II* aired its program were absolutely unparalleled in terms of speculation and outright lies about what really occurred at Abu Ghraib. So Bremer suggested that he and I make a personal presentation to the Council to apologize for the abuses, and provide an overview of actions under way to remedy the situation and bring the perpetrators to justice. Bremer's suggestion was very appropriate and I immediately agreed to do it.

As my convoy approached the Green Zone that morning, we were all unusually cautious, because the previous day at 6:45 a.m. we had suffered an IED attack. Two vehicles had been disabled and the Suburban I was riding in took some shrapnel. Fortunately, no one was harmed and we managed to make it inside without any further fire.

A couple of hours later, Bremer and I walked into the Governing Council meeting room. President Ezzedine Salim sat at the head of a long rectangular conference table with the twenty-five council members seated around him. They were dressed in a wide array of clothing. Some were in conventional Western business suits, some had on their traditional Arabic garb, and others wore business casual. After exchanging greetings, we sat down at the other end of the table and Ambassador Bremer began the presentation.

"The abuses at Abu Ghraib prison were unacceptable and outrageous behavior from a few individuals," he said. "We apologize to you for the entire incident, and we want you to know that we have taken steps to ensure that it does not happen again. Those individuals involved will face justice under the American system."

During his remarks, Bremer held up a copy of *Time* magazine's 2003 Person of the Year issue, which featured three 1st Armored Division soldiers on the cover, representing "The American Soldier." "What happened at Abu Ghraib is a contradiction of everything we are about," he said. "We've been here for a year, and have had thousands of soldiers working very hard to reconstruct your country and bring you security. These soldiers on the cover of *Time* are the real face of America, not the few individuals who perpetrated the abuses."

After Bremer's overview, I gave about a ten-minute synopsis of the findings of the investigation. Generally, I talked about the specifics of what went on inside Abu Ghraib, including what had happened to the prisoners, how many Americans were involved, and actions that had been taken to correct the problems. Next, I went through the charges against those undergoing court-martial, I apologized for what had happened, and I assured the Governing Council that we viewed the abuses as unconscionable. Then Ambassador Bremer asked the members for their comments.

"Thank you very much for that presentation and for your apologies," said Ezzedine Salim, president of the Governing Council. "But why isn't the American press talking about the tortures of Saddam Hussein and his regime? These abuses took place at Abu Ghraib prison, the site of Saddam's worst crimes. In this case, the prisoners were just being humiliated. It is not significant. It does not compare to what happened to us. Why don't you get on to something important? When will you show the world the evidence of Saddam's tortures? That is what you should be doing."

President Salim's statement took me back to Kosovo, when I had to deal with the rape and murder of a twelve-year-old girl by an American sergeant. Back then, I had called a meeting of local leaders, expressed my condolences, and stated that we were going to be aggressive in bringing the individual responsible to justice. But their immediate response was to say that the incident wasn't all that bad, that it was the girl's destiny, and then they tried to use the situation as leverage to get some of their people released from custody. In this case, it seemed that President Salim's general attitude was similar. "Yes, this is bad, but let's get on to something really important, like justice for our own people." It was a difference in value systems, and perhaps it reflected a greater tolerance for man's inhumanity to man. Either way, it was very clear that Salim was sincere in his statement, because he became quite emotional when speaking about what he described as "the heinous acts of torture perpetrated by Saddam."

As each of the Governing Council members took his turn speaking, however, I realized there was a vast range of reactions to Abu Ghraib. "A crime is a crime and this came from the United States Army," said one council member. "It is inhumane and it is unacceptable."

"Do you have Israeli interrogators at Abu Ghraib?" asked another.

"No, sir," I replied. "We do not."

Other council members wanted to know about specific CIA and Special Forces methods of interrogation, and how we were going to ensure that our new plans were implemented properly. They also wanted to know when the Iraq government would be able to run the prisons. One individual said that he had recently visited Abu Ghraib, spoken with some of the detainees, and saw for himself that there were no abuses currently taking place. "The television reports coming out of the United States are clearly contradictory to what is really happening on the ground," he said. "I've seen it with my own eyes."

That particular council member's statement was right on the mark. Actually, nearly everything the U.S. media had been reporting was incorrect. In the wake of the *60 Minutes II* report, it seemed like every media entity, whether print or television, had come up with a new idea of who did what at Abu Ghraib. That led to the creation of numerous rumors and a rush to judgment.

Much of the media focused on the picture of one hooded prisoner as proof that American forces tortured detainees while they were being interrogated. Of course, *60 Minutes II* had already inferred as much during its original broadcast. From that followed the assumption that the administration's interrogation policies caused the Abu Ghraib abuses. And during the first two weeks of May, Abu Ghraib was elevated to a watershed event that exemplified what people wanted to believe was a Bush administration regime of torture. And from that portrayal, prognosticators and pundits predicted that Muslim jihadists would issue a call to arms around the world. That did, in fact, happen.

As soon as I heard the false links to interrogations, I thought to myself, "Okay, here it comes. This is just the beginning." And sure enough, additional outrageous allegations were made, many of which were specifically directed at me. Some of them included: "The entire chain of command was involved and complicit." "Sanchez had control over every entity in Iraq that was conducting operations." "Sanchez had been present during the picture-taking sessions at Abu Ghraib, and had personally approved the interrogations." "Sanchez had personally been involved in questioning and torture." When the September 14, 2003, and October 12, 2003, memorandums became public, it was stated that I had expanded the interrogation techniques, that I was advocating interrogation approaches in violation of the Geneva Conventions, and that I had been directed to do so by Secretary Rumsfeld. It was all untrue.

Regrettably, the Department of Defense had not put in place

any kind of public relations plan to deal with all the negative publicity. So we had no ability to counter the speculation. Moreover, legal standards and judicial rules of evidence prevented us from stating what we already knew to be true. Because both criminal investigations and courts-martial were under way, we could not make public that some of the soldiers charged had already confessed, nor could we reveal the content of those confessions. Meanwhile, we were being decimated by the media onslaught.

The press blitz had far-reaching consequences. Democrats in Congress immediately jumped on Abu Ghraib and made it the linchpin of their attacks on Republicans for the upcoming presidential election. They ferociously attempted to discredit, embarrass, and pin some sort of wrongdoing on leaders at the highest levels of the Bush administration. "Look!" they said. "These pictures are proof that the administration both condones and orders torture." Some congressional Democrats refused to accept any warnings, facts, or otherwise truthful information that they were headed down the wrong path. In my mind, there was an absolute belief that they thought they could connect all the dots from Abu Ghraib right up the line to the intel community, Rumsfeld, and the President.

Administration leaders, on the other hand, were doing everything they could to defend themselves from the partisan Democratic attacks. Public calls for President Bush to fire Donald Rumsfeld reached a fevered pitch, but neither Bush nor the Republicans in Congress would place any blame on the Secretary. It was just a matter of time, however, before others would be exposed for creating an unconstrained interrogation environment. Leaders in the Pentagon were especially aware of this fact. Many of them were guilty of dereliction of duty for not putting in place proper standards, training, and resourcing programs before sending soldiers into battle. The Pentagon's strategy, however, was to keep quiet as long as possible, hoping not to expose the greater problems our soldiers had been struggling with for years.

All these factors were lining up to create an environment that would never allow full-scale accountability in the matter of the prisoner abuses. Personally, I felt like I was in the middle of a perfect storm. I was at the confluence of the high-pressure attack system coming out of Congress, the low-pressure "we're not saying anything" doldrums of the military establishment, and the Bush administration's blame-shifting tornadoes—all swirling in and around the inflated winds of Hurricane Abu Ghraib. Any one phenomenon, individually, wouldn't have been so bad. But taken collectively, there was just no stopping the havoc.

Back in Iraq, I was not too concerned about detention and interrogation operations, because we had literally done everything we could to fix the problems that had existed. By the end of the first two weeks in May, for instance, we were nearly ninety days into corrections at Abu Ghraib. We had suspended the chain of command, moved out problem people, taken disciplinary and punitive actions, and levied formal charges. Multiple investigations, both administrative and criminal, had been initiated against the perpetrators. We understood issues voiced by Red Cross personnel and had incorporated their concerns into the investigations. We had requested and finally received additional help from Washington, and the resulting Miller and Ryder reports had validated our concerns. MP training teams had arrived in mid-March 2004 and detention operations training was at its peak by May. It had taken since the summer of 2003, but interrogation training fixes were finally put in place when GTMO Tiger Teams came into the country.

In addition, we knew we had ongoing problems with abuse during tactical questioning. So, through memorandums and commanders' conferences, we reiterated that all Iraqis were to be treated with dignity and respect, and in accordance with the Geneva Conventions. Written updates had been sent to all levels of leadership up to and including the Army's Chief of Staff and the Secretary of Defense. Based on the survey of other theater-wide potential problems, we had also identified a number of field-related detention and interrogation issues that were being handled properly by the appropriate division and brigade commanders.

In another significant move, MG Geoff Miller was assigned as our deputy commanding general for detention operations, and immediately took steps to eliminate the possibility of abuse occurring in the future. After conducting a comprehensive review of our theater-wide procedures, Miller suggested that we modify my October 12, 2003, memorandum, which listed certain approved interrogation techniques. "Sir, we're never going to use these tougher techniques," said Miller. "It'll be too hard for us to ever get the legal concurrences to use them and they would probably yield little intelligence. So we should remove them. That will also make the list more acceptable in terms of a printed document."

COL Marc Warren also stated his opinion: "Sir, there is nothing wrong with the policy we have now," he said. "These techniques are not in violation of the Geneva Conventions. Such a change would only be more aesthetically appealing with the items off the list. However, I have no problem if you wish to follow MG Miller's advice."

"Okay, if it'll help us with perception, let's go ahead and reissue the guidance with those techniques off the list," I said. So on May 13, 2004, we issued a new memorandum to that effect.

I next turned my attention to all of the other problems affecting Iraq. One of my highest priorities was to address the morale of our men and women in uniform. They were upset with all the media attacks on our military leaders. I received, for example, many questions from young soldiers, midgrade officers, and senior commanders. "Sir, why are they doing this?" many of them asked. It wasn't a question of whether or not I did these things. They knew the charges were not true.

"Don't you worry about it," I told them. "Focus on doing your job, especially in combat. Your judgment, your leadership, your ability to perform will make a difference in the lives of the people under your command."

During those first two weeks of May, I was heartened to see that both President Bush and Secretary Rumsfeld also expressed concern for troop morale. In a video teleconference, the President asked me to pass on the message that "we're thinking hard about protecting the reputation of those on the ground," and that he "wanted to make it clear that the troops are our top priority." Bush was very supportive of our military efforts during this conference and, I believe, he was genuinely concerned about how Abu Ghraib was affecting both the troops and our mission in Iraq.

Secretary Rumsfeld paid a visit to Baghdad shortly after testifying to the Senate Armed Services Committee about Abu Ghraib. Shortly before, Stephen Cambone sent us a list of questions that we had to be prepared to answer, and told us that the Secretary was coming in to skewer us. But it didn't turn out that way. Accompanied by Chairman of the Joint Chiefs of Staff GEN Richard Myers, Rumsfeld walked into my office, plopped himself down in a chair, and said, "Well, with all the crap being thrown around, we thought it would be a good idea to come and talk to the troops."

"That's great, Mr. Secretary," I responded. "They need to hear from you."

"You know, this is all just too bad," he said, referring Abu Ghraib. "Why isn't the press paying more attention to what Saddam Hussein did?"

"Sir, that's the same question the president of the Governing Council asked," I replied. "He was very emotional about describing the torture inflicted by Saddam on his people."

The Secretary expressed deep concern about the detainee issue and said he wanted to go to Abu Ghraib and assess the current situation himself. So later that day, MG Miller gave Rumsfeld and Myers a tour of the prison, walked them through all the procedures, and showed them all the changes that had been made. Before leaving Iraq, Rumsfeld felt fairly comfortable that we were doing all the right things and that we had solved previously identified problems. He also spoke to, reassured, and thanked as many soldiers as he could while he was in the country.

While in Iraq, Rumsfeld and Myers also expressed interest in the status of the new four-star command headquarters we would be opening in mid-May. They were now convinced that restoring the command system—that had been so quickly, and so mistakenly, shut down the previous year—was the right thing for the theater. GEN Abizaid and I had been fighting for its reestablishment from the beginning. What's more, I had maintained that had McKiernan's command headquarters stayed active and been given even a minimum of economic and political capacity, the military could have done a very credible job of bringing stability and security to Iraq during that first year of occupation.

It was a lesson I had learned in Kosovo. In the initial aftermath of that war, nobody but the military had the capacity to calm the region. It had taken a year to eighteen months of work to improve the political and economic elements of power on the ground. And compared to Iraq, Kosovo was a fairly benign environment.

Without the four-star command in Iraq, CJTF-7 had assumed the entire burden for the strategic political-military interface *and* for the tactical warfighting aspects of the mission. It was simply too much of a burden for an Army corps headquarters to bear. With a dedicated four-star headquarters, however, ample attention could be given to both strategic and operational issues while the corps-level headquarters focused on fighting the insurgency. It would also provide the credibility and stature necessary to communicate directly with the highest levels of the Pentagon, the Department of Defense, the National Security Council, and the United Nations.

So, on May 15, 2004, after a year of concerted efforts led by GEN John Abizaid, CJTF-7 split into two components. Multi-National Force Iraq (MNFI) was the new four-star headquarters with responsibility for the strategic international, political, and military interface that would direct the Iraqi theater of operations. It would also coordinate with CENTCOM and other national and international agencies. Multi-National Corps Iraq was now a subordinate unit of MNFI that would focus on operational and tactical warfighting operations. As for the formal chain of command, I remained in command of the country and assumed leadership of MNFI (even though I was still a three-star general). LTG Tom Metz assumed command of Multi-National Corps Iraq and reported directly to me. In addition, LTG David Petraeus was named commander of Multi-National Security Transition Command with the responsibility of rebuilding the Iraqi military and security forces. He also reported to me, and I still reported to John Abizaid at CENTCOM. At this point, the military command relationships in Iraq were similar to the way it was in May 2003 when GEN Tommy Franks commanded CENTCOM and LTG David McKiernan was in charge of ground forces in Iraq. They had defeated Saddam's army with that setup. Perhaps now we could finally stabilize Iraq and win the war.

Overall, our actions focused on returning to a counterinsurgency posture after the offensive operations in Fallujah and southern Iraq. We strove to achieve conditions that would allow for a smooth transfer of sovereignty. In that vein, I wanted to know if there were repercussions from the Abu Ghraib media blitz that might be impacting our efforts on the ground. So I asked our division commanders to have the troops keep their eyes and ears open and report back. Shortly, I received word that there were no noticeable adverse reactions from Iraqi citizens relating to Abu Ghraib.

With regard to Najaf, we were still engaged in discussions about the way ahead. We knew that Muqtada al-Sadr was reorganizing his militia forces, even though sporadic fighting continued at a sustained pace. Still, the CPA and the Iraqi Governing Council kept sending delegations back and forth in hopes of working out some sort of peaceful accommodation. The political process, however, just never seemed to be sufficient to close on the issues of Muqtada al-Sadr's arrest, his future in the political structures of the country, and the status of his militia.

In the wake of Fallujah, we were trying to fix breakdowns in the Iraqi Army and the ICDC. The Army itself had to be completely reconstituted in terms of training and equipping, because they had deserted as soon as the fighting got tough. So we had some serious rethinking to do about how we would train, man, and equip the ICDC, and whether it would become the Iraqi Reserve, the National Guard, or simply go away. And finally, we were in the process of reconstituting the police and security forces in the southern provincial capitals that had been recently taken back from al-Sadr's militia.

Our coalition partners continued to be a constant challenge, especially because we had to scramble to redistribute the missions of the Spaniards and Nicaraguans after they left the theater. On the other hand, the Republic of Korea was in the process of coming into Iraq as a new addition to the coalition. We had been working their deployment over the previous two or three months and found that they were very specific about their conditions for deployment. For example, they adamantly requested that they be placed in a safe sector that did not involve combat. They were sending in several thousand people, most of whom were engineers with the intent, they said, to help rebuild Iraq. So we cut the Koreans a sector up north and put them in a fairly benign area where they performed a lot of good work.

Actually, the northern sector of Iraq was one of the few places we could place the Koreans, given the fact that they wanted to avoid violence. After Fallujah, word went around that three or four hundred fighters had held the mighty U.S. military at bay, which only added to the confidence of enemy fighters like Abu Musab al-Zarqawi. In turn, al-Zarqawi continued to stage a full spectrum of attacks, including IEDs, mortars, suicide bombings, and murder-kidnappings. One of the most publicized and worst incidents occurred when American contractor Nick Berg was beheaded by al-Zarqawi in retaliation for "the treatment of Iraqi prisoners." That awful execution, which was filmed and released on the Internet, validated the brutality of the enemy we were facing. For that reason, in part, we reiterated to our forces that they needed to do everything possible to keep from being captured by the enemy. I recall Barry McCaffrey telling his subordinate officers one time that "sometimes the worst thing that can happen to you in a firefight is to not get killed." It seemed to me that we were now operating in that type of an environment. Sadly, there were very few instances where we were successful in finding a kidnapped civilian or soldier alive. And that was despite the fact that we focused on every single one of these incidents, and acted on any bit of credible intelligence in an effort to save the individual who had been taken.

Another very publicized attack that Abu Musab al-Zarqawi likely perpetrated was a mid-May suicide car bomb explosion just outside the Green Zone. Seven Iraqis were killed and five wounded. One of the dead was Governing Council President Ezzedine Salim. It was just a few days earlier that President Salim had spoken so emotionally to Bremer and me about the abuses committed by Saddam Hussein. Now he himself had been killed in apparent retribution for cooperating with the Americans in trying to create a free Iraq.

In the wake of the Berg and Salim murders, and the overall increased levels of violence, rumors surfaced that Baghdad was in chaos and the city about to fall. Employees of the CPA seemed especially upset. Actually, I believe their panic never completely subsided from the moment that full cafeteria menus were cut off during the Fallujah and southern Iraq offensives. In that context, Ambassador Bremer began making his own military assessments on the extent of the threat. I constantly reassured him that Baghdad was in no danger, because after all, we had more than enough forces in the city (over 38,000). We also had more than 150,000 American troops in Iraq due to the force rotation overlap. Actually, for the very first time, I had an operational reserve force (elements of 1st Armored Division) that could reinforce from Najaf.

Ever since the Abu Ghraib story on *60 Minutes II*, Bremer had been distancing himself from the military. He never told me as much, but I knew he and his people were very angry about the entire affair. As a result, the CPA stopped engaging with our command headquarters, which quickly became problematic. When members of my staff brought the matter to my attention and asked me to intervene, I discussed the issue with Bremer. He responded by stating that there were simply some things the CPA needed to work internally, without military involvement.

About this time, Bremer reacted to the panic that Baghdad might be coming under siege, and asked me what I would do if I had one or two more divisions (about 40,000 more soldiers). "Well, sir, I'd keep them right here to secure the city and I would constitute a reserve to handle crises as they occur," I replied. "But at this point, we've got things well under control, both here and in the south. And besides, the Pentagon can't send us any more divisions right now. They just don't have them."

I don't think Bremer ever completely believed what I was telling him. And we left it at that.

ON MAY 17, 2004, my fifty-third birthday, I received a call from CENTCOM placing me on standby. "Sometime here in the next couple of hours, we'll know whether you may have to fly back to the States," was the message.

So I picked up the phone and called GEN Abizaid. "What's this all about, sir?"

"We've just been informed that you, Miller, Warren, and I may have to testify in front of the Senate Armed Services Committee in fairly short order," replied Abizaid.

"But, sir, we haven't had any time to prepare."

"Yeah, I know, Ric. But that's the way it is. If we're called, we'll just have to do the best we can."

Three hours later, I received word that the scheduled testimony was on. So I boarded a plane and flew to Washington.

CHAPTER **21**

Keeping the Lid on Pandora's Box

"Don't fight among yourselves."

"Speak plain English."

"If you don't know something, just say you don't know."

"If you can't remember the exact contents of a memorandum, say, 'May I provide that for the record?' "

"If you need to correct a mistake, respond to the chairman and say, 'Mr. Chairman, I misspoke. Please let me correct my statement.' "

"If you can't discuss something in open session, say, 'Mr. Chairman, we can cover this in a closed session.' "

"Take notes."

"Listen to the news and read *The Early Bird* [the Department of Defense news synopsis]. Make sure you know what's in the newspapers before you walk in there."

"Oh, and don't forget to let the senators pontificate, because they're all going to want to pontificate. Success is if you spend only 20 percent of the time talking and they take the other 80 percent making their political statements."

It was May 18, 2004. GEN John Abizaid, MG Geoff Miller, COL Marc Warren, and I were in the Pentagon listening to a member of the Department of Defense congressional affairs staff give us pointers on how to testify the next afternoon before the Senate Armed Services Committee.

I wasn't looking forward to it. None of us were. Some of the Democrats were going to be rough on us. Besides, there were simply too many erroneous and outrageous speculations that had been presented in the press, ensuring that the congressional hearings would be a spectacle.

The next day, at 2:02 p.m., Senator John Warner (R-VA) convened the hearing of the Senate Armed Services Committee. Thirteen Republicans and twelve Democrats served on the committee with Warner as chairman and Senator Carl Levin (D-MI) as the ranking Democrat. After Abizaid, Miller, Warren, and I were sworn in, GEN Abizaid made his opening statement, and then it was my turn.

"I am fully committed to thorough and impartial investigations [of Abu Ghraib events] that examine the role, commissions, and omissions of the entire chain of command, and that includes me," I told the committee. "As the senior commander in Iraq, I accept responsibility for what happened at Abu Ghraib, and I accept as a solemn obligation the responsibility to ensure that it does not happen again."

From there, I pointed out to the committee that we had already initiated courts-martial in seven cases, and that there were ongoing criminal investigations. In that regard, I had to be circumspect in my testimony. "I cannot say anything that might compromise the fairness or integrity of the process or in any way suggest a result in a particular case," I said.

During my opening statement, I also recounted my actions after learning of the abuses in January 2004, including ordering investigations, suspending the chain of command, and reassigning the individuals implicated in the abuses. I recounted MG Miller's assessment of detention and interrogation operations, and I asserted, "The laws of war, including the Geneva Conventions, apply to our operations in Iraq. This includes interrogations."

Senator Warner began the questioning, and in his first query he asked about my reasoning behind placing COL Pappas in charge at Abu Ghraib. This order had been a cause for concern in the Taguba report. Senator Levin also brought up the issue, and I testified that Pappas's authority was limited to tactical control for the security and defense of Abu Ghraib. I had issued the order, I said, because "we had been receiving significant amounts of direct and indirect fire . . . and during my visits, I had found force protection and defensive planning seriously lacking. I needed to get a senior commander in charge of the defense of that forward operating base."

Both Senator Levin and Senator Hillary Clinton (D-NY) brought up a document, titled "The Interrogation Rules of Engagement," that had been posted by CPT Carolyn Wood at Abu Ghraib, and had been presented at a previous hearing. I stated that I had not seen the document prior to it being presented to Congress and that I had no role in preparing or approving it. COL Warren subsequently explained to the committee that CPT Wood had prepared the document to reflect measures contained in the Army *Field Manual* on interrogations. "Items in the left column were authorized; items in the right column required commanding general approval," said Warren. "The intent of the posting was to remind interrogators that anything not authorized had to go to the commanding general. CPT Wood prepared it with all good intentions."

In my opinion, CPT Carolyn Wood was a good military intelligence officer who was doing her duty at Abu Ghraib. She had originally been stationed in Afghanistan where serious prisoner abuses had occurred, and she had made several requests to her chain of command for guidance. She had posted the document "The Interrogation Rules of Engagement" in an attempt to make certain that proper guidelines were followed at Abu Ghraib. Every one of the techniques listed fell within the limits of the Geneva Conventions—both those that needed approval and those that did not. This was a fact that the press and some of the Democrats on the committee did not want to accept.

A little while later, Senator Robert Byrd (D-WVA) turned to me, held up an article from *The New York Times*, and read the title. "Officer Says Army Tried to Curb Red Cross Visits to Prison in Iraq," said Byrd. "Is that allegation correct?"

"Sir, I never approved any policy or procedure or requirement to do that," I responded.

After questioning COL Warren about interrogation techniques that fell within the Geneva Conventions, Senator Jack Reed (D-RI) quoted a false statement made in *USA Today* that I had ordered specific interrogation methods against a certain prisoner. His implication was that this particular method was severe.

"Sir, I have never approved the use of any of those methods [during] the 12.5 months that I've been in Iraq," I replied.

The next senator to speak was Mark Dayton (D-MN), who referred to an article out of that morning's *New York Times* regarding Red Cross complaints. Dayton also mentioned another unnamed newspaper report that implied that I had issued an order specifically placing key cellblocks under the authority of COL Pappas, where he was then supposed to have committed an abuse.

"I never issued such an order," I responded.

When Senator Dayton next implied that I had approved certain harsh interrogation methods via various memorandums, all I could do was again simply tell him the truth. "Such a specific request never got to my level," I said. "And I never approved any interrogation methods other than continued segregation."

The open session, which was televised live, lasted five hours, with questions evenly directed toward Abizaid, Miller, Warren, and me. In general, I was amazed at the steady references to inaccurate newspaper reports. There was also a constant pounding from some of the Democrats on the committee that we had very little regard for the Geneva Conventions, in spite of our testimony to the contrary. As soon as the hearing recessed to move into closed session, both Senators Warner and Lindsey Graham (R-SC) came up to me and said, "General, thank you for your service. Don't take any of this personally."

"Well, that's pretty hard not to do," I replied.

In the closed hearing, Senator Jack Reed was obsessed with finding an intelligence stovepipe "shadow" chain of command that circumvented the normal flow of orders. Senator Reed embraced virtually every improper conclusion made by the rest of the committee, and concluded that I was being circumvented as commanding general in Iraq when it came to interrogations, intelligence, and detention operations. He was absolutely flabbergasted that he could not link Secretary Rumsfeld and the administration to the Abu Ghraib scandal via a "smoking gun" memorandum. Senator Reed also expressed absolute disbelief when I told him that no such guidance had been issued from higher headquarters.

The entire Senate hearing was stressful, but I thought Abizaid, Miller, Warren, and I accomplished a lot. We clearly stated the context and rationale for decisions made and, I believe, we were all transparent on policies and directives. In retrospect, it was clear to me that some of the committee's members did not like the truth. They would not accept it, because it did not fulfill their political agendas.

Unfortunately, Senator Jack Reed continued his erroneous statements about me after the hearing. As a distinguished graduate of West Point, he should have known better, because cadets there are taught that integrity and honor are part of a warrior's ethic. He had apparently been designated the Democratic point man, because he repeated reports that I was in command of everything going on in Iraq, including the CIA, Special Forces, and all interrogations. And then he and other Democrats on the Senate Armed Services Committee portrayed my memorandums of September 14, 2003, and October 12, 2003, as having opened the door to aggressive interrogation techniques that went far beyond those listed in the Army *Field Manual*. However, this was not an issue of aggressive techniques. It was an issue of what the Geneva Conventions allowed.

In responding through various communication channels (such

as follow-on investigations and responses to congressional questions for the record), I tried to explain the context and reasons behind the publication of the memorandums—that there were no standards or guidance from the Army or anyone else in the Department of Defense as to what interrogation procedures should be used; that, as a result, absolutely no boundaries existed; that the Army *Field Manual* does not establish any controls or safeguards, and leaves the entire universe of techniques available for use; that I had given up on the Department of Defense ever issuing guidelines, and subsequently had exercised leadership by instituting specific standards so that we could get some controls into the Iraq interrogation environment; and most important, that every technique listed in the memorandums fell within the guidelines of the Geneva Conventions, and therefore could not be labeled torture. No one, however, wanted to listen. My explanations fell on deaf ears.

The members of the Senate Armed Services Committee were convinced that they knew my true intent. These two simple memorandums, which I issued to ensure the security of detainees and to synchronize force protection elements, became, according to the committee, the root of all evil that caused young soldiers at Abu Ghraib prison to become criminals. Even more disconcerting to me personally, however, was the fact that no experts at the highest levels of the Armed Forces stepped forward to explain to either congressional Democrats or the American public that I had done nothing illegal, and that I had, in fact, done the right thing when I issued these memorandums.

ON THE MORNING OF May 20, 2004, GEN Abizaid and I gave a closed-door Iraq operations update to a few members of the Senate Armed Services Committee, which was a breeze compared to the previous day's open and closed hearings. Afterward, at the request of Secretary Rumsfeld, we proceeded over to the White House for a meeting with President Bush.

Rumsfeld met us in the waiting room outside the Oval Office and, a few minutes later, National Security Advisor Condoleezza Rice opened the door and invited us in. President Bush, who was already standing, stepped forward and shook my hand. "Hi, Ric," he said. I barely noticed as a photographer snapped our picture. Abizaid and I greeted several other presidential advisors in the room and then sat down on the couch to the left of Bush.

GEN Abizaid began the conversation. "Mr. President, Ric's convoy was hit by an IED about ten days ago," he said. "When I called him up to ask how he was, his immediate response was, 'Hey, sir, no big deal. A couple of our vehicles were disabled, but none of our soldiers were wounded. Everybody is okay.' That's who this guy is, Mr. President. He doesn't think about himself. He thinks about his soldiers."

President Bush smiled and nodded. "That's good, that's good," he said.

Secretary Rumsfeld spoke next. "Mr. President, I just received a close-hold memorandum from Ambassador Bremer requesting that two additional divisions be deployed to Iraq."

Then turning toward Abizaid and me, he asked, "Have you guys seen this?"

"No, sir," replied Abizaid.

"Never heard of it," was my response.

Bush then addressed Condoleezza Rice, to whom Bremer reported. "Did you know about this?" he asked.

"No, sir," she responded. "I'm not sure why Jerry's doing this."

"Well, why didn't he go through the military?" asked Bush, who seemed visibly upset. "What are we going to do about it?"

"Mr. President, you ought to be glad he didn't send it to you, because now you don't have to respond," said Rice. "Bremer is ready to leave. He'll be writing his book. He needs to go."

"Well, this is amazing," said Rumsfeld, shaking his head negatively. "Mr. President, you don't have to do anything. He addressed it to me. I'll take care of responding to him."

Over the next hour or so, we discussed how the congressional testimony went, Abizaid gave a broad overview of his theater of operations, and I spoke about the current ground situation in Iraq. President Bush supported our efforts, and paid close attention to what we were saying.

When the meeting finally broke up, Bush stood up, grasped my hand warmly, and said, "Good job, Ric. Thanks for everything you're doing."

"You're welcome, Mr. President," I replied.

As GEN Abizaid and I walked out of the Oval Office, Secretary Rumsfeld asked us to wait for him in the Situation Room. "I'm going back in to see the President for just a second, and then I need to talk to you," he said.

"Any idea what this is about, sir?" I asked Abizaid when we got downstairs.

"No. Not a clue."

A minute or two later, Rumsfeld came into the Situation Room and closed the door behind him. "The President has approved the following personnel moves," he said. "He can't send General Craddock to Iraq, because it would formalize the shadow chain of command links that the Democrats are tying to prove. Therefore, the choices are Abizaid, Casey, and McKiernan. McKiernan would be a good choice if it were a warfighting assignment. Abizaid has to stay focused on the CENTCOM theater as a whole. Therefore, he's sending General Casey to Iraq."

Then the Secretary of Defense looked directly at me. "Ric, he's afraid to send your nomination [for a fourth star] forward at this time, because it's likely to get mired in the ongoing political debate. He's decided to keep you in V Corps and send General Craddock to Southern Command [SOUTHCOM]. We'll keep you on hold, let this thing die down, and renominate you later. So hang in there."

John Abizaid immediately spoke up. "Mr. Secretary, I don't understand why we're doing this," he said. "Ric has been told he was going to command SOUTHCOM."

"Well, the political conditions are not right," replied Rumsfeld. "We've got to let this blow over."

"But, sir, this is wrong!" protested Abizaid.

"Timing isn't right. We just can't go forward."

I was stunned to learn that I was being replaced in Iraq, sent back to Germany, and my nomination for a fourth star had been rescinded. All I could manage to say was, "I understand, Mr. Secretary."

"Okay, that's all I have for you," said Rumsfeld, effectively ending the meeting.

When exiting the White House, I ran into Paul Wolfowitz, who was just coming in. Under the awning, he came up to me, looked me in the eye, and shook my hand. "Ric, you're a great American hero," he said in a sincere, almost regretful tone. "It's been nice knowing you."

When Abizaid joined me, I mentioned Wolfowitz's remark, and said, "Something is just not right."

"Aw, Ric, you're reading too much into it," Abizaid replied with a nervous laugh.

I had been scheduled to testify in front of the House of Representatives Armed Services Committee, but Speaker of the House Dennis Hastert had questioned the need for me to be present. "What is General Sanchez doing in Washington?" he wanted to know. "With so much going on in Iraq, why isn't he back there?" So when Hastert said I didn't need to appear at the second congressional hearing, I headed to the airport for my return flight to Baghdad.

I had arrived in Washington a few days earlier believing I enjoyed the support of both the administration and my military chain of command. Now I was leaving without any clear knowledge of where I stood or who really supported me. All I knew was that I was being replaced in Iraq, sent back to Germany to continue with my command of V Corps, and that my nomination for a fourth star had been pulled back. I felt deeply disappointed and terribly betrayed. In boarding the airplane, I turned to my aide and said, "Boy, am I glad to be leaving Washington. At least in Iraq I know who my enemies are and what to do about them."

Waiting on the runway for our turn to take off, I jotted down some thoughts in my notebook:

> Putting two and two together, this decision was probably made before meeting with the president. Prior to this, there was never any concern about a loss of confidence in me. Things shifted when the story broke in the press. Now it's clear to the administration that Abu Ghraib is a big problem for them. Politically, they now have a huge challenge as they face the elections.
>
> The congressional hearings are not about objectivity or fact-finding for the purpose of taking care of the problem. This issue will continue to be pushed to the forefront until the elections. There will be a lot of collateral damage, with many political casualties and scapegoats along the way in order to appease the politicians. Both parties are hungry for someone to be held accountable at the highest levels, but for different reasons. Democrats are unrelenting in their attacks. Republicans are trying to cut their losses. Either way, somebody has to hang for this. I guess that's me. It may be that I need to retire.
>
> What a setback! There must be some reason for it. I will have to place my trust in the Lord. But it is awfully difficult to accept after being told that the nomination was at the level of the president for approval. People will speculate and talk. Few, if any, will know. Many will demand perseverance. The Hispanic kids of America and other Latin American countries deserve my continued struggle. Can't let a political issue defeat me. I have overcome too many obstacles to get to this point.

Less than a week after my return to Baghdad, an anonymous Pentagon official orchestrated a coordinated press release. On

May 25, 2004, a flurry of articles appeared in major newspapers around the country, which were subsequently picked up by national and international media outlets. *The New York Times* pretty much summed up the substance of the released information in an article headlined "The Struggle for Iraq: Army Shifts; No. 2 Army General to Move In As Top U.S. Commander in Iraq." Among other things, the article stated:

> The top American officer in Iraq, Lt. Gen. Ricardo S. Sanchez, will leave his command this summer, to be replaced by the Army's second-ranking general, senior Pentagon officials said Monday . . .
>
> Pentagon officials said that replacing General Sanchez with the Army Vice Chief of Staff, Gen. George W. Casey Jr., in no way reflected on General Sanchez's handling of the widening prisoner-abuse scandal at Abu Ghraib prison, outside of Baghdad, which was under his authority . . .
>
> His intended new assignment, which was to lead the United States Southern Command in Miami, may now have been given to Defense Secretary Donald H. Rumsfeld's senior military assistant, Lt. Gen. Bantz J. Craddock . . .
>
> . . . [A] leading plan was to promote General Sanchez to four-star rank, making him the Army's senior-ranking Hispanic officer and rewarding his work in Iraq by giving him the Southern Command, which has responsibility for most of Latin America . . . But something happened in the past few days to derail that plan . . . [D]efense officials would not say Monday night what caused the plan to change.

This article, like the others released on the same day, created the perception that I was being held accountable for the abuses at Abu Ghraib, regardless of what the pending investigations might conclude. It also cemented a perception among my peers that I was now a wounded general and likely would not survive the scandal.

GEN Abizaid, upset by the whole event, placed a call to Paul Wolfowitz and demanded to know who orchestrated the press release. My understanding, after speaking with Abizaid, was that it had been done to set up a firewall between Secretary Rumsfeld and the Abu Ghraib abuse issue. And I suspected that the whole thing had been coordinated by Larry Di Rita, Rumsfeld's press secretary.

A short time later Thom Shanker, one of the coauthors of the *New York Times* article, traveled to Baghdad and asked me for an interview, which I granted. After discussing the current situation in Iraq, Shanker asked, "How did you feel about the way your reassignment was released to the press?"

"Well, Thom, I was not particularly happy about it," I replied.

"General Sanchez, it wasn't appropriate the way it was orchestrated with all the major news agencies across the country," said Shanker. "It was clearly a well-thought-out, calculated action that was meant to serve a political purpose. The impression was clear: that you are being relieved, and have lost a fourth star, for your involvement in the Abu Ghraib scandal. Why are they doing this to you, General?"

"I don't know," I replied. "I really have no idea."

As things began to settle back into a normal routine in Iraq, GEN Abizaid called to inform me that John Negroponte, the newly designated Iraq ambassador (set to take over from Bremer), and GEN George Casey, the four-star set to replace me as Multi-National Force commander, would both assume their positions at the same time, coincident with the transfer of sovereignty. "Well, that makes sense," I said to Abizaid. "I understand."

I also received a call from GEN Casey who, at the time, was the Vice Chief of Staff of the Army. "I've just issued guidance that we're going to have your headquarters manning level up to 90 to 95 percent by mid-June," said Casey.

"Well, that's great," I thought to myself. "It's about time."

For most of the first thirteen months I was in Iraq, my CJTF-7 staff was left manned at about 45 to 50 percent. During the overlap of forces, it peaked at just above 60 percent. But now that the Vice Chief of Staff of the Army was being transferred in, proper staffing was suddenly a priority. Before that, all of our pleas for help had fallen on deaf ears. GEN Casey knew he was assuming command and, rightfully so, he wanted his staff fully manned. So the Army made it happen.

A couple of weeks after I spoke with Casey, I was interviewed via video teleconference by the investigators for the Schlesinger panel. In May, Secretary Rumsfeld had appointed the panel to investigate the abuses at Abu Ghraib. To head the group, Rumsfeld had handpicked James R. Schlesinger, former secretary of defense under Presidents Richard Nixon and Gerald Ford.

Expecting to be speaking with them for hours, I blocked out most of the evening. But the questioning was both abrupt and limited in scope. For the most part, they questioned me about why I did not relieve BG Karpinski. After I explained the reasoning that MG Ryder and I went through during our outbriefing discussion, they asked me about my order placing COL Pappas in charge of security at the prison. "What were you thinking when you did that?" they asked "What did you not do that you were contemplating? What could you have done better?"

In response, I mentioned three things—all of which the final Schlesinger report would later use against me. "First, given more time, I would have become more personally involved in staff supervision of our interrogation and detention operations," I responded. Second, I would have been more adamant about solving the command relationship problem that existed with the 800th MPs. And third, I would have been more aggressive in getting CENTCOM, the Joint Chiefs, and the Department of Defense to respond to our requests for assistance.

After about forty minutes, the Schlesinger panel closed the interview. I had answered all their questions in a straightforward and honest manner.

When I got back to my office, somebody asked me how it went.

"It was superficial," I replied. "They didn't want the truth. The Schlesinger panel is nothing more than a cover-your-rear move on the part of the Secretary of Defense. It appears preordained that when the final report is issued, it'll blame me, some others, and lay some minor responsibility on Rumsfeld. But it'll protect him in the end."

Shortly after my testimony to the Schlesinger panel, Vice Admiral (VADM) Albert T. Church III showed up in Baghdad as part of his charge to conduct a comprehensive review of Department of Defense interrogation operations. Secretary Rumsfeld had asked Church to conduct the investigation on May 25, 2004, the same day as the Pentagon's coordinated press release. VADM Church's mission was to investigate the entire spectrum of interrogation operations, including those in Afghanistan, Iraq, and GTMO. My personal meeting with Church and his team was brief. Rather than have me testify under oath, they only asked me to submit written testimony in response to a series of detailed questions.

With time, I began to think more about Thom Shanker's simple question, "Why are they doing this to you, General?" Now, I thought I understood what was really happening. In my mind, there was clearly both self-protection and a political purpose involved. Unfortunately, there was also a deeper, darker purpose that had to be kept hidden. A meaningful and unlimited investigation, which the Bush administration adamantly opposed, would result in an unmitigated disaster. It would open up Pandora's box and let out a world of evil.

The administration didn't want Donald Rumsfeld's 2003 memorandum (or the administration's related trail of memorandums and decisions) to get out, because it advocated an interrogation policy with few constraints. They didn't want details of their treatment of Guantánamo Bay inmates released. And they certainly wanted to contain the fact that both torture and murder had occurred in Bagram, Afghanistan. The Bush administration also could not afford to have information released about the CIA's practice of having "ghost" detainees at Abu Ghraib, or the agency's torture and murder of "Ice Man." And at this point, the American public did not know about the CIA spiriting people out of the country and taking them to secret prisons in countries that were not subject to the Geneva Conventions.

The Pentagon did not want revealed the military's tremendous deficiencies in interrogation and detention operations. And the Pentagon's lack of action to fix known deficiencies and their repeated refusal to answer the calls for help from ground commanders wouldn't look good, either. The Army, in particular, did not want its problems with military intelligence and the military police exposed. And most important, key senior leaders knew that the Army was guilty of dereliction of duty in providing doctrinal guidance for the conduct of interrogations. We had sent improperly trained soldiers into battle and had failed to respond to their pleas for training, standards, and resources.

Overall, there were early moves to contain the damage being done by the ongoing investigations and hearings, which is probably why Secretary Rumsfeld appointed the Schlesinger panel in May 2004. It was no coincidence that that report was released the day before the Army's Fay-Jones report, which had the potential to be more comprehensive in nature, and therefore more likely to arrive at the truth. In addition, I believe there was a concerted effort to keep the political and media focus on the specific abuse cases at Abu Ghraib. Eventually, it would be determined that the primary cause of Abu Ghraib was nothing more than a bunch of soldiers screwing around. But if the focus of investigation grew broader and deeper, the skeletons in the closet would pose tremendous risks to people in power.

My nomination for promotion was sent to the President on May 11, 2004. It was withdrawn on May 20, 2004, the day after my Senate Armed Services Committee testimony. I repeatedly communicated to senior administration officials that I was going to truthfully answer each and every question put to me in a future confirmation hearing. It did not matter how committed I was to undergoing the rigors of the nomination process. In their minds, it was not in the best interests of the administration, the Department of Defense, or the Army to allow me to testify. And they were right. I was committed to tell the truth to Congress.

As the May 25, 2004, article in *The New York Times* said, "Generals Sanchez and Craddock are both three-star officers who would have needed Senate approval for promotion to a higher rank, and either might have faced a lengthy confirmation process. General Casey is already a four-star officer, and presumably could be installed in the new position more rapidly."

So the strategy of the senior administration officials was to keep me on hold. They would send me back to Iraq, and then on to Germany after the transfer of sovereignty. They would tell me to "hang in there," to "wait until this thing dies down," and then "we'll renominate you later." They would try to give me just enough hope so that I would not retire too soon, but still stay in the service where I was bound to my oath—and therefore prevented from speaking out. All they had to do was get past the November 2004 election. After Bush was reelected, it wouldn't matter.

Sure enough, President Bush would later win reelection, and, in the ensuing years of his second term, many of the administration's evils were released to the public as, from time to time, the media managed to pry open Pandora's box. Unfortunately, just as in Greek mythology, Pandora closed the lid before hope could emerge.

ONCE BACK IN BAGHDAD, I tried not to think about the political shenanigans going on in Washington and instead focused on my job. After all, there was still a lot of fighting that hadn't subsided. And then there were all the preparations for the upcoming transfer of sovereignty. As the CPA rushed to the finish line, a great deal still had to be accomplished, and with each passing day, fewer and fewer people were left to do the work. While we were assisting Ambassador Ricciardone and retired LTG Kicklighter in setting up the new embassy, I implored them, the CPA, and higher headquarters to issue orders prohibiting people from leaving the country until properly relieved. "You just can't allow everybody to get on planes and go back home before the job is done," I said. They took note of my concern, but didn't really do anything to stop the mass exodus.

On June 1, 2004, the Interim Governing Council chose Ghazi al-Yawar (a Sunni Muslim) to serve as Iraq's new interim president. Then the council dissolved to make way for the new thirty-three-person cabinet that was to be stood up at the end of the month. With that event, my mind flashed back to the day we stood up the first Interim Governing Council and one of the ministers came into my office and asked, "General, where is my office?" So I asked my staff to find out what the transition plan was for the incoming new ministers. Sure enough, they came back and told me that there really wasn't a plan. "Okay, that's the bad news," I said. "The good news is that we've got four weeks to put one together and make it happen." So over the next month, our staff put a tremendous effort into helping the new prime minister, the ministers of interior and defense, the Iraqi Joint Chiefs, and their army commanders establish basic capacities, such as offices, people, a fully functioning command and control center, and the beginnings of a process to efficiently handle national security functions.

During this time, Ambassador Bremer was not only in the process of transferring control of all Iraqi monies over to the incoming minister of finance, he was also finalizing spending obligations for the $18 billion U.S. congressional supplement. At the same time, CJTF-7 was attempting to secure a continued flow of Iraqi funds for ongoing projects. These were things like equipment for police and security forces, and construction projects for building medical clinics and schools. The funds I was looking for amounted to about $10 to $15 million, which was really not much compared to the entire $18 billion. There were also other major projects that had to be funded directly from the congressional supplement. So I went to the Ambassador, explained the problem, and urged him to make sure the processes and agreements were in place to continue funding these projects after the transfer of sovereignty. But when I approached Ambassador Bremer for the money, he threw a fit. "This was supposed to have been taken care of," he said. "You should already have accomplished these things."

"No, sir," I replied. "These are projects that are under construction and will be accomplished over an extended period of time."

"Well, I'm not going to fund them," he said.

"Okay, sir, then these projects will have to stop. But I suggest that you take a look at the rest of your organization, because I am certain that the CPA has committed to these projects and many more like them."

Sure enough, Ambassador Bremer came back a couple of days later and informed me that he did, indeed, have some huge funding requirements of which he had not been aware. And those projects called for much larger amounts than I was requesting. So in the end, CJTF-7 received the necessary funding to continue our initiatives. As for the rest of the $18 billion supplement, I do not know where all that money ended up.

As the publicized date for the formal transfer of sovereignty approached, we went through numerous rehearsals for the elaborately planned ceremony, and made detailed security arrangements for what was sure to be an appealing target for the insurgents. However, Ambassador Bremer decided at the last moment to hold the transfer ceremony two days early to avoid any potential violence. So on June 28, 2004, a brief formal ceremony was held in the new Iraqi prime minister's office with just a few people in attendance. I stood behind the photographer as he took a few pictures to mark the occasion. And then it was all over.

Ambassador Bremer left later that same afternoon. At the airport, we shook hands and I thanked him for his leadership and for the political strategy he had left behind for Iraq. Our personal relationship had remained intact, but professionally we had encountered some significant differences on how to implement the comprehensive strategy. Still, Bremer deserved a lot of credit for his strategic vision, the political strategy he was leaving behind in Iraq, and for pulling off the transfer of sovereignty in such a short period of time. In response, Bremer warmly shook my hand, thanked me for all my help, and wished me the best.

The new American Ambassador to Iraq, John Negroponte, came in shortly after Bremer left. Just as when Bremer replaced Jay Garner the previous year, there was little overlap. GEN Casey, who was already in the country, immediately began engaging with Negroponte. I stayed a couple of more days to finish some reports and administrative work. After the formal change of command ceremony with Casey, I left the country in the early morning hours of July 4, 2004.

When I stepped onto that final outward-bound plane, I had mixed feelings. I felt guilty, because I was walking away from my soldiers on the ground. Even if I was no longer Multi-National Force commander, I was still the leader of V Corps, and we had two full divisions deployed in Iraq. My duty was to make certain they were well looked after. On the other hand, I felt a tremendous sense of relief. No longer did I have to worry about the tremendous complexities of getting Iraq back on its feet. That was now the job of GEN Casey and Ambassador Negroponte. When I arrived in Germany later that day, Maria Elena and I joked that I had returned home just in time for the July 4th Independence Day holiday. It was true freedom at last.

This was the first time I had been home in quite a while. I had indeed made it back for my son Daniel's high school graduation in late May, though. Overall, our children were very upset about all the media coverage and questioning of my actions in regard to Abu Ghraib. Lara and Rebekah urged me to retire and get away from the unfairness of it all. But Maria Elena never wavered. "Whatever you want to do, I'll support you," she said.

And she kept telling the children that their father did not do the awful things that were being said. "Just wait," she told them. "The Lord is going to set things right. In the end, the truth will come out." I would never have been able to cope as well as I did had it not been for Maria Elena's blessing and support. It meant everything to me.

Sadly, I was only able to spend about a day with my family. I had to fly straight back to the United States to appear before the Fay-Jones investigative panel. And I decided to leave a little early so I could prepare and get a good night's sleep before testifying.

My plane arrived in Washington, D.C., at 2:00 p.m. on July 5, 2004. I checked into my hotel, changed into my desert camouflage uniform, and headed straight over to Walter Reed Army Hospital to visit the wounded soldiers who had served under my command in Iraq. They were scattered about on several floors. Some were in regular hospital rooms, some were in the intensive care unit, and others were in rehab. I saw some severely wounded soldiers and Marines, some with burns, some with amputated limbs, and some with various parts of their bodies bandaged. When I walked in and they recognized me, most started to stand up and salute. "No need to do that," I said. "You've got to conserve your energy. Just relax."

I asked how they were doing and tried to engage them in a little conversation. "Are you comfortable talking about your situation and how this happened?" I asked. Just about every one of them said they were, so we discussed their various situations. Many of them talked about wanting to go back to Iraq and rejoin their buddies. If there were family members present, I pulled them aside and asked how they were being treated. None of the family members or the soldiers themselves ever expressed any anger at the Army or the nation. And all of them said they were being treated well. Before leaving, I thanked each soldier for his or her sacrifice and service to the nation. Then I handed each of them one of my commander's coins.

One of the last people I saw was a young sergeant who had

a terrible injury to the left side of his face. He immediately recognized me when I walked in, and was on his feet with a salute before I could tell him to rest easy. "Sir, it's great to see you," he said. "Thank you for your leadership. I want you to know that I'd do it all over again. It was an honor serving in Iraq under your command."

The next morning, at 8:00 a.m. on July 7, 2004, I walked into a conference room at Fort Belvoir to undergo questioning for the Fay-Jones investigation about Abu Ghraib. The first thing the interviewer said to me was, "General Sanchez, you have the right to remain silent."

CHAPTER 22

Hang In There

After testifying to the Fay-Jones panel at Fort Belvoir, I turned right around and flew back to Germany to be with my family. Officially, I had to report to the office for seven days. But after that, as a soldier returning from combat duty, I was eligible for a thirty-day leave—and I took it all. I needed it. When I returned to duty in the fall of 2004, V Corps still had two divisions in Iraq, 1st Armored and 1st Infantry, and I was back to commanding in a normal garrison environment with a focus on training, manning, equipping, and deploying forces.

As the corps headquarters began retraining, we scheduled an internal exercise called Victory Start, which was nothing more than just turning on the lights, so to speak, in our command and control centers. We had left not only our soldiers behind (both divisions were due to return in February 2005), but almost all of our equipment. And because most of the corps had been gone for about a year and a half, garrison systems and procedures had completely atrophied in all areas of supply, maintenance, safety, personnel, and training. Meanwhile, in order to meet the operational needs and demands in two theaters of war, the Army was continuing to dismantle units by picking and choosing all the way down to the platoon level.

As fall turned to winter and 2004 into 2005, V Corps kept busy by preparing to receive our troops back from Iraq. We implemented the U.S. Army Europe plan called "R-4: Redeployment, Reintegration, Reconstitution, and Retraining."

Redeployment was mostly mechanical. The 1st Armored and 1st Infantry Divisions came back over a period of three weeks, with their last full day in Iraq on February 19, 2005. Moving these divisions required extremely detailed timetables. We had to get them on the planes, off the planes, and back to quarters. Upon arrival, we held very brief welcome-home ceremonies and then the soldiers went home. Redeployment was as important as anything we do for our men and women in uniform, because we are finally getting them out of harm's way, back with their families, and into a peacetime routine.

Reintegration consisted of two phases. First, everybody reported to duty for seven days. It was the chain of command's responsibility to assess each soldier's physical and psychological condition. At the end of those seven days, we released them to go on an extended leave to be with their families (second phase). After their return, we gave them a proper welcome home with parades, entertainment, community parties, and various celebrations. Part of the festivities included an awards ceremony for the family readiness groups. While some military functions may have declined during combat deployment, the family functions had remained strong, robust, and effective. GEN B. B. Bell had seen to it.

Our reconstitution period lasted ninety days. During this time, we tried to get the soldiers back into their routines, and keep them busy. A normal duty day was from six thirty in the morning until five o'clock in the afternoon. There was no extra duty and minimal individual training. At the end of the day, everybody went home to their families. Early on, I had town hall meetings with every community to ensure the spouses knew it was my responsibility to fix any problems they might have. It was critical that I warn the families about post-traumatic stress disorder and to ask them to be aware that, after such a long time in a combat zone, their loved ones would be different people and might have problems reintegrating into the community. It was okay to ask for help, I told them. And that help included additional psychiatrists, psychologists, chaplains, doctors, and dentists. Once restrictions were lifted and our soldiers were able to change jobs, go to school, or transfer back to the United States, we helped line up transportation and move household goods to get them settled in their new locations. Overall, GEN B. B. Bell's theater-level family support and readiness program became a standard across the rest of the Army.

After reconstitution came retraining—with the goal of recertifying units as warfighting organizations within 180 days after returning home. By the time we began extensive training operations, the Army was already executing its third rotation into Iraq, and it was clear to us that V Corps would eventually be going back into combat, either in Afghanistan or in Iraq. After all, there were only three Army corps that could serve in these theaters (3rd Corps, 18th Airborne Corps, and V Corps). Third Corps had already been deployed for six months. The 18th Airborne was in training to go in and replace 3rd Corps. And V Corps was the only corps available to come in behind the 18th Airborne. It was the standard 3-to-1 rotation cycle, and the only way we could sustain a permanent presence.

At this point, we were only a year and a half into the war, but training and deployment were already experiencing big problems Army-wide. The overall process hadn't really changed very much. A commander on the ground identified a requirement that was forwarded to the combatant commander, such as CENTCOM. After validation, it went up to the Joint Chiefs in Washington and then (as a necessary part of civilian control of the military) to the Secretary of Defense, who was the sole and final approval authority for deployment orders. From the Secretary's office, the request proceeded to the operations office at Army headquarters. The service then issued the deployment orders. Of course, after a unit was approved, and before it could actually be deployed, it had to be certified (through training) as being combat-ready. It was at this stage that we were beginning to see problems that stemmed from the summer of 2003 when GEN Abizaid and I insisted that any unit leaving Iraq had to have a replacement. The Army adapted to the new situation by expediting transformation to build additional brigades, modifying training requirements to shorten predeployment timelines by thirty to forty-five days, and by deploying "in lieu of" units. Adding to the difficulty was the fact that the Army had continued with the drawdown of forces in Europe, including the transformation of installations, bases, and general infrastructure. As a consequence, some units were programmed to go back to the United States and/or to completely deactivate.

We faced our first real challenge when our 18th Military Police Brigade was placed on deployment orders to Iraq. Because of previous multiple sourcing and rapid deployment timelines, they had been completely fractured as an organization. Our task was to try to get them back to a brigade-level command and put them through training as a cohesive unit. One impact of the Army's multiple sourcing strategy was that some units were together for the first time when they stepped onto the battlefield; only then could they build relationships within the new command structure. And that violated one of the very fundamental principles that we, as an army, had always embraced and cherished—to train as a team before going to war.

When we learned that our MP brigade was going to be filled with Reserve "in lieu of" units from the United States, we began to ask questions about the certification of those units. To my surprise, I learned that not only were they placed on an accelerated two-week training schedule for quick deployment, but many of the soldiers were being thrown together at the last minute to fill the units. Even more disturbing was that in some cases, people were put into the units, underwent training, and departed the unit. The permanently assigned personnel arrived later and received no unit-level training at all. From our perspective in V Corps, it appeared that many "in lieu of" units were not receiving sufficient training for their mission, and were being filled in a haphazard manner, without any regard for long-term effectiveness. Accordingly, I decided to intervene in the process. On at least three occasions, I sent letters to higher headquarters concerning the deployment of active-duty and Reserve component units from the United States that would be fighting under the command and control of V Corps. I made it very clear in my letters that, as senior leaders, we had a responsibility to ensure that our soldiers went off to war properly trained. Not doing so would create problems and, most certainly, result in increased casualty rates.

My letter put Forces Command into a tailspin and, while I did not receive any direct complaints, I was informed through separate channels that some people were very angry. However, I was not about to back down. I had been on the ground in Iraq and had seen firsthand the problems caused by poorly trained soldiers and units. Abu Ghraib was the prime example. Now that I was V Corps commander back in Germany with responsibility for training, there was no way in the world I was going to allow any element under my command to go to Iraq or Afghanistan without being properly trained. It just wasn't going to happen.

IN THE LATE SUMMER of 2004, both the Schlesinger and Fay-Jones panels released their investigative reports into the abuses at Abu Ghraib. The ninety-two-page Schlesinger report was the first one out, on August 25, 2005. I was in Europe at the time and I initially read the findings in the press. Nobody from Washington called to talk to me about it.

Essentially, the Schlesinger panel placed most of the blame for the abuses at Abu Ghraib on me. According to them, I failed to ensure proper staff oversight of detention and interrogation operations; I should have ensured that my staff dealt with the command and resource problems; I should have ensured that urgent demands were placed for appropriate support and resources through CFLCC and CENTCOM to the Joint Chiefs of Staff; the fact that I delegated responsibility for detention operations at Abu Ghraib led to the damaging result that no single individual had responsibility for overseeing operations; I could have set in motion the development of a more effective alternative course of action; I was responsible for establishing the confused command relationship at the prison; I should have taken more forceful action; I failed to report the abuses up the chain of command in a timely manner with adequate urgency; although the abuses were known and under investigation as early as January 2004, the gravity of the abuses was not conveyed up the chain of command to the Secretary of Defense; even though the Taguba report was transmitted to both CENTCOM and CJTF-7, the impact of the photos was not appreciated; CJTF-7 determined that some of the detainees held in Iraq were to be categorized as unlawful combatants; the memorandums I issued allowed for interpretation in several areas, did not adequately set forth the limits of interrogation techniques, and I should have relieved BG Janis Karpinski sooner.

The Schlesinger report also leveled criticism at Karpinski and COL Pappas. It stated the Joint Chiefs of Staff both underestimated the need for personnel and neglected to provide troops once the need became apparent. And it noted that Secretary Rumsfeld contributed to confusion over permissible interrogation techniques. "There is both institutional and personal responsibility at higher levels," the report stated. But then it noted that senior military commanders bore a greater share of the responsibility than did Rumsfeld.

In general, the report read just as I suspected it would after I came out of my cursory forty-minute interview. It was clearly a report designed to protect Secretary Rumsfeld.

On August 26, 2004, the Army's Fay-Jones final report was released to the public. Shortly afterward, GEN Paul Kern, commander of the U.S. Army Material Command at Fort Belvoir (and the four-star officer responsible for the report), testified before the Senate Armed Service Committee on the findings of his investigation. Once again in Europe, I was not informed about GEN Kern's investigative findings. After scrambling to obtain a copy, I was finally sent an executive summary of the Fay-Jones report.

Overall, GEN Kern's congressional testimony supported my position. Among other things, he stated:

- From the time V Corps transitioned to become Combined Joint Task Force 7 (CJTF-7), and throughout the period under investigation, it was not resourced adequately to accomplish the missions of the Combined Joint Task Force.
- The military police and military intelligence units at Abu Ghraib were severely underresourced.
- CJTF-7 had to conduct tactical counterinsurgency operations while also executing its planned missions. That is the operational context in which the abuses at Abu Ghraib took place.
- The primary causes are misconduct (ranging from inhumane to sadistic) by a relatively small group of soldiers and civilians.

The substance of the written Fay-Jones report, however, placed some blame on CJTF-7. In general, it stated that: There was a lack of clear command and control of detainee operations at the CJTF-7 level; CJTF-7 failed to ensure proper staff oversight of detention and interrogation operations; CJTF-7's relationship with 800th MP Brigade resulted in disparate support from the CJTF-7 staff; a lack of aggressive oversight by the CJTF-7 leadership resulted in a lower priority for resources needed for detention operations; a lack of one person on the staff to oversee detention operations and facilities complicated the coordination among the CJTF-7 staff; I was wrong in assigning COL Pappas responsibility for force protection at Abu Ghraib and that, in fact, doing so was doctrinally incorrect; our interrogation policy memorandums led indirectly to some of the nonviolent and nonsexual abuses, which led to the belief that additional interrogation techniques were condoned in order to gain intelligence, and contributed to the confusion about what techniques could be used.

The Fay-Jones report included some statements that indicated to me that Reserve component investigators did not have a complete grasp of military doctrine or combat-command relationships. And while not as damaging to me personally as the Schlesinger report, the Fay-Jones report was still not good.

Usually, when the Army does an after-action investigation and report, both the details of the incident and its causes are examined. In these two reports, however, the institutional problems that *led* to the abuses at Abu Ghraib were barely mentioned—and they were certainly not addressed in a comprehensive manner, in part because Kern did not have the authority to investigate the actions of the Army staff. Abu Ghraib was not a problem involving only CJTF-7 and its soldiers on the ground. To a very large degree, it was created by institutional negligence and, in some cases, by individual dereliction of duty. GEN Kern, in his congressional testimony, and the Fay-Jones written report clearly stated that CJTF-7 had not been manned adequately enough to perform the missions assigned. So it came as no surprise that shortly after Kern testified to the Senate Armed Services Committee, the Secretary of Defense summoned me to Washington.

Rumsfeld scheduled me for a lunch meeting at the Pentagon at 12:30 p.m. on Thursday, September 16, 2004. In his office, there was a couch along the left wall, a regular desk in the center of the room, and a standing desk where he did most of his work. At the far end was a conference table where we sat down to eat. Also present were Larry Di Rita, his press secretary, and VADM James G. Stavridis, the senior military assistant to the secretary of defense.

The first subject Rumsfeld brought up was my status as commander of V Corps in Germany. "Mr. Secretary, my concern is that I am due to rotate out of command of V Corps as part of the normal rotation cycle," I said. "That would place me at risk of reverting back to a two-star on active duty or being forced to retire as a two-star, because I will not have completed my three years time in grade."

"That will not happen," Rumsfeld quickly said. "I am the decision authority for those moves, and I will not allow you to be placed in the position of having to retire as a two-star. Besides, you have a big supporter in Pete Schoomaker [GEN Peter J. Schoomaker, Chief of Staff, U.S. Army]."

"That's good to hear, Mr. Secretary," I replied.

"We're committed to keep you in the current command billet as long as necessary until we can nominate you for a fourth star," Rumsfeld said. "But you must be patient. We need to wait until things settle down. Closer to the election, I'll start working the political issue to allow for a favorable outcome in the Senate."

The Secretary then started talking about a report he had just been given by the Joint Chiefs of Staff about the manning levels of CJTF-7 headquarters in Iraq. He grabbed a pen and started drawing a chart on his napkin. Rumsfeld drew the *X* and *Y* axes, and then two lines moving from left to right, separated by about a half inch. "Here is your requirement number way up here," said Rumsfeld, pointing to the top line. "And here is your actual fill percentage down here for the entire time you were in Iraq. Your headquarters never came close to meeting the fill rate. In fact, it was below 50 percent almost the entire time. Why didn't you tell somebody? This is exactly why I had argued for having standing joint task forces. It's supposed to be the core for a much larger organization, with fully trained personnel across the spectrum of operations."

Rumsfeld's voice rose and he became more animated, waving his hands up and toward me in an emotional manner. "How could this happen, General," he said. "Why in hell didn't you tell somebody about it?"

"I did, Mr. Secretary," I responded. "Every senior leader in the Pentagon knew the status of CJTF-7. We were constantly arguing for support. I personally gave this message to every leader that came into Iraq, including the Chairman of the Joint Chiefs, the Vice Chairman, the Army Chief of Staff, the Vice Chief of Staff, and members of congressional delegations. General Abizaid was working on it constantly, sending reports on our status into your office. Everybody knew, sir. Everybody."

"Oh, well, this is not right," said Rumsfeld. "Why didn't the services fill you up?"

"Sir, I can't answer that question. That's a question you're going to have to ask the service chiefs. I do know that the Army didn't establish this as a priority until the Vice Chief of Staff was sent to Iraq in my place."

"What the hell was wrong?" thundered Rumsfeld. "Why couldn't they get people?"

"Mr. Secretary, I don't know."

"Well, we're going to fix this," he said. "We're going to fix this."

Rumsfeld then startled me by saying, "I understand that General McKiernan and CFLCC left the country in the summer of 2003. How did that happen? That's unbelievable!"

"Sir, he left on orders from General Franks," I responded.

"Well, I just can't believe they did that. I didn't know about it."

"You didn't know that McKiernan's command had left Iraq?" I asked, incredulously.

Rumsfeld didn't respond. Rather, he started talking about Marine LT GEN James Conway, who had just returned from Iraq and given a newspaper interview in which he discussed the administration's indecisiveness surrounding the Fallujah operation, especially the fact that we had committed the forces and had no resolve to finish the job.

"What was Conway talking about?" asked Rumsfeld.

"Well, he was upset," I responded. "The issue was that we launched the attack on Fallujah and, within a few days, we decided to completely reverse our decision—even though, as you well know, sir, that we had plenty of forces on the ground and could have accomplished the mission. From his perspective, Conway saw that as indecisiveness on the part of our political leadership."

"What the hell was going on between you and Bremer during this time frame?" asked Rumsfeld.

"Sir, everything you've heard about my relationship with Ambassador Bremer has been as result of some very heated discussions about what we had to do in Fallujah, and in Najaf when we were trying to take out Muqtada al-Sadr," I replied. "Bremer was pushing very hard for us to unilaterally disengage from Fallujah. I refused to do that, because it would have resulted in a strategic defeat, or at least that's what the perception would have been on the part of Al Jazeera."

"Absolutely," said Rumsfeld. "It was bad enough the way it was done."

"Sir, I believe it would have been even worse if we had unilaterally withdrawn all our forces under fire."

"I agree with you," Rumsfeld replied. "But why didn't I know about this?"

The look on my face must have reflected some puzzlement, because, the Secretary quickly answered his own question. "I know, I know," he said. "You had to go through the chain of command."

"Right," I replied. "I communicated these issues to General Abizaid. In fact, he was present during some of these discussions."

"The Army headquarters structures are so byzantine," said Rumsfeld. At that point, we ended the meeting. I said goodbye to Larry Di Rita and VADM Stavridis, both of whom barely said a word during the entire forty-five-minute meeting.

"We'll work this promotion for you, Ric," said Rumsfeld, as he walked me to the door. "Just hang in there."

In the outer office, I ran into GEN Richard Myers, who was surprised to see me. "What are you doing here?" he asked.

"The Secretary called me in, sir. We just had a lunch meeting."

"Well, Ric, you did a fantastic job for our country," said Myers. "We're going to get you through this. Hang in there."

I walked out of the Pentagon wondering exactly why I had been called all the way back from Germany. And I thought long and hard about what Secretary Rumsfeld had just said to me.

He didn't know about the staffing levels at CJTF-7? I guess that was possible. The Joint Chiefs and the other service leaders were gun-shy about going to him with problems, because of the tremendous beatings they took in the run-up to the war.

He didn't understand why LT GEN Conway was upset about pulling out of Fallujah? Not likely, I thought.

But was I really supposed to believe that the Secretary of Defense didn't know about CFLCC leaving the Iraq theater of operations? He was the "decision authority" for all unit movements and he didn't know about LTG McKiernan's command moving back to the United States? Frankly, that was impossible. And while I tried to act professionally, I did everything short of saying, "You had to know, Mr. Secretary."

Rumsfeld had summoned me to Washington right after GEN Kern testified before Congress that CJTF-7 hadn't been properly supported while I was in Iraq. It crossed my mind that this meeting might have been set up so that he could deny knowing about any of these findings. And he had two of his own people in the room who could witness his claim of ignorance.

In subsequent discussions with the leadership of the Army, I brought up parts of my conversation with Secretary Rumsfeld. With regard to staffing, it was their contention that Rumsfeld was the problem all along—not them and not the Joint Chiefs of Staff. I didn't know whom to believe. So I said, "Look, this whole thing is a big mess and it's dragging itself out far too long. Maybe I just need to retire." But they were all very supportive behind closed doors.

"No, no, don't do that," I was told. It's not an option. "Everything will be all right. Just hang in there, Ric."

Upon my return to Germany, Maria Elena and I had some long talks about what course of action to take. Do we stick it out? Do we trust in them to ride out the storm? Do we accept their commitment to promoting me?

The truth is that I wanted to believe them. And I was absolutely not ready to retire at that point. I wanted to continue to serve my country. So Maria Elena I made the decision to "hang in there" and see what happened.

On October 12, 2004, barely a month after my meeting with Secretary Rumsfeld, I was notified that the Inspector General of the Army was directed to conduct an inquiry into the allegations contained in the Schlesinger and Fay-Jones reports. The IG was going to determine whether or not I was derelict in the performance of my duties pertaining to detention and interrogation operations in Iraq, and if I had improperly communicated interrogation policies. At this point, I was devastated. After all I had been through, I could not believe it was me they were investigating for dereliction of duty.

After the two investigative reports were issued, there had been a big push in Congress for another comprehensive review of what happened at Abu Ghraib. Some individuals were absolutely convinced that I was somehow responsible for the abuses, and that they could start with me and follow the bread-crumb trail all the way up to the White House. There was a never-ending call from the Senate Armed Services Committee to hold the highest levels of leadership accountable for the events at Abu Ghraib. Therefore, the Army's Inspector General had been directed to conduct a more comprehensive investigation designed to fill the gaps in the previous reports.

It took me a while to regain my balance, but when I finally did, I thought about my own experience as an investigator in the Inspector General's office some eleven years earlier. I was intimately familiar with the process of how an investigation like this would proceed—and that comforted me. At least now, I believed, I stood a fair shot at vindication. But being under a continuing cloud of uncertainty from my contemporaries in the general officer corps disturbed me. And of course, it was now apparent that Secretary Rumsfeld was not going to be approaching Congress about any possible nomination for a fourth star—not with a pending investigation by the Army Inspector General's office.

A short time later, I spoke about my situation with one of my former mentors, GEN Barry McCaffrey. I respected him deeply, and knew he was in my corner. "Ric, you are too honest for your own sake," he said. By that, GEN McCaffrey meant that I had impeccable integrity and would stick by the truth to the end. And in such a political environment, it might result in my eventual downfall. "The best way to describe you, Ric, is that you're like a wounded zebra in the Serengeti," McCaffrey said. "The herd is mildly interested in your survival. You're injured. If you survive, they will take you back into the herd. If not, then so be it."

I knew he was right. But the words "so be it" stuck in my mind. "So be it." "If God wills." *In cha'Allah. Si Dios quiere.*

ON NOVEMBER 2, 2004, George W. Bush was reelected President of the United States. The margin was narrow—only 3 million votes out of 121 million cast in the popular vote. The final tally in the Electoral College was 286 to 251. The Bush administration now had four more years in power.

CHAPTER 23

The End of the Line

During the latter part of 2004, after meeting with Secretary Rumsfeld, and after the Army Inspector General began his investigation, I had some serious discussions with the Pentagon leadership about my situation. We talked about the fact that my two-year billet was up in June, about my time-in-grade problems, and about whether I would receive a new assignment or be forced to retire as a two-star. "Don't worry," they reassured me. "If need be, the Secretary will approve your waiver and you'll be able to retire as a three star."

By early 2005, I began to press the issue with my boss, GEN B. B. Bell, and Army Chief of Staff GEN Schoomaker. "What are you going to do?" I said. "A decision has to be made one way or the other. We need to get a decision on my replacement. The new V Corps commander should be on the ground by June in order to give him time to build the team and to train the headquarters before deployment. The corps begins deploying in August and will be back in Iraq by December."

I could see why everybody was dragging their feet on this decision. It was a tough one to make. Politically, they could not allow me to go back to Iraq with the corps. And they couldn't simply keep me as V Corps commander, deploy the unit colors to

Iraq, and leave me back in Europe without a billet. The question was, "What the hell do we do with Sanchez?"

The ability to come up with a solution might have been made easier on March 2, 2005, when VADM Albert T. Church III released the executive summary of his investigative report regarding Department of Defense interrogation operations in Afghanistan, Iraq, and GTMO. *[The bulk of the 368-page report remained classified.]* Regarding Iraq, the Church commission found that my policy memorandums did not play any role in the detainee abuses or lead to the use of illegal or abusive interrogation techniques. "It should be noted," the report read, "that none of the techniques contained in either the September or October CJTF-7 interrogation policies would have permitted abuses such as those at Abu Ghraib." The Church commission also confirmed that the offenses at Abu Ghraib did not occur during interrogations nor were they connected in any way to official interrogation policy. Our key purpose, the commission determined, was to regulate interrogations in Iraq by specifying approved techniques, mandating oversight and safeguards, and requiring adherence to the Geneva Conventions. Overall, I was very pleased with Admiral Church's findings. However, few media outlets took note of the executive summary or published it.

Not long after the report's findings were released, the Army designated MG John Batiste, commander of the 1st Infantry Division, to be the new deputy V Corps commander. Although unannounced, it was apparent to everyone that Batiste would assume command of the corps, at some point. As the deputy, he would train the troops and deploy to Iraq with the headquarters staff. Under this arrangement, I agreed to step back and act as a senior mentor to help with training at the headquarters, while the Army sorted out a future role for me.

GENs Schoomaker and Bell finally came up with a plan to split and deploy the headquarters staff element of V Corps and leave the unit colors in Germany. This solution required the creation of an additional three-star position, but they justified it as a wartime billet. I would become deputy commanding general for U.S. Army Europe, still reporting to GEN Bell, and whoever assumed command of the Multi-National Corps in Iraq would deploy there with the V Corps staff. I would be left in Europe with a small reserve component staff, but would also be able to tap into the staff at U.S. Army Europe. I'd retain my command responsibilities to train, man, and equip the remaining V Corps forces for combat, and would maintain my reporting relationship with the division commanders. The plan amounted to a very complicated and unusual way to keep me in command.

However, in May 2005, two days before he was due to relinquish command of the 1st Infantry Division, MG John Batiste threw everything into chaos by abruptly resigning from the Army, because of his disgust with the Bush administration. *[Batiste was a big supporter of GEN Eric Shinseki, who had been forced into retirement back in 2003 for bucking the administration. Batiste would later testify before Congress about Rumsfeld's poor leadership and general mishandling of the war.]*

MG Peter Chiarelli became the designated Multi-National Corps commander later that summer, but in the meantime, I had the duty of training the V Corps staff for combat in Iraq. Of course, I had done the job before, and didn't mind doing it again. But it was not an ideal situation, because it forced the troops to train with a commander who was not going to be on the ground with them in combat.

It became obvious during the training program that we had mortgaged a lot due to the wars in Iraq and Afghanistan. For instance, the entire training proficiency of the Army was sliding away from high-intensity conflict and moving down to the lower end of the spectrum, which involved training only for the specific mission at hand. And professional development had been reduced to the point that we were now developing leaders who had exceptional counterinsurgency skills, but had never actually been trained in other critical, high-intensity warfighting tasks.

I couldn't control the missions that would be assigned once our troops were sent into combat. But I certainly could control how I was going to get them ready, and I wasn't going to sacrifice a single standard. Right from the very beginning, I determined that we would train as a corps, and part of that training would include a high-intensity conflict scenario. Moreover, I was not going to allow soldiers to be deployed unless they were fully trained. That guidance, in and of itself, caused some significant problems for us, because the Pentagon on several occasions waited until the last minute to issue deployment orders.

In one case, MG Doug Robinson came to me in the summer of 2005 with a typical problem. *[Doug left Iraq in July 2003, was promoted to major general, and was sent to the Army staff in Washington where he oversaw the sourcing of forces. So he was intimately familiar with the Army's problems in this area. He had returned to Germany in 2005 as commander of the 1st Armored Division.]* "Sir, we have a UAV [Unmanned Aerial Reconnaissance Vehicle] platoon that has been ordered to deploy immediately," Robinson told me. "I know your guidance is that we must certify units ready for combat before they deploy, but I can't do it in this case. They haven't completed their training."

I thanked Doug for informing me, and instructed him to continue their training, but to also determine the absolute minimum required time needed. Then I fired off a letter to GEN B. B. Bell. "We can't certify these soldiers for war within the timelines of their deployment orders," I wrote. "There are two choices: delay their deployment or, if we insist on deploying them, then I will need to receive an order from higher headquarters ordering me to send them into combat without being properly trained and certified."

That letter set off some fireworks in the G-3 operations office of the Army staff. I received a call demanding that we deploy the platoon and they, in turn, would complete the training in either Kuwait or Iraq. "No, I refuse to do so," I said. "We do not send soldiers off to war without proper training."

"But we need them on the ground now, General Sanchez."

"We don't need them bad enough to put them at risk," I said. "If you want them that bad, then some four-star is going to have to send me an order telling me I need to compromise on their training."

Of course, no senior leader would ever be willing to do such a thing, and I knew it. Finally, they relented. "Okay, sir, we'll delay their deployment. But we really need to make it as quickly as possible. How much time do you need?" I then turned around and asked Doug Robinson to work nonstop to complete the UAV platoon's training, but not to sacrifice any standards.

As our troops began to deploy for Iraq and Afghanistan, I went out to speak with them and, invariably, received questions about why I was staying behind. "Sir, why is this happening?" they asked. "Why aren't you going with us? We need you."

"Well, I want to go with you," I said. "But they won't allow me to deploy. It's politics."

Of course, as soon as the public learned that I was not deploying to Iraq with my soldiers, a number of media outlets made a big deal about it. "Sanchez is now staying behind" and "His Corps is Leaving" ran the headlines. Many speculated about the reasons for this unusual situation. For the most part, I did not comment. Responding to all the media hype, I reasoned, would only fan the flames of negativity.

On the whole, in 2005 there was a big shift in national media focus—from the specifics of Abu Ghraib to the war in Iraq in general. People were now beginning to realize that the entire first year of the war had been lost. But the reasons presented by media pundits and analysts were mixed as to who exactly caused the mess. Some placed the blame on civilian leadership in the Bush administration. Others placed it squarely on my shoulders. After all, they reasoned, I had been the military commander on the ground. It must follow, then, that I did not adapt properly to the changing conditions, I had not recognized the insurgency, and I was completely unprepared for the responsibilities. After all, I was the youngest three-star in the Army at the time.

A panel discussion on NBC's *Meet the Press* on April 28, 2005, was representative of the school of thought that placed responsibility on the administration. Former GENs Wesley Clark, Barry McCaffrey, Montgomery Meigs, and Wayne Downing participated. When asked by moderator Tim Russert if it was a mistake going into Iraq, GEN Clark responded, "I think it was a strategic blunder. First, it wasn't connected to the war on terror, at least not to the people that struck us [on September 11, 2001]. Secondly, it has proved a huge recruitment tool for al-Qaeda . . . Seeing American soldiers engaged there just raises the temperature and the blood pressure throughout the Islamic world." GEN Downing later commented, "We wasted the first twelve months in Iraq, because we didn't plan for postwar hostilities."

Representative of those placing blame on my shoulders was Andrew J. Bacevich, a retired colonel and now Boston University professor and writer who wrote a blistering column in the June 28, 2005, issue of *The Washington Post*. "Lieutenant General Ricardo Sanchez . . . can best serve his country by retiring forthwith," wrote Bacevich. "His advancement would do untold damage to the military professional ethic . . . Historians will remember [him] as the William Westmoreland of the Iraq war—the general who misunderstood the nature of the conflict he faced and thereby played into the enemy's hands."

This new round of attacks in 2005 was punctuated when the ACLU (American Civil Liberties Union), in coordination with Human Rights First, filed a major lawsuit against Rumsfeld, Karpinski, Pappas, and me on March 1, 2005. Brought on behalf of former prisoners, the suit charged me, personally, with direct responsibility for the torture and abuse of prisoners at Abu Ghraib. It also stated that I disregarded warnings about the abuse and authorized the use of illegal interrogation tactics that violated the constitutional and human rights of prisoners.

Additionally, ACLU Executive Director Anthony D. Romero urged Congress to initiate a formal inquiry against me, because, as he said, I had lied in front of the Senate Armed Services

Committee during my testimony. His proof, he said, were the memorandums of September 14, 2003, and October 12, 2003. Romero also sent a letter to U.S. Attorney General Alberto Gonzales requesting that the Justice Department open an investigation into possible perjury charges against me. "Lieutenant General Sanchez's testimony, given under oath before the Senate Armed Services Committee, is utterly inconsistent with the written record . . ." he wrote. "This clear breach of the public's trust is also further proof that the American people deserve the appointment of an independent special counsel . . ."

Of course, Romero refused to accept my testimony to Congress about either the context in which the memos had been written or their specific contents. I had issued the directives to limit rather than expand interrogation techniques, and none of the techniques listed were in violation of the Geneva Conventions. Overall, I was very upset by this development. And when asked by a reporter what I thought of the ACLU, I responded that they were "a bunch of sensationalist liars, I mean lawyers, that will distort any and all information that they get to draw attention to their positions." I was deeply angry at the time.

Just when I thought things were looking as bleak as they could get, I received a call from the Department of the Army Inspector General's Office. They had completed their report, they said, and were releasing findings to the Senate Armed Services Committee and then to the public. The actual written report would come out in another week or so. In general, I was told that none of the allegations against me were substantiated and that none of my staff officers had been found complicit in anything. However, the IG had substantiated the allegation of dereliction of duty against BG Karpinski. As for me, personally, I'd been cleared of any wrongdoing in relation to the Abu Ghraib abuses, the Iraq detention and interrogation problems, and every other allegation the IG had investigated.

When I finally got hold of a copy of the written report, the words jumped off the pages at me. Among other things, this is what the investigation found:

- Sanchez properly delegated authority [and] properly provided oversight.
- CJTF-7 leadership provided routine oversight [and] Sanchez was directly involved in providing such oversight. He proactively sought assistance and additional resources to address shortfalls in both the detention and interrogation operation.
- Sanchez requested assistance in detention operations. [He] also recognized that CJTF-7's interrogation operations were not configured to produce the actionable intelligence required to fight the insurgency, and identified this as a shortfall to his chain of command.
- Sanchez was . . . proactive in responding to identified shortcomings by requesting appropriate assistance.
- Sanchez properly exercised his responsibilities at the strategic level of leadership. [He] appropriately focused on supporting the CPA (Coalition Provisional Authority), interfacing with DoD, the National Command Authority, and contributing to the effort to rebuild Iraq's infrastructure.
- Sanchez, as a strategic leader, was not responsible for the direct supervision of soldiers operating at AGP (Abu Ghraib Prison). Command and staff failures cited by the Schlesinger report and attributed directly to detainee abuse were failures in leadership at the BDE (brigade) and BN (battalion) levels. These failures were not attributable to a lack of oversight by Sanchez.
- CJTF-7 was never fully resourced in terms of personnel, either in raw numbers, experience, or grade-level. The Joint Manning Document reached no more than a 60 percent fill and much of the available staff's effort was directed toward support of the CPA.

The Inspector General's report completely cleared me of any wrongdoing and I was glad to read it. However, my emotions were mixed. I was relieved that the truth had finally come out. But I was also angry that I had ever been accused in the first place. Being tried in the media had taken its toll on me both personally and professionally. The more frequently negative information was released by the press, the more it became conventional wisdom. And after a period of time, most people lost interest in the subject and did not take note of the final outcome. The military newspaper *Stars and Stripes* did run a front-page story that I had been cleared. But the news did not receive anything close to the amount of exposure that the original accusations did.

I knew that the majority of the American public still believed I had done something wrong. And, at this point, I wasn't sure whether the Inspector General's report was going to make any real difference at all. The political climate was still far too volatile and polarized.

A few days after learning of the IG's findings, I received a message from President Bush inviting me to the Cinco de Mayo celebration at the White House. At first blush, I was startled. "I wonder what this means?" I thought. "Should I fly back to Washington for this event or not?"

After thinking about it, I placed a call to the Army Chief of Staff's office, and asked if they were aware of the President's invitation. "No, sir, we didn't know about it," I was told.

"Well, I don't know whether I should accept this thing or not," I said. "I don't want you all to find out on CNN that I'm at the White House with the President."

"It'll be all right," came the response. "Go ahead and go."

"Okay, but somebody needs to tell the Secretary of Defense and other key leaders that I'm coming in to Washington for this event. Could you please make sure that happens?"

"Sure, General Sanchez. We'll take care of it."

With the Army's blessing, Maria Elena and I flew back to Washington, D.C., and on May 5, 2005, we joined a few hundred other people on the White House lawn for a moving celebration of our nation's Hispanic heritage. I remember seeing the Washington Monument in the background as we approached President and Mrs. Bush in the receiving line. And for a moment, I believed that I might, indeed, be able to survive the vicious politics that had distorted reality for over a year. As soon as the President saw me, he snapped to attention and saluted me. Then, when I returned the salute, he grasped my hand warmly. "Hey, Ric," he said. "I'm glad you could make it."

OVER THE SUMMER AND early fall of 2005, I concentrated on training our troops for deployment to Iraq and Afghanistan. Professionally, I was working hard, and personally, I was feeling better about the Army. I had received a number of congratulatory calls from friends and family about the outcome of the IG report. And by the time the new nomination cycle came around, it appeared that I was finally going to be nominated for a fourth star. GENs Schoomaker and Bell were leading the charge to make it happen, and they informed me that I was under consideration for a number of possible billets, including Vice Chief of Staff, Army; Commanding General, U.S. SOUTHCOM; and Commanding General, U.S. Army Europe.

By early October, Chief of Staff GEN Schoomaker informed me that my nomination for the position of Commanding General, U.S. Army Europe (a four-star billet) to replace B. B. Bell was being forwarded to the Secretary of Defense. But before it would actually be submitted to the President and the Senate, Rumsfeld wanted to discuss the chances that my nomination would be received favorably by the Senate Armed Services Committee.

Earlier in 2005, a couple of trial balloons had been sent out by the Pentagon (I assume from the Secretary's office) that Rumsfeld was considering nominating me for promotion. They immediately drew flack from the press, which issued such responses as "The administration has no desire to hold senior leaders accountable"; "Rumsfeld is focused on holding only junior soldiers accountable"; "No Senior officers held accountable for Abu Ghraib"; "Now they want to promote the guy responsible for Abu Ghraib." Given that the IG report had been issued, I wondered if it would make any difference.

During the second week of October, Secretary Rumsfeld went to the Senate with this latest Sanchez trial balloon. Later that very same week, Senators Levin and Warner began raising new calls for accountability hearings. "This hasn't been settled yet," they said. "There's a lot more to follow. We're going to schedule an accountability hearing in two weeks."

As soon as I heard that, I called the Pentagon. "Look, if there is an accountability hearing on Abu Ghraib," I said, "I've got to be there personally or I must submit a statement for the record."

"Okay, General Sanchez. We'll let the Secretary know."

I didn't hear anything from anybody for a couple of weeks. But on the evening of October 31, 2005, I had just landed in Kansas City, where I had been scheduled to speak at a meeting of the Army's new general officers, and was walking out of the airport when my cell phone rang. It was GEN Peter Pace, the new Chairman of the Joint Chiefs of Staff. Pace, who had been the Vice Chairman, had been promoted the previous month upon the retirement of GEN Richard Myers.

"Bad news doesn't get better with age," said Pace. "I've got to inform you that we've made the decision not to nominate you for a fourth star or another three-star assignment. It is not in the best interests of the Department of Defense, the Army, or you. A confirmation hearing would be too contentious."

I paused a moment to absorb the shock of Pace essentially telling me that I had to retire. "Well, sir, you all have betrayed me," I finally said. "I understand. Thank you for calling." And that was the extent of the conversation.

The next morning, at Fort Leavenworth, I met with the Chief of Staff, who had also flown in for the conference. As soon as I walked into his office, GEN Schoomaker asked, "Did Pace call you, Ric?"

"Yes, sir, I talked to him last night," I responded.

"Ric, I'm very sorry this didn't work out," he said. "They wanted me to deliver the message to you and I refused. 'This is wrong and I will not do it,' I told them. 'Get someone else.'"

All I could manage was a nod and a "Thank you, General."

"I'd like to help you in any way I can to accommodate whatever timelines you want to set up for your retirement," continued Schoomaker. "The only stipulation is that you'll have to be out of command before the corps headquarters comes back. Otherwise, you can do it as early or as late as you want, and, of course, wherever you want."

"Thank you, sir," I said again.

The Chief of Staff and I chatted for a few minutes. He tried to make me feel better by telling me it was all politics. And I really got the feeling that he was deeply disappointed about this outcome. Pete Schoomaker was a good man. He had done a lot for me and I knew there was really nothing else he could have done. So when I left, I shook his hand and thanked him for his support.

After the conference ended, I walked out of Bell Hall, the age-old heart and soul of the Command and General Staff College, and stood on the building's steps for a few moments. Fort Leavenworth is perched on a high plateau overlooking a spectacular stretch of the Missouri River.

As I paused to take one last look around, it occurred to me that, a year and a half earlier, right after my nomination for a fourth star had been withdrawn the first time, I was on a flight out of Washington, and the view of the Potomac River below had reminded me of the Rio Grande in Texas. Now I was looking at a magnificent view of the Missouri River Valley. Only this time, the flowing waters below really were leading me home.

My military career was over. It was the end of the line.

CHAPTER 24

Hail and Farewell

After the phone call from GEN Pace, it didn't take me long to come to a realization about what really had happened. Obviously, key members of the Senate Armed Services Committee had told Secretary Rumsfeld that it didn't matter what the Inspector General's report concluded. If my nomination was submitted for a fourth star, I would still have to endure a contentious hearing and be subjected to some tough questioning. In addition, by threatening to hold new "accountability" hearings, the Senate Democrats were intentionally putting pressure on the administration not to send my nomination forward. Combine those two factors and my fate was sealed. The Bush administration just could not take the risk of having me appear before a Senate committee determined to resurface the issues of interrogations, suspension of the Geneva Conventions, and torture.

On several occasions, I had wondered what Secretary Rumsfeld's real intentions were in regard to me and my career. Had he really been interested in taking care of me? Or was it that he was stringing me along, trying to keep me in the service, so I would have to stay silent? In the end, I think it was a combination of both, with heavy weight on the latter.

My comment to GEN Pace, that I had been betrayed, stemmed

from my sense of values—in particular, that a man's word is his bond. Every step of the way, right up until two weeks before Pace called me, I had been reassured that everything was going to be okay. GEN Myers was especially supportive. He understood all the complexities of what I had gone through in Iraq, and all the realities of Abu Ghraib. Every time I saw him, he said something encouraging, like "Hang in there, Ric," or "We're going to do the right thing." Of course, when Myers retired, I may have lost a stalwart supporter, but would it really have made any difference? Pace was new to the job, aligned with Rumsfeld and the administration, and didn't have the depth of commitment that Myers did. When Schoomaker refused to give me the news, Pace had to do it. I felt not only betrayed but deeply hurt. My superiors had walked away from me—the President, the Secretary, the Chairman, even my beloved Army. It was a very tough thing for me to swallow.

My other two direct superior officers, John Abizaid and B. B. Bell, were unaware of the decision until it was too late. When Abizaid called me a short time later, he had no idea of what had really happened. "Ric, I'm hearing that you are thinking of retiring," he said.

"Now, sir, you know me better than that," I said with a laugh. "I wouldn't give up this fight if I had any choice at all."

"Oh, no," he replied. "You mean they didn't give you a choice?"

"That's right, sir." After I recounted the whole story, Abizaid expressed both anger and frustration.

"This is unbelievable!" he said. "Unbelievable! But . . . but I guess it's a done deal, huh? Maybe I'll talk to the President about you. I'm going to see him in a few days."

"Yes, sir. But I don't believe there's anything anybody can do."

THE NEXT SIX MONTHS went by in a blur. Maria Elena and I made the decision to retire in the fall. That way our youngest son, Michael, could graduate from high school in Germany and begin to make plans for college. I would also be able to get past the three-year time-in-grade requirement that would allow me to retire with three stars. I kept my mind off the unpleasantness by burying myself in my work. And there was much to do. We were in the middle of intensive training, deployment, and support operations for the troops in Iraq. And with the Army's unit sourcing solutions, there was always something going on with regard to training and deploying units, whether it was to Iraq or to Afghanistan.

In early April 2006, I received a telephone call from Secretary Rumsfeld. He asked if I would come in to see him the next time I was in Washington. What that really meant was that I needed to visit him, but that I didn't have to drop everything and leave. After trying to find out what the Secretary wanted to talk to me about, and learning that the rest of the chain of command had absolutely no idea, I made an appointment with Rumsfeld during the Army three-star commander's conference in Washington, which would take place a couple of weeks later.

I walked into Rumsfeld's office at 1:25 p.m. on April 19, 2006. He had just returned from a meeting at the White House, and the only other person present in the room was his new Chief of Staff, John Rangel.

"Ric, it's been a long time," Rumsfeld said, greeting me in a friendly manner. "I'm really sorry that your promotion didn't work out. We just couldn't make it work politically. Sending a nomination to the Senate would not be good for you, the Army, or the department."

"I understand, sir," I replied.

Then we walked over to his small conference table. "Have a seat," he said. "Now, Ric, what are your timelines?"

"Well, sir, my transition leave will start in September with retirement the first week of November."

"That's a long time. Why so long?"

"I want to have my son graduate from high school in June. After that, I'll have forty-five days to hit my three years' time in grade, so I can retire as a three-star without a waiver."

"Oh, yes, I remember now. That's why we kept you in Germany in your current job."

"Right."

"Ric, I wanted to tell you that I'm interested in giving you some options for follow-on employment as a civilian in the Department of Defense." Rumsfeld then talked about a possibility with either the Africa Center for Strategic Studies or the Center for Hemispheric Defense Studies. There was a director they were thinking of moving to make room for me, he explained.

"Well, I'll consider that, sir, but I'm not making any commitments. I have some other opportunities I need to explore."

Secretary Rumsfeld then pulled out a two-page memo and handed it to me. "I wrote this after a promotion interview about two weeks ago," he explained. "The officer told me that one of the biggest mistakes we made after the war was to allow CENTCOM and CFLCC to leave the Iraq theater immediately after the fighting stopped—and that left you and V Corps with the entire mission."

"Yes, that's right," I said.

"Well, how could we have done that?" he said in an agitated, but adamant, tone. "I knew nothing about it. Now, I'd like you to read this memo and give me any corrections."

In the memo, Rumsfeld stated that one of the biggest strategic mistakes of the war was ordering the major redeployment of forces and allowing the departure of the CENTCOM and CFLCC staffs in May–June 2003. "This left General Sanchez in charge of operations in Iraq with a staff that had been focused at the operational and tactical level, but was not trained to operate at the strategic/operational level." He went on to write that neither he nor anyone higher in the administration knew these orders had been issued, and that he was dumbfounded when he learned that GEN McKiernan was out of the country and in Kuwait, and that the forces would be drawn down to a level of about 30,000 by September. "I did not know that Sanchez was in charge," he wrote.

I stopped reading after I read that last statement, because I knew it was total BS. After a deep breath, I said, "Well, Mr. Secretary, the problem as you've stated it is generally accurate, but your memo does not accurately capture the magnitude of the problem. Furthermore, I just can't believe you didn't know that Franks's and McKiernan's staffs had pulled out and that the orders had been issued to redeploy the forces."

At that point, Rumsfeld became very excited, jumped out of his seat, and sat down in the chair next to me so that he could look at the memo with me. "Now just what is it in this memorandum that you don't agree with?" he said, almost shouting.

"Mr. Secretary, when V Corps ramped up for the war, our entire focus was at the tactical level. The staff had neither the experience nor training to operate at the strategic level, much less as a joint/combined headquarters. All of CFLCC's generals, whom we called the Dream Team, left the country in a mass exodus. The transfer of authority was totally inadequate, because CENTCOM's focus was only on departing the theater and handing off the mission. There was no focus on postconflict operations. None! In their minds, the war was over and they were leaving. Everybody was executing these orders, and the services knew all about it."

Starting to get a little worked up, I paused a moment, and then looked Rumsfeld straight in the eye. "Sir, I cannot believe that you didn't know I was being left in charge in Iraq."

"No! No!" he replied. "I was never told that the plan was for V Corps to assume the entire mission. I have to issue orders and approve force deployments into the theater, and they moved all these troops around without any orders or notification from me."

"Sir, I don't . . ."

"Why didn't you tell anyone about this?" he asked, interrupting me in an angry tone.

"Mr. Secretary, all of the senior leadership in the Pentagon knew what was happening. Franks issued the orders and McKiernan was executing them."

"Well, what about Abizaid? He was the deputy then."

"Sir, General Abizaid knew and worked very hard with me to reverse direction once he assumed command of CENTCOM. General Bell also knew, and he offered to send me his operations officer. In early July, when General Keane visited us, I described to him the wholly inadequate manning level of the staff, and told him that we were set up for failure. He agreed and told me that he would immediately begin to identify general officers to help fill our gaps."

"Yes, yes," replied Rumsfeld. "General Keane is a good man. But this was a major failure and it has to be documented so that we never do it again." He then explained that he would be tasking ADM Ed Giambastiani, Vice Chairman of the Joint Chiefs, to conduct an inquiry on this issue.

"Well, I think that's appropriate," I said. "That way you'll all be able to understand what was happening on the ground."

"By the way," said Rumsfeld, "why wasn't this in the lessons-learned packages that have been forwarded to my level."

"Sir, I cannot answer that question," I replied. "But this was well known by leadership at multiple levels."

After the meeting ended, I remember walking out of the Pentagon shaking my head and wondering how in the world Rumsfeld could have expected me to believe him. Everybody knew that CENTCOM had issued orders to drawdown the forces. The Department of Defense had printed public affairs guidance for how the military should answer press queries about the redeployment. There were victory parades being planned. And in mid-May 2003, Rumsfeld himself had sent out some of his famous "snowflake" memorandums to GEN Franks asking how the general was going to redeploy all the forces in Kuwait. The Secretary knew. Everybody knew.

So what was Rumsfeld doing? Nineteen months earlier, in September 2004, when it was clearly established in the Fay-Jones report that CJTF-7 was never adequately manned, he called me in from Europe and claimed ignorance, "I didn't know about it," he said. "How could this happen? Why didn't you tell somebody about it?" Now, he had done exactly the same thing, only this time he had prepared a written memorandum documenting his denials. So it was clearly a pattern on the Secretary's part, and now I recognized it. Bring in the top-level leaders. Profess total ignorance. Ask why he had not been informed. Try to establish that others were screwing things up. Have witnesses in the room to verify his denials. Put it in writing. In essence, Rumsfeld was covering his rear. He was setting up his chain of denials should his actions ever be questioned. And worse yet, in my mind, he was attempting to level all the blame on his generals.

But why now? Why was he doing it in September 2006? I wasn't completely sure. I knew it had been a hectic week. The media was hounding Rumsfeld, because a number of former generals had staged something of a revolt and were calling for his resignation. Perhaps he wanted to set up this link in his chain of denials before I left the service, or gauge how I was going to react to his position. Or Rumsfeld might have been anticipating a big political shift in Congress after the midterm November elections, which, in turn, might lead to Democratic-controlled hearings. I didn't know exactly why it happened at this particular time. I just know that it did happen.

Upon returning to Germany, I had some very long discussions with my wife, especially about Rumsfeld's offer of a possible high-paying job in the Department of Defense. "I'm not sure I want to pursue something like that," I said. "But given my reaction to Rumsfeld's memorandum, he now knows that I'm not going to play along. So I don't think he'll pursue it."

"Ricardo, they are just trying to buy you off and keep you silent," said Maria Elena. "I don't think we should mess with them anymore."

My wife had hit the nail right on the head. "I believe you're right," I replied. And sure enough, no one from the Department of Defense ever followed up. So at that point, I closed out all options of doing anything with the DoD after retirement.

On my first day back in the office, I received a phone call from ADM Giambastiani, who had obviously talked to Rumsfeld. "Ric, what happened in that meeting?" he asked. "The Secretary was really upset."

"Well, sir, I essentially told him that his memorandum was wrong," I said. "I guess he didn't like that."

"Well, no, I guess he didn't. Anyway, he's asked me to make this study happen, so we'll get right on it."

Giambastiani assigned the task to the Joint Warfighting Center and gave them a pretty tight timeline. So it wasn't long before I was giving the investigative team a complete rundown of everything that had happened in Iraq between May and June 2003. I later learned that GEN Tommy Franks, however, had refused to speak with them.

A few months later, I was making a presentation at the Joint Warfighting Center and ran across several of the people involved with the study. "Say, did you guys ever complete that investigation?" I asked.

"Oh, yes sir. We sure did," came the reply. "And let me tell you, it was ugly."

"Ugly?" I asked.

"Yes, sir. Our report validated everything you told us—that Franks issued the orders to discard the original twelve- to eighteen-month occupation deployment, that the forces were drawing down, that we were walking away from the mission, and that everybody knew about it. And let me tell you, the Secretary did not like that one bit. After we went in to brief him, he just shut us down. 'This is not going anywhere,' he said. 'Oh, and by the way, leave all the copies right here and don't talk to anybody about it.'"

"You mean he embargoed all the copies of the report?" I asked.

"Yes, sir, he did."

From that, my belief was that Rumsfeld's intent appeared to be to minimize and control further exposure within the Pentagon and to specifically keep this information from the American public.

Continuing the conversation, I inquired about the "original twelve- to eighteen-month occupation deployment," because I wasn't sure what he was talking about. It turned out that the investigative team was so thorough, they had actually gone back and looked at the original operational concept that had been prepared by CENTCOM (led by GEN Franks) before the invasion of Iraq was launched. It was standard procedure to present such a plan, which included such things as: timing for predeployment, deployment, major combat operations, post–major combat operations, staging for major combat operations, and redeployment. The concept was briefed up to the highest levels of the U.S. government, including the Secretary of Defense, the National Security Council, and the President of the United States. And the investigators were now telling me that the plan called for a Phase IV (post–major combat) operation that would last twelve to eighteen months.

To say I was shocked would be an understatement. I had never seen any approved CENTCOM campaign plan, either conceptual or detailed, for the post–major combat operations phase. When I was on the ground in Iraq and saw what was going on, I assumed they had done zero Phase IV planning. Now, three years later, I was learning for the first time that my assumption was not completely accurate. In fact, CENTCOM *had* originally called for twelve to eighteen months of Phase IV activity with active troop deployments. But then CENTCOM had completely walked away by simply stating that the war was over and Phase IV was not their job.

That decision set up the United States for a failed first year in Iraq. There is no question about it. *And I was supposed to believe that neither the Secretary of Defense nor anybody above him knew anything about it? Impossible!* Rumsfeld knew about it. Everybody on the NSC knew about it, including Condoleezza Rice, George Tenet, and Colin Powell. Vice President Cheney knew about it. And President Bush knew about it. There's not a doubt in my mind that they all embraced this decision to some degree. And if it had not been for the moral courage of GEN John Abizaid to stand up to them all and reverse Franks's troop drawdown order, there's no telling how much more damage would have been done.

In the meantime, hundreds of billions of taxpayer dollars were unnecessarily spent, and, worse yet, too many of our most precious military resource, our American soldiers, were unnecessarily wounded, maimed, and killed as a result. In my mind, this action by the Bush administration amounts to gross incompetence and dereliction of duty.

In the summer of 2006, I was asked to speak at a number of military forums about the war in Iraq. I had no problem relating the truth as I had experienced it. But I did not speak to the press, because I did not believe it was the right thing to do for a serving general officer.

I gave a presentation at the Command and General Staff College in Fort Leavenworth, Kansas. LTG David McKiernan, who commanded CFLCC during major combat operations, spoke first. McKiernan talked about the great war we fought, but he ended his presentation at May 1, 2003. Then I got up and spoke to the same audience about the near catastrophic failures of the United States.

In August 2006, the Center for Army Lessons Learned sent me its preliminary history of the Iraq occupation period (May 2003 to July 2004) with a request to make comments and review it for accuracy. I had not been interviewed nor had any of the CJTF-7 general officers. In reading the report, I found that it appeared to be nothing more than an effort to protect the Army and push all the blame for what went wrong off onto CENTCOM and the Joint Task Force. I was being asked to provide comments, because the publication date was in mid-September. The Army had refused to tackle this issue for a long time. And now, when they finally did address it, they were in a defensive mode. So I blasted them. "This is *not* what happened," I said. "*Our* Army made many of these disastrous decisions. *Our* Army didn't provide proper training and guidance. We have to embrace and recognize the fact that we screwed this thing up miserably. Only then can we possibly get back on the right track."

With time, the Department of Defense did make changes to fix some of the problems. By 2007, the concept of a standing joint task force headquarters had been embraced. It would be designed to prevent what happened to CJTF-7 in Iraq. A U.S. Army Detainee Operations Training Update was published in March 2005. And on the day I formally left command in Europe, the Army finally published a pamphlet called "Interrogation Guidance," which specifically outlined approved interrogation techniques. Everything mentioned in that set of guidelines complied with the Geneva Conventions.

In early September 2006, Maria Elena and I attended what the Army calls a "Hail and Farewell" social event. It was a casual affair with members of the corps staff and subordinate unit commanders, where everybody said goodbye, presented some mementos, and gave me a bit of a roast. They poked fun at my training methods, noting that I was adamant we had to have all our gear on, and that we had to train under expeditionary conditions. Some of the jokes were very funny and we laughed a lot while remembering the good times in command. There were also some very serious statements made. Training soldiers the way I did, they said, was the right thing to do, because it saved lives on the battlefield. They told me that I was the most feared but also the most respected general they'd ever served with, that I was always very fair, that I took the time to listen, and that I cared. And they thanked me for my leadership.

When it came time for me to say a few words, I thanked them for everything they had done for me, for the Army, and for the nation. And one piece of advice I offered was not to compromise their personal lives for the possibility of career advancement. "I had a problem with that early on," I told them. "I was too focused on the Army and the mission. I was too willing to sacrifice my family. But then my nine-month-old son, Marquito, was killed in a horrible automobile accident. After that, I got things back in balance. Don't wait until some awful event jars you into realizing what is truly important in life."

I urged them to go home for dinner every night, to go to the youth soccer games, and to the teacher conferences. And then I related the advice given to me by MG Dick Boyle. "Here was a man who had served three decades, had retired as a two-star, and said that it was as though he had never been in the Army. In the end, he said, the only thing you'll have left is your family, your friends, and your faith."

Three days after the Hail and Farewell event, on September 7, 2006, I formally relinquished command. I had been the longest serving commander in V Corps history. Normally, a commander serves for two years. I was there for three years and four months. Maria Elena, the children, and I left Germany the next day.

When the Army asked me where I wanted to hold my retirement ceremony, I chose Fort Sam Houston in San Antonio where I could be back home in South Texas with my friends and family. We held the event on November 1, 2006, and everybody showed up. Of course, Maria Elena, and our four children, Lara, Bekah, Daniel, and Michael, were there. So were my mother, my brothers (Mingo, Robert, Leo, David), and my sisters (Maggie and Diana)—and our extended family on both sides. Some friends that I had not seen in thirty years also showed up to share the moment. We caught up on old times and rekindled old friendships. Even my childhood pals David Saenz and Chuy Trevino were there, along with eight members of the King's Rifles. John Abizaid was the highest-ranking military officer present. I was deeply moved that he had taken the time to attend, and even more appreciative when he said to me, "Ric, history is going to show that you and your soldiers at CJTF-7 held that mission together for over a year. It'll be a while before it comes out, but it will be recognized."

In my farewell address, I again thanked all the important people in my life, especially my family, friends, and colleagues. But they all knew I was not ready to leave the Army. Here is a bit of what I said:

> It has been thirty-three years, three months, and eleven days since I stood at the ROTC Cadet Club at Fort Riley, Kansas, and took my oath of office as an officer in the United States Army. On that day I embraced the value systems and the ethics of a warrior, which I will carry to my grave. These values have served me well, especially when confronted by politicians and pundits who will do anything to protect themselves and to preserve their hold on power.
>
> Retirement is a very difficult thing for me to accept, because my soldiers have been fighting and dying for four years now and we have no end in sight for this war. To walk away while engaged in battle is not the ethic of a warrior. However, I had no choice.
>
> A few weeks ago, a friend of mine asked me what I would miss most about the Army. Instinctively, without thought or hesitation, I replied, "my soldiers." As I reflected on that answer I thought of shared experiences that only soldiers appreciate, and of feelings only soldiers can understand. We have shared the blood, sweat, and tears that warriors shed when they go to war. As Shakespeare wrote, "For he who sheds his blood with me this day shall be my brother."
>
> Soldiers must always be guided by our values. The value system that we embrace is the toughest of any profession in our society and it is often not understood by the average American. It is our sense of duty, integrity, and honor that must guide every action and every decision we make as leaders. There is no place in our warrior ethos for compromising our integrity. The soldier must always do what is

right, knowing that many will question and second-guess his actions. The soldier cannot afford to hide behind policies, nuances, or rhetoric. Regardless of personal attacks from the media, pundits, and others with their own selfish agendas, a soldier must never leave the moral high ground. He must never waver from the truth.

I now join the endless ranks of those who have served our country. I do not know what lies ahead. What I do know is that, as the days go by, the Sanchez family will continue with the same commitment to service, willingness to sacrifice, and dedication to duty that our military career always reflected.

Praise be to the Lord, my rock,
who trains my hands for war,
my fingers for battle.
He is my Loving God and my Fortress.

Thank you very much.

SIX DAYS AFTER MY retirement ceremony, on Election Day, November 7, 2006, the American people swept the Democratic Party to power in the U.S. Congress. Democrats captured majorities in both the House of Representatives (233 to 202) and the Senate (51 to 49). As a result, there would be a dramatic shift in the chairmanships of all committees in Congress.

The very next day, on the afternoon of November 8, 2006, President Bush removed Donald Rumsfeld from his position as Secretary of Defense. "The timing is right for new leadership at the Pentagon," the President said in making the announcement. Donald Rumsfeld had been in his position for five years and ten months. He was George W. Bush's longest-serving original cabinet member.

. . . .

TWO WEEKS LATER, MY family and I shared the first Thanksgiving of my retirement. All four of our kids were with us and we began reminiscing. "Say, do you remember when you kids were growing up and the whole family would go to Thanksgiving lunch at the base dining facility?" recalled Maria Elena. "We haven't done that in seven or eight years. Why don't we go?" Everybody got all excited and we decided to go over to San Antonio's Brooke Army Medical Center, a regional active-duty facility that is also one of the finest major burn centers in the country.

As soon as we walked into the dining room, I saw a severely burned young man sitting in a wheelchair at a table with his mother. I could tell by the extent of his injury that he had probably been there for at least a year. "I'm going to go over and talk to that soldier," I said to Maria Elena. "Be right back." So I walked over and introduced myself.

"Young man, I'm General Sanchez," I said. "Thank you for all of your sacrifices. I really appreciate your service to our country."

When he looked up at me, tears started to roll down his cheeks. "Sir, I was with you in Mosul in '03," he said. "I saw you a couple of times when you came to visit us. This is a great honor, sir. Thank you so much for coming over." Then his mother started to cry.

I leaned down and put my hand on the young man's shoulder. "Son, believe me when I say that the honor is all mine," I said. "I owe you a tremendous debt for what you've sacrificed. We all do. God bless you, son. God bless you."

Epilogue

On March 20, 2003, U.S.-led coalition forces invaded Iraq. Six months earlier, the Bush administration had begun a national media campaign to persuade the American people that Saddam Hussein was a threat to the security of the United States. In September 2002, Vice President Dick Cheney, National Security Advisor Condoleezza Rice, Secretary of Defense Donald Rumsfeld, and other key officials announced that Iraq had long-standing ties to al-Qaeda and, by inference, was partly responsible for the 9/11 attacks on the United States. They also stated that Saddam was capable of inflicting death on a massive scale through its pursuit and/or possession of weapons of mass destruction (WMDs). Believing these statements to be true, on October 10, 2002, both houses of the U.S. Congress passed resolutions authorizing President Bush to use U.S. armed forces to defend the national security of the United States against the continuing threat posed by Iraq, and enforce UN Security Council resolutions.

In his January 2003 State of the Union address, President Bush asserted that Saddam was in possession of "the material to produce as much as 500 tons of sarin, mustard [gas] and VX nerve agent . . . , more than 38,000 liters of botulinum toxin . . . and upwards of 30,000 munitions capable of delivering chemical agents." He also stated that Iraq had attempted to purchase high-strength aluminum tubes suitable for nuclear weapons production, and had sought significant quantities of uranium from Africa. A week later, in February 2003, Secretary of State Colin Powell spoke before the UN Security Council, reasserted the link between al-Qaeda and Iraq, and presented satellite photos and illustrations that he said showed chemical weapons bunkers and mobile biological weapons factories. Secretary Powell also stated that Iraq's persistent denials of these U.S. charges were "all a web of lies."

None of it was true. Iraq had no links to al-Qaeda, no nuclear weapons program, and no stashes of chemical or biological weapons. I was on the ground in Iraq and I know. We never found anything. And there was no link between al-Qaeda and Iraq, or a major presence of al-Qaeda in Iraq, until after the aborted battle of Fallujah. These facts were verified (in part or in whole) by three independent government study groups: the U.S. Senate Report of Pre-war Intelligence on Iraq (July 9, 2004); a British study, the Butler Review (July 14, 2004); and Iraq Survey Group (September 30, 2004), which was initially led by David Kay before he resigned. To use Colin Powell's phrase, then, which was really the "web of lies"—Iraq's denials or the Bush administration's assertions?

More than three decades ago, in the jungles of Vietnam, the U.S. Army was micromanaged by the White House, was forced to fight incremental battles, and got stuck in a military quagmire. Back then, the Army became almost totally focused on Southeast Asia to the exclusion of everything else. It took over a decade to fix the "broken Army," as it was called back then.

The Iraq War, too, became a national nightmare with no end in sight. The initial plans of the military were micromanaged by the Bush administration, as were many of the individual battles, troop deployments, and strategic operations. In fact, that micromanagement took place at the total exclusion of the political and economic aspects of rebuilding Iraq. The U.S. Army and the Marine Corps became almost totally focused on the Middle East, and great damage was done to its full-spectrum readiness. It will take at least a decade to repair the damage.

Moreover, there has been no comprehensive investigation by a full independent commission, or any other entity, to explore the full truth behind why we went to war in Iraq, how the suspension of the Geneva Conventions led to putting America on the path to torture, and why the political, economic, and military elements of power were not properly coordinated as part of a grand strategy during the first year's occupation of Iraq. Until such an investigation is completed, we will never know the extent of our government's actions—nor will we ever be able to learn from the entire debacle.

For America, the cost of the war has been high—approximately 4,000 killed and 30,000 wounded. In October 2007, the Congressional Budget Office estimated that the monetary costs of the Iraq War could reach $2.4 trillion through 2017. As an analogy for the future, officials in the Bush administration have used South Korea, where we have had U.S. forces stationed for more than half a century.

IN THE SPRING OF 2006, defense attorneys asked that I be made available to testify at a number of trials for individuals charged with abuses at Abu Ghraib. Several of the defendants alleged that they had been ordered by higher authority to use harsh detention tactics to soften up prisoners for interrogation. But in each case, after defense lawyers either spoke to me in advance, or heard I was present, I ended up not testifying. They wouldn't use me, because my testimony would have refuted their client's allegations and supported the prosecution's case.

In the end, seven soldiers were either convicted or pled guilty to charges related to the abuses at Abu Ghraib prison. SPC Charles Graner was sentenced to ten years in prison, SSG Ivan Frederick, eight years, and PVT Lynndie England, three years.

The four others convicted in the abuses received various penalties ranging from six to ten months in prison and/or fines and dishonorable discharges. COL Tom Pappas admitted approving the use of unmuzzled dogs during interrogations (without my approval), specifically to intimidate a prisoner after the capture of Saddam Hussein. Pappas subsequently received nonjudicial punishment via a written reprimand and was fined $8,000. He was then granted immunity from prosecution by the commanding general, Military District of Washington, and ordered to testify at the courts-martial of the MP defendants. MG Geoff Miller decided to remain silent and invoked his Fifth Amendment rights against self-incrimination during preliminary trial proceedings against two soldiers. After I heard that he was planning to retire, I placed a call to Miller to wish him well. "They've stopped my retirement," he told me. "They won't let me go." The Army gave him a small office in the Pentagon and told him to show up every day. He had no assigned duties. Miller eventually retired on July 31, 2006, in a private ceremony at the Pentagon. To this day, a debate continues about Miller's involvement at GTMO and Abu Ghraib.

In a way, the Abu Ghraib prison scandal was a grotesque blessing for our country. When the pictures of the abuses were broadcast worldwide via the media, it forced America to walk away from the uncontrolled interrogation environment that had been established back in 2002 when the Bush administration suspended the Geneva Conventions. Bush's doctrine led to a totally unsupervised environment where anything was allowed. Torture, which did occur in a number of incidents, should never, ever be permitted in operations conducted by the United States of America.

Before my retirement, and while still in Europe, I made a trip to Kosovo to see what the province looked like. It had been five years since I left command, but when the chief of the Kosovo Protection Corps, soon to be prime minister, learned I was coming, he went out of his way to visit with me. I was amazed at the pace of economic progress in the sector I had once commanded. There were new buildings all over the place, including hotels, sporting arenas, service stations, schools, and municipal buildings. Major roads had been constructed, a variety of European corporations had moved in to do business, and there had been only one major violent incident in a year. Overall, Kosovo had flourished. When my host and I dined together, he made a point of thanking me for my leadership and everything I had done for his country. It was pretty satisfying to look back and realize that I had a little bit of a hand in helping set Kosovo on the right path. At the same time, I wondered what Iraq might have looked like if, during that first year of occupation, we had been able to synchronize all of our efforts.

The U.S. Army has an after-action process in the wake of every success and failure so that, the next time out, we will be wiser in battle. In much the same way, I believe that America must assess itself and understand where we have gone wrong in Iraq. And we should do so in a nonpartisan environment, free of self-interest or political influence. Our national leadership and the public, in general, must understand and accept the fact that there were major mistakes made in the early days of this war. We must also acknowledge that there was never a synchronized strategy to bring together all elements of power in Iraq. Actually, because the Bush administration ignored and/or handed off de-Baathification, reconciliation, and most other critical issues during the occupation period to the Iraqis, a synchronization of the political, economic, and military elements was made impossible once the transfer of sovereignty occurred in June 2004. Even my beloved Army must use the Iraq war as a forum for training future military leaders—not just in all aspects of Phase IV, counterinsurgency, and occupation operations, but also in coalition and joint operations. If we do not learn and embrace these fundamental lessons, our nation will probably go down this path again—and it will be our warriors in uniform who suffer most from the consequences.

In my thirty-three years as a warrior, I learned not only how to be wiser in battle, but how to be wiser in life. I always prepared for my next encounter by leaning on my instincts, experiences, and history. I came from a poverty-stricken town on the desolate banks of the Rio Grande in South Texas, where my soul is still anchored. From my parents, I learned values that have lasted a lifetime, including hard work, keeping my word, and always telling the truth. And when I joined the U.S. Army, my personal values were reinforced and sustained by the Army's values. L-D-R-S-H-I-P became not only my professional code of conduct, but my personal mantra: Loyalty, Duty, Respect, Selfless Service, Honor, Integrity, and Personal Courage. These values have sustained me during some of my darkest hours, as has my faith. When my son Marquito died, rather than turning away from God, my belief was strengthened—to the point where I was never afraid in combat. Since that terrible moment, the phrase *"Si Dios quiere"* has always held a very deep spiritual meaning for me.

In the wake of my son's death, I sat down night after night reading the Bible, looking for comfort, searching for sustenance. And when I read Psalm 36, I was moved by the words and metaphors of "the children," "the river," and "life." In my travels around the world, there was always a river nearby—its flowing waters providing nourishment to the plants, animals, and people along its path. Sometimes the river was an international boundary separating poverty from prosperity. But always, the people on the other side of the river were just like the rest of us. They wanted freedom, friendship, and hope for a better future. Each river I encountered, whether in the dusty desert of South Texas or in the cradle of civilization in the Middle East, had an associated valley. These are the valleys that made me. The Tigris, the Euphrates, the Han, the Kačanik, the Missouri, and the Potomac all reminded me of my roots and the values that had been instilled at an early age—and that comforted and sustained me in my toughest moments. In the end, the flowing waters led me home to the Rio Grande Valley.

In the town where I grew up, the little house on Roosevelt

Street is still standing, the people still speak Spanish, and Fort Ringgold is still the hub of public school activity. As a matter of fact, there is a new school there named in my honor. In 2004, I went to the dedication of General Sanchez Elementary School in my full dress uniform. The children came out of their classrooms and lined the halls. Some stood at attention. Some saluted. It reminded me of when I was a little boy and my big brother Mingo used to come home on leave from the Air Force and inspect my brother Robert and me. I walked through every wing of the new school, greeted the teachers, and shook hands with every single child.

Outside, after the dedication ceremony, I was chatting with some ROTC cadets when I noticed a familiar face near the building, kind of peeking around the corner. It took me a moment, but I suddenly recognized him to be my old friend Santos Gonzalez, who was now the janitor at the school. As I walked toward him, he started to walk off. "Santos! Santos!" I shouted. "Wait! Wait!"

I went up and gave him a big hug. "Santos, how the heck are you doing? It's great to see you."

"Hello, Ricardo," he said with a smile.

"Why did you walk off?"

"Well, I was afraid," he said. "I didn't know if you would recognize me or would even want to talk to me."

"But we grew up together," I said. "We're friends. We'll always be friends."

I was glad to see Santos, but his comment made me feel a little sad. We both had come from the same place and grown up under the same conditions. But in our early teens our paths diverged. He left school and began working full-time as a migrant worker. I stayed in school and joined the ROTC. And now, more than thirty years later, here we were in the same place again, but under very different circumstances.

Seeing Santos Gonzalez made me think of his father, Benito, who had served in the Army during World War II. When I remembered Mr. Gonzales, it made me think about my heritage, and all the other Hispanic military veterans who had honorably served our country. And I hoped and prayed that, during my own career, I had not let down their legacy. Thinking about Mr. Gonzalez also made me remember the young Hispanics I had encountered in Iraq on Christmas Day 2003. When I asked them what I could do for them, they surprised me by revealing that they were caught up in the bureaucratic tangle of becoming U.S. citizens. By the end of my deployment, we had created a citizenship process for our warriors in uniform, and were planning citizenship ceremonies for them. To my astonishment, it wasn't just Hispanics who had signed up to fight for freedom. Also taking the oath of U.S. citizenship were Bosnians, Poles, Ukrainians, Koreans, and individuals from several other nations.

Ever since my experience in Operation Desert Storm, whenever I sent soldiers into battle, I always thought back to the order I gave a young lieutenant who was leading his scout platoon into dangerous territory. "Have I done everything possible to ensure that the warriors under my command are properly trained so they will return safely?" I asked myself. "Better not lose a soldier due to a failure in training, leadership, or discipline."

During the year and a half I served as commander of coalition forces in Iraq, 843 soldiers and Marines were killed in action. Back then, the first thing I did each morning as I drove to my headquarters in the Green Zone was to review the casualty reports. Even though I thanked God every day that the numbers weren't higher, I was still only looking at numbers. And even though I wrote sympathy letters to every family of a soldier killed in action, it was still a fairly impersonal process. But when I returned from Iraq and was on my thirty-day combat leave, I knew I had to look at their faces and see who they were. And so, for that month, I rose early each morning, went down to my computer, and found a website with biographies of every young man and woman who had died. In chronological order, I read their biographies, learned about their interests, and took note of the loved ones they had left behind. I also looked at each of their pictures for the longest time. Now, they were no longer numbers to me. They were people with hopes and dreams and families.

Before my last tour in Iraq, I used to whistle a lot. Maria Elena knew that meant things were good. But she is still waiting to hear me whistle. Before the war, I had also been a sound sleeper. But when I returned home this time, Maria Elena would wake up in the middle of the night to find me gone from the bedroom. She'd go downstairs and discover me in the living room, swaying back and forth in the rocking chair. She said I had a very peculiar look on my face—a pained look, a grieving look. On those nights, my wife left me alone, because she knew this was something I had to work out for myself. She was right. I did have to work it out for myself. And I *was* grieving. I was grieving for all those fine young men and women who had died under my command. Their loss is a heavy burden to bear. I carry it to this day.

So now, I have come full circle. I am back home with my family, my friends, and my faith—my head held high, my honor and integrity intact. In many ways, I find I had never really left the Rio Grande Valley. Older now, and hopefully wiser from all my battles, I believe I now understand my home better than I ever did before.

I am grateful for my blessings. And I *will* whistle again.

Si Dios quiere.

Acknowledgments

We'd like to thank David Hirshey, vice president and senior editor at HarperCollins, for his hands-on masterful guidance of this project from its inception. Bob Barnett, the best agent in the business, turned the book idea into reality, and provided expert and personal advice during the course of the project. Thanks also to GEN Barry McCaffrey for introducing Bob Barnett into the process and for his mentorship, encouragement, and support through the years.

Also at HarperCollins, our thanks go to Josh Baldwin for his astute editorial suggestions and shepherding of the manuscript through production with unflappable grace; and to Kate Hamill for her outstanding assistance and professionalism. Maria Elena, Lara Marissa, Rebekah Karina, Daniel Ricardo, and Michael Xavier Sanchez all reviewed parts of the manuscript and provided important and valuable insights. Ismael Garza provided computer technical expertise whenever we needed him, and we thank him for his unwavering support.

This career story would not have been possible without the mentorship, friendship, and encouragement of MG (Ret) Richard Boyle, GEN (Ret) Montgomery C. Meigs, GEN (Ret) Wesley K. Clark, GEN (Ret) John P. Abizaid, GEN William Scott Wallace, and LTG (Ret) Randolph W. House.

Ricardo S. Sanchez
Donald T. Phillips

Index